Introducing
and Implementing
Revit® Architecture

Introducing
and Implementing
Revit® Architecture

LAY CHRISTOPHER FOX
JAMES BALDING

autodesk Press

THOMSON

DELMAR LEARNING

Australia · Canada · Mexico · Singapore · Spain · United Kingdom · United States

autodesk Press

Introducing and Implementing Revit® Architecture
by Lay Christopher Fox and James Balding

Autodesk Press Staff

Vice President, Technology and Trades ABU:
David Garza

Director of Learning Solutions:
Sandy Clark

Managing Editor:
Larry Main

Senior Acquisitions Editor:
James Gish

Senior Product Editor:
John Fisher

Marketing Director:
Deborah S. Yarnell

Channel Manager:
Kevin Rivenburg

Marketing Coordinator:
Mark Pierro

Production Director:
Patty Stephen

Production Manager:
Stacy Masucci

Production Technology Analyst:
Thomas Stover

Content Project Manager:
David Plagenza

Art Director:
Benj Gleeksman

Editorial Assistant:
Sarah Timm

Library of Congress
Cataloging-in-Publication Data:

Fox, Lay Christopher
  Introducing and implementing Revit Architecture / Lay Christopher Fox and James Balding. -- 3rd ed.
      p. cm.
   Rev. and expanded ed. of: Introducing and implementing Autodesk Revit Building. 2006.
   Includes index.
   ISBN-13: 978-1-4283-1944-8
   ISBN-10: 1-4283-1944-1
1. Architectural drawing--Computer-aided design. 2. Architectural design--Data processing. 3. Computer-aided design. I. Balding, James J. II. Fox, Lay Christopher. Introducing and implementing Autodesk Revit Building. III. Title.
   NA2728.F695 2007
   720.28'40285536--dc22

                          2007039903

              ISBN-13: 978-1-4283-1944-8
              ISBN-10: 1-4283-1944-1

NOTICE TO THE READER

CONTENTS

CHAPTER 9 ILLUSTRATING THE DESIGN—GRAPHIC OUTPUT FROM REVIT ARCHITECTURE 503

INTRODUCTION

WELCOME

"Get the right tool for the job."

—Mr. Natural

The preceding words entered the U.S. popular lexicon in the free-wheeling 1960s, spoken by a metaphoric "underground" cartoon character who combined confident disregard of hidebound conventional thinking—a common commodity in those days—with a wise, practical outlook, rare enough in any generation. His advice is timeless, not old school.

In today's design world, the tools of the trade are computerized, and applications or releases appear every couple of years. How do we recognize the right tool amidst today's clamor of constant change? What marks a real advance in technology that will benefit designers and the firms that employ them? How do we take advantage of changes in tools while constantly involved in the rigors of meeting expectations and deadlines? Are there definitive answers to these questions?

This book has a double purpose:

- To introduce Autodesk Revit Architecture as the next generation in architectural design, a fully three dimensional building modeler of exceptional power, depth and ease of use, that functions well at all phases of the design process, from concept sketches all the way through construction documentation.

- To prepare individuals, firms and schools for the process of implementing this new tool into their current workflow and processes. Successful implementation will require planning, training and a transition phase from current Computer Aided Design applications. We will use AutoCAD, the worldwide standard, as a reference.

CAD AND REVIT ARCHITECTURE: WHAT ARE THE DIFFERENCES?

CAD stands for Computer Aided Design (or Drafting). This acronym explains succinctly the approach that mid-priced design software (what individuals and small firms can afford) has embodied until the last few years: the computer has been used as an aid to an age-old design process that hasn't changed substantially even as the tools became electronic. CAD software still requires the user to design line-by-line for the most part, and lines representing a wall in a plan do not create or appear in elevation, section or 3D views of that wall.

Revit Architecture was first called simply Revit. Autodesk has expanded that single application into a platform; Revit Architecture, Revit Structure and Revit MEP (for Mechanical, Electrical and Plumbing) are three separate programs using the same engine and sharing the same modeling format.

Revit Architecture is a robust, affordable and efficient software application created specifically for architectural designers. It is not based on the drafting language of lines, arcs and circles. Revit Architecture is a building modeler database, with drafting tools included for those inevitable cases where simple linework is the most efficient way to illustrate a design concept or detail. Revit Architecture differs fundamentally from CAD; it is not a drafting application that has been expanded to attempt limited 3D content and views. Revit Architecture models consist of objects that are completely definable in size, structure, appearance, data content and relation to other model components. All possible views of the model, including sheets and schedules, update with each command. Much of current design process routine—especially propagating changes throughout a set of construction document pages—disappears as a result. Designers can spend the bulk of their time designing, not transcribing.

An example of the effectiveness of the "right tool for the job" comes from newspaper headlines. The *Freedom Tower* project for New York City, possibly the most complicated skyscraper proposal in the world, is being developed in Revit Architecture, from subway interchanges beneath the foundation to an electricity-generating wind turbine matrix at the top of the tower. On May 1, 2005, very late in the design process, the New York City Police Department announced that it was dissatisfied with security measures in the site and building design, and put the entire project on hold. Public outcry and consternation included "Where do we go from here?" editorials in the *New York Times*. On June 8, 2005, only five weeks later, the Police Department announced its satisfaction with design changes put in place by the architect, and the review board certainly did not base its decision on napkin sketches. If you can imagine yourself sitting in the hot seat on a project of that size, scope and public scrutiny, are you convinced that your present design software tools would allow you to respond to a major last-minute challenge as quickly and as well?

WHAT DOES PARAMETRIC DESIGN MEAN?

Good design of any kind is composed of harmonious relationships, which are based on rules, or parameters. Revit Architecture objects are placed in relation to all other objects in a model, and parameters govern the relationships. Designers can now establish relationships among building components that will hold through design revisions unless specifically removed. Suppose that an array of windows in a wall works best if the windows are equidistant from one another, the end windows are six feet [1.8m] from the corners and the sills are two feet [600mm] from the floor. Revit Architecture will recognize and apply those constraints, and adjust the positions of the windows if the wall length or floor height changes. The potential use of constraints flows throughout a model.

Revit Architecture carries parametric specifications much further than the previous simple example of windows in a wall. All types of components can carry information (job address, object size or placement specifications, supplier, cost, department, electrical requirements, etc.) according to supplied or user-defined parameters. This information can be collected in schedules or diagrams and exported to other applications.

Revit Architecture's use of parameters underlies every aspect of the user interface. There is no drafting-oriented organization such as layers to control the appearance of objects; much of the learning curve for CAD applications therefore does not pertain to new Revit Architecture users.

WHO SHOULD READ THIS BOOK?

If you currently work in architectural design and your firm is considering or moving toward Revit Architecture, this book will explain the basics of the Revit Architecture tools that you, your co-workers or your employees will be using every day. An important part of implementing Revit Architecture in an active design company's workflow is blending this powerful new tool with the often vast store of valuable information in existing drawing files. The first chapters of the text deal with importing AutoCAD information into Revit Architecture and exporting DWG files from Revit Architecture, so that the new technology can be transparent to clients, collaborators and other departments or offices in your firm. We mention throughout the text (see particularly Appendix A) the effects Revit Architecture will have on workflow, process, personnel and attitudes in organizations of nearly any size making this particular technology transfer.

If you are a student of architectural design, this book will show you the very latest software tool in the marketplace. You will be among the first to design buildings parametrically, entirely in three dimensions, without having to first learn the mechanics of representational lines, arcs and circles. A few words of caution that your professors will no doubt echo: you will never become a good designer if you don't understand and practice good drafting techniques. Architects who can't successfully create clear sketches or readable detail pages do not win prizes, or even commissions.

FEATURES OF THIS TEXT

WHAT YOU WILL FIND IN THIS BOOK

The third edition of *Introducing and Implementing Revit Architecture* has been extensively revised and expanded from the 2004 first edition.

Each chapter in the book covers specific, logically sequential skill development using Revit Architecture in a planned progression from basic modeling through more complicated refinements, data extraction and illustration. All work is based on tutorial exercises, with conceptual explanations, tips, notes and cautions where appropriate. Each chapter ends with review questions suitable for quizzes.

Chapter 1 provides an overview of Revit Architecture's rapidly growing position in the market of architectural design software, and contains a tutorial exercise showing how to create, revise and document a simple building model. Chapter 2 covers template files, their uses and pre-loaded content. Chapter 2 also shows how Revit Architecture can import information from AutoCAD during the creation of a custom titleblock. These chapters should be of particular interest to company principals, supervisors or CAD managers.

Chapters 3 through 10 use a hypothetical multi-building, multi-phase project on a college campus as the framework for their exercises. Chapter 3 covers the creation of a site plan with building shell, using walls, floors and other building element objects combined with imported information, and also shows how to export information from a Revit Architecture project file to AutoCAD. Chapter 4 covers sketching, massing objects, transformation of massing into building components and file linking. Chapter 5 develops modeling techniques in the context of multi-user design teams, and introduces project phasing. Many of Revit Architecture's most powerful modes and tools are covered in these exercises.

Chapters 6, 7 and 8 concentrate on various aspects of annotation, within the context of the hypothetical project workflow. Chapter 6 covers customization and management of the appearance of annotation, from tags to titleblocks. Chapter 7 works through Revit Architecture's extensive and powerful scheduling capabilities. Chapter 8 moves from schedules into area and room plans, with customizable color fills. These chapters show—in broad strokes, concentrating on design development—how to create, manage and enhance the "information" side of Revit Architecture, including streamlined data export to external applications.

Chapter 9 covers graphic output: printing, perspective views, rendering and animations. Chapter 10 wraps up with Revit Architecture's tools for organizing and managing design options, and a look at Revit Architecture family creation, itself a topic worthy of its own book.

WHAT YOU WON'T FIND IN THIS BOOK

This text is not a Revit Architecture help manual or command dictionary. Its scope extends to all aspects of design in Revit Architecture—sketching, mass modeling, conceptual design, design development, construction documentation, data export and illustration. Space limitations prevent us from covering every tool available in the software, or all the possible control options for the tools and methods we do illustrate. Where appropriate, the text shows or mentions alternate ways of accessing tools or completing tasks.

STYLE CONVENTIONS

Text formatting and style conventions used in this book are as follows:

Text Element	Example
Step-by-Step Tutorials	1.1 Perform these steps
Menu Selections	Select File>Import/Link>RVT
Keys you press are in SMALL CAPS	Press CTRL or ESC
User input is in **bold**	Change the Eye Elevation value to **55' 6" [16650]**
Metric Units are contained in brackets	[1800]
Files and Paths are shown in *italic*	*Content\Imperial Library\Doors*

HOW TO USE THIS BOOK

The order of chapters has been planned to follow normal design firm flow and process. Chapters 1 and 2 introduce information of particular value for instructors, firm principals and CAD managers, although the exercises presume no extensive design experience. If you are new to Revit Architecture, we recommend that you complete the exercises in order. If certain chapters do not pertain to your work or your firm's work, feel free to skip topics as you see fit.

FILES ON THE CD

Certain tutorial exercises use startup files. These files are on the CD enclosed with this book. Users can load files from the CD into convenient locations on their own systems, or as specified by their instructor(s). Revit Architecture project files are on the CD in unzipped *rvt* format; Family files are in *rfa* format. Additional supplement files are also in native format (*doc*, etc.).

WE WANT TO HEAR FROM YOU

We welcome your suggestions and comments regarding *Introducing and Implementing Autodesk® Revit® Architecture*. Please send your correspondence to:

The Revit Architecture Team

c/o Autodesk Press

Executive Woods

5 Maxwell Drive

P.O. Box 15015

Clifton Park, NY 12065

You can also visit us on the web at:

www.autodeskpress.com
www.archimagecad.com

REVIT ARCHITECTURE RELEASES AND SUBSCRIPTIONS

Autodesk has announced plans to make substantive new releases of Revit Architecture roughly once a year, with an interim (.x) release halfway through the cycle. The latest build of Revit Architecture will be available as a download from the Autodesk website, or you may request a free CD. Without a license, Revit Architecture works for a 30-day demonstration period. After that time, you need to purchase a subscription to save work or print.

The exercises in this book were tested with Revit Architecture 2008. While the authors have tried to coordinate exercises and illustrations with the latest available information, there may be inconsistencies between this book and the behavior of the software you have installed. Exercise files on the CD are in Revit Architecture 2008 format. Check the CD or websites for additional information not available at press time.

ABOUT THE AUTHORS

Lay Christopher Fox is an independent architectural drafter/illustrator, author and educator. He has co-authored two books on Architectural Desktop® (now AutoCAD Architecture) and one on Revit® Architecture prior to this text. He is the Revit Technical Editor for *AUGIWORLD Magazine,* the bi-monthly newsletter of Autodesk User Group International (AUGI®). Chris is now a resident of Australia, where he conducts university and corporate classes on Revit Architecture and Revit Structure, and leads implementation projects for firms of all sizes. He has written Autodesk Official Training Courseware for Revit Architecture and Revit Structure. He has been a speaker at Autodesk University (2003). He received his undergraduate degree in literature at Harvard, and has pursued technical training in Rochester, NY, and Adelaide, South Australia.

Jim Balding is a licensed architect with more than 17 years of experience integrating technology into the architectural field. He is currently employed with Wimberly Allison Tong & Goo (WATG) in Newport Beach, CA. Jim earned his Bachelor of Environmental Science degree from the University of Colorado, Boulder. He has been a member of the Autodesk® Revit Architecture® Client Advisory board since its inaugural meeting and is currently serving as the Revit Architecture Product Chair for AUGI®. He is also currently serving as the South Coast Revit Architecture Users Group (SCRUG) president. Jim has spoken at several technology conferences and is one of the top-rated speakers at Autodesk University. He has developed a successful Autodesk Revit Architecture implementation strategy and is currently bringing the seven offices of WATG up to speed in its use.

ACKNOWLEDGEMENTS

The authors would like to thank all of those who have contributed to this project and to our personal and professional development over these many years.

First, thanks to the editors at Delmar/Thomson Learning/Autodesk Press: Jim Devoe, John Fisher, Sharon Popson, MaryBeth Vought, MaryEllen Martino, and Katherine Bevington. Particular thanks to John Fisher for his patience and understanding.

Special thanks to those who reviewed the text and exercises:

Michael Gatzke—Des Moines Area Community College, Ankeny, Iowa

John Knapp—Metropolitan Community College, Omaha, Nebraska

Robert Mencarini—Autodesk

The authors would also like to acknowledge the help of others who took the time to review the book in its formative stages and assist with feedback. You went above and beyond the call of duty:

Scott Davis, Greg Cashen, Justin Kelly, Carl Walls

The international Revit Architecture community of users is lively and committed to mutual growth. Recognition should go to Chris Zoog for starting a thriving web discussion group where users new and experienced share comments, concerns and content. This forum has since moved to the site of Autodesk Users Group International (www.augi.com), along with the Revit Users Group International. Revitcity (www.revicity.com) is another source for shared content, forums and general information regarding Revit Architecture. Many discussion groups exist at Autodesk's discussion group site, news.autodesk.com. Blogs on Revit are appearing all the time.

Jim Balding not only wrote a significant part of the book's text, but he worked as Technical Editor on the rest of the book. Little did he know what he was getting into—many thanks from Chris.

Chris would like to make particular acknowledgement to John Clauson and David Harrington of AUGI, who encouraged me to keep writing newsletter articles for years, and Elise Moss (also of AUGI), who generously and graciously let me assist her on a number of book projects.

Jim would like to thank the many people that have had an influence on his career as an architect and leader in architectural technology. Many thanks to: Chuck Cutforth, Ed Conway, M. G. Barr, Gerald Cross, Curt Kamps, Charlie Wyse, Jim Grady, Perry Brown and last, but certainly not least, Larry "Figurehead" Rocha. Tom (Buck) and Collette, thank you very much for all of your love, support and understanding while I was under your wings. I love you both very much.

Finally, we both would like to recognize the Revit Architecture Team—Leonid Raiz, Irwin Jungreis, Marty Rozmanith, Steve Burri, Dave Heaton, Rick Rundell, David Conant, the development and support staff and the hard workers too numerous to name at the Factory. Your pride and inspiration shine through, and do you honor.

DEDICATIONS

This book is dedicated to my wife, Diane, and sons, Judson and Luke. Without their understanding and patience there would be no book at all. Thank you for saying, "That's okay, we can do it another time," when there was too much work and too little time to do everything. I love you.

—Jim

This book is dedicated to my wife, Sally Goers Fox, a bright-eyed, intrepid bird of passage, with all my love and strength forever. Coming in on a wing and a prayer...

—Chris

CHAPTER 1

The Very Basics

INTRODUCTION

This chapter addresses the concerns of architectural design firms and technical schools or university architectural departments that seek to meet the current and future needs of the architectural design marketplace through the adoption and implementation of Revit® Architecture.

Students and designers new to Building Information Modeling will find the summary and "quick tour" exercise in this chapter a helpful introduction to the basic modeling and documentation tools in Revit Architecture, in preparation for more advanced concepts in later exercises.

OBJECTIVES

- Understand Revit Architecture's position as an architectural design tool
- Prepare for successful implementation of Revit Architecture in your current work process
- Experience an overview of the software

BUSINESS SUMMARY

WHAT'S REVIT ARCHITECTURE?

Autodesk, Inc. is the world's premier producer of mid-price design software. Its general-purpose AutoCAD drafting package is the world standard. It has sold millions of copies worldwide, and there is an entire industry of third-party developers dedicated to creating task-specific or industry-specific design applications based on the AutoCAD engine. Autodesk has developed and successfully marketed "Desktop" vertical applications to push AutoCAD into mechanical, architectural, civil and land-development specialties.

Architectural Desktop now AutoCAD Architecture has been renamed AutoCAD Architecture, with approximately 250,000 licenses worldwide. In April of 2002, Autodesk acquired Revit Technologies, Inc. (formerly Charles River Software), whose sole product was Revit, a parametric building modeler package. "Revit" is taken from architecture's slang for "revise it," a nearly constant refrain in building design. Autodesk has established Revit as its parametric 3D building design and documentation system and its strategic platform for software serving the building industry. The Revit platform now includes Revit Architecture, Revit Structure and Revit MEP (for Mechanical, Electrical and Plumbing). Why the change from AutoCAD, and what does it mean for architectural design firms using AutoCAD and Architectural Desktop (now AutoCAD Architecture)?

Autodesk's move away from the DWG format (AutoCAD's output file) for architectural models is consistent with a previous transition the company made in the mechanical design field. Autodesk developed Inventor, a parametric modeler of 3D solids that uses file formats incompatible with the DWG, and has used it to replace the previously successful and widely accepted Mechanical Desktop. With Revit, Autodesk purchased a product (and company) rather than attempt to replicate the effort that had been put into Revit by its developers. Here again Autodesk has been consistent. AutoCAD Architecture itself is an outgrowth of a purchase—Autodesk bought out Softdesk software in the mid-1990s. Softdesk had developed AutoArchitect, an add-on package to AutoCAD for architectural documentation. Autodesk reworked AutoArchitect and released it as Architectural Desktop now AutoCAD Architecture; other packages Softdesk had developed turned into Autodesk's Building Systems.

Autodesk executives have recognized the limitations inherent in the DWG format, whether for coordinating large mechanical assemblies or the complexities of building design and construction documentation. The results of their aggressive search for the next phase of design software have been the development of Inventor and the purchase of Revit.

Firms using AutoCAD or AutoCAD Architecture, that may have a huge inventory of DWG files and an ongoing commitment to developing/purchasing/applying peripheral applications to enhance parts of their work process such as schedules or space planning, will undeniably face a transition period while they come to grips with this change in the tools of the trade. This book is designed to show what changes actually lie in store for firms and institutions that decide to embrace the new shape of architectural design, how to manage and make best use of the parametric capabilities of Revit Architecture, and what job skills, office practices and client benefits are likely to develop in answer to the new conditions.

If you are reading this book because you have come to the conclusion that maintaining your company or school's current investment in AutoCAD is not as important as learning to make effective use of the next generation of tools, then you understand the costs of doing business and are willing to pay the price of keeping ahead of the pack rather than the price of falling behind.

SIMPLICITY MEANS SPEED

One of the biggest differences between Revit Architecture and any other CAD architectural design software currently in widespread use lies well under its surface, but shows up in the entire interface and organization of its information handling. Revit Architecture is a complete architectural design software application with drafting components, not drafting software that has been modified or expanded to make architectural design possible. This means a number of major differences from CAD packages, AutoCAD in particular.

- Every object created in Revit Architecture is classified according to its function (walls, doors, structural members, components such as furniture, etc.); the relations between objects are also based on architectural function. Walls "host" doors and windows and the necessary openings, for example. Geographic positioning, size and other properties of objects can be explicitly controlled by constraints or made subject to parameters, so that relationships can be established which will survive changes. A door can always center on a given wall, and windows on either side can be equidistant from one another, no matter how long the wall or the size of the windows. This eliminates many steps checking and rechecking dimensions during design iteration (as shown in Figures 1.1 and 1.2).

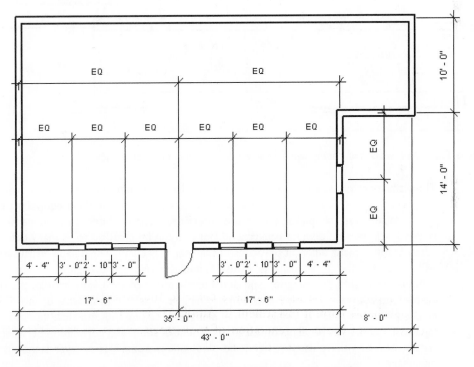

Figure 1.1   *Doors and windows spaced equally in walls of certain lengths*

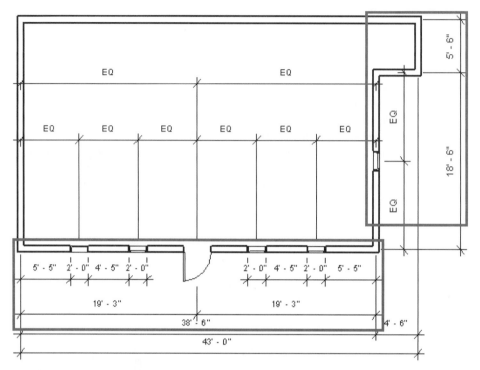

Figure 1.2 *Drag walls and change window size—the equal spacing still holds*

- The visibility of objects is controlled in views (plan, elevation, section, 3D) according to properties of the views and the objects. There are no layers or their equivalent used, which means vast simplification of the interface. Revit Architecture's appearance is largely black lines on a white background, but this can be changed to as many colors or patterns for as many different purposes as the user may desire. The default visual settings will suffice for the great majority of users, and the visible lineweights correspond to printed lineweights. The function and detail level setting of each view control the detail visible—no material symbols or section hatching in walls at coarse scale or in reflected ceiling plans, but hatch is visible in floor plans at medium and fine scale (as shown in Figures 1.3 and 1.4). This eliminates a large amount of setup and correlation between the screen view and the plotted results—no plot configuration files or layer standards are necessary.

Parameter	Value
View Scale	1/8" = 1'-0"
Scale Value 1:	96
Display Model	Normal
Detail Level	Coarse
Visibility	Coarse
Model Graphics Style	Medium
Advanced Model Graphics	Fine

☐ Coarse
☑ Medium
☒ Fine

1/8" = 1'-0"

Figure 1.3 *The view detail level control in the View Properties dialogue, and on the View Control Bar*

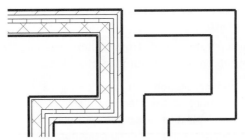

Figure 1.4   *The same walls shown in plan—medium detail level in the left view, coarse detail right*

- The views in Revit Architecture are extremely powerful—a fundamental organizational principle. All come from the existing model, so changes in the model are automatically propagated throughout all possible views of it, as shown in Figure 1.5. An elevation will always contain exactly the number of windows in the exact location along a wall that the plan shows, and a wall section will always show the correct components of the wall no matter how many times the wall type changes. No manual updating procedure requires the user's attention. This saves significant amounts of design time and eliminates chances for error over the course of a project.

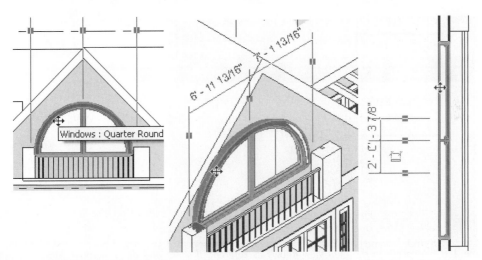

Figure 1.5   *Elevation, 3D and plan views. A change to the model in one view (note the window currently selected) shows immediately in the other views.*

- Revit Architecture includes "traditional" line-arc-circle drafting tools and editing commands, which are used, as one would expect, for sketching custom components or 2D detailing. There is no visible coordinate system in Revit Architecture, but a system of work planes on which one drafts or sketches. These work planes utilize reference planes that can be governed by parameters when creating components, so part of a column can always be twice as wide (or as long) as another part, or the column will always run from its base level to the one above (plus or minus an offset value if desired) even as floor heights change. Design intent can be specified to carry through the iterative process that is architectural design.

Revit Architecture also separates itself from other available architectural applications, including AutoCAD Architecture, in a number of ways:

- Plotting sheets and annotations are smart—Figure 1.6 shows how a plan tag will renumber itself according to the sheet and detail number of its reference as the sheet package develops, eliminating a source of much error and time spent checking annotations in the latter stages of documentation.

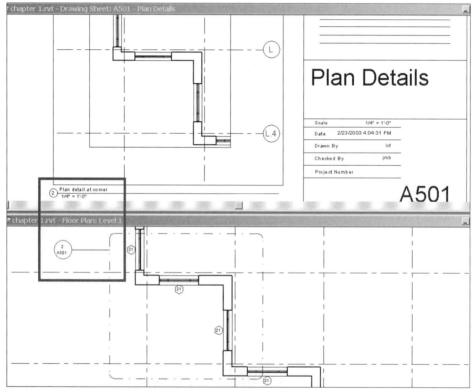

Figure 1.6 *A callout in the plan view knows its referenced sheet and detail number. Move the detail to another sheet, and the plan symbol updates immediately wherever it is visible.*

- Nearly any object with parametric properties can be scheduled, and nearly all objects are or can easily be made parametric. Schedules (which are views) are live and reflect changes in quantities or properties.

- Room and area tags are easy to create and apply, and feed directly into space planning and energy analysis calculations, schedules and color fill views.

- Revit Architecture includes a rendering engine that makes lighted, colored views, with material maps applied, possible at any time in the design process, for any part of the model. Still views and animated walk-throughs are quick to generate. Early visualizations of conceptual models are so easy some Revit Architecture users have resorted to subterfuge so as not to respond "too quickly" to client requests for renderings.

All this boils down to greater ease of use for more users than design systems based on drafting methods. There have been totally object-based architectural design packages available before, but they were not as easy to use, nor did their interfaces provide the ready visual understanding and ease of plotting that Revit Architecture does, and those applications are no longer viable in the market.

QUICK PRODUCTIVITY

Revit Architecture's entire system has been carefully crafted to provide a design environment that architects will readily understand and find easy to use—all software strives for the elusive goal of providing an intuitive experience for the user, and Revit Architecture, in its basics, comes very close to the ideal.

As a result, designers who may not be CAD-literate can expect to create models, edit them, change their appearance, and create specifications in considerably less time than it takes to learn a drafting package and use that drafting package to create a building design. AutoCAD Architecture requires facility with AutoCAD—Revit Architecture does not, and much productive time can ride on that simple difference.

Experienced CAD designers who are not disconcerted by the lack of layers to control, or by the lack of X-Y-Z coordinates to enter when defining the location of an object, can find themselves productive in a surprisingly short training cycle. Competent architectural drafters who are not simply CAD jockeys will find their drafting skills as much in demand as ever. There is a world of 3D content yet to create for this growing program.

SIZE AND STRENGTH MATTER

Revit Architecture is a robust application, capable of generating complex and complete building models of considerable size. Its hardware requirements are substantial but fully in line with those of current CAD workstations. Revit Architecture is designed to hold a single building model in each file. It contains a number of mechanisms (Worksets, Groups, Copy/Monitor) that allow teams to parcel out and work on parts of a project and correlate their updates. File linkages (corresponding to external references in AutoCAD-based products) allow for assemblies of models or campus layouts combined in a single file location.

Since Revit Architecture contains a rendering module, architects can produce detailed presentation images without leaving the application. The controls of the AccuRender module within Revit Architecture are easy to learn, production and capture of useful images or animations is quick, and no updating of the model or file linking is necessary. For firms with a separate graphics department apart from building design teams, Revit Architecture models export readily into DWG format for import into VIZ. The latest release of VIZ now recognizes most Revit Architecture object properties.

Autodesk's position in the global market as a premier supplier of design and visualization applications provides a stable development and support environment for Revit Architecture. The original Revit application has become a worldwide platform standard. Three compatible programs (Architecture, Structure and MEP) now cycle yearly releases together.

Architecture is a notoriously conservative industry. As late as the mid-1990s nearly half of the architectural firms in the United States had next-to-no investment in CAD technology in their offices. When AutoCAD was Autodesk's flagship design product, each new release of the software occasioned hand-wringing and consternation as firms agonized over the decision whether to keep CAD stations and standards current or not. Now that the software decision before today's firms involves a new file format, different design techniques, changing office roles and fluid client relationships, the basic question is actually simpler than before.

Autodesk stated from the moment it acquired Revit that this application is its building modeling solution of the future. The marketplace is responding; sales have doubled consistently for years. The transition is underway from the only CAD standard many architectural firms have ever known, and the myriad third-party applications developed by an entire industry devoted to extending AutoCAD and the DWG, to a parametric 3D modeling environment.

Autodesk has substantial resources to help firms making a commitment to Revit Architecture understand and manage the transition. Autodesk is a company dedicated to developing and maintaining a growing stable of industry-leading design products. The company takes its leadership position seriously enough to know the dangers of out-pacing the market. Revit Architecture is Autodesk's proposition to those architectural firms that also take leadership seriously.

NO LONGER THE NEW KID ON THE BLOCK

Whether your firm or institution is an aggressive front runner or a wait-and-see trend adopter, you should take some issues into account in your transition and implementation planning.

Tip: Autodesk makes Revit Architecture available as a free download from the www.autodesk.com/revit website, and will send a CD on request. Revit products will work in demo mode without a license, and for a limited time as a full function trial version. You will be unable to save work or print after the trial period expires until you purchase a subscription.

FILE COMPATIBILITY

Given the relative user base, the issue of file compatibility will be daunting to some firms, though it need not be a reason to avoid or delay exploring Revit Architecture's capabilities. Revit Architecture is built to interface easily with AutoCAD products. "Interoperability" is the word Autodesk uses to emphasize the two-way connectivity. Revit Architecture can link to AutoCAD files as easily as AutoCAD does. Revit Architecture imports and links to 2D DWG files with full layer, linetype and color information, and handles a growing variety of 3D solid information. Revit Architecture outputs 2D or 3D DWG files that are tidy and complete, with layers logically assigned.

TRAINING

Executives facing the decision to begin using Revit Architecture will need to budget for training as part of the implementation process. Revit Architecture ships with a series of tutorials and the company provides free live training sessions online, so for those individuals who can successfully self-teach using the provided lessons there will be no cost for materials or instruction. Autodesk will provide a training team for larger firms making an investment in Revit Architecture. Resellers and Autodesk Training Centers have begun providing scheduled classes and on-site instruction sessions by certified instructors.

PROJECT SIZE

Large building projects are difficult to manage no matter what software you use to design and document them, and with Revit Architecture's single-file building model, larger projects have proportional challenges. Your first projects in Revit Architecture should be relatively modest, perhaps in the 100,000 square foot range. With careful configuration of template files, work sharing and model linking, firms are presently doing projects in Revit exceeding a million square feet. Each release of Revit has improved performance in project size, and with Revit Architecture an experienced team can confidently tackle just about any building design. As we mentioned in the Author's preface, the Freedom Tower project for New York City, possibly the most complicated skyscraper proposal in the world today, is being developed in Revit Architecture, from the subway interchanges beneath the foundation to the electricity generating wind turbine matrix at the top of the tower.

Revit Architecture's file is based on a single building model that carries all the relevant information about the building structure and contents, rather than a system of individual files created per floor, for instance, and brought back by reference into a shell or assembly file. Revit Architecture's system of Worksets allows for a growing team to work on a model as a design develops by "checking out" parts of the model with a central file holding and coordinating data. This system has the advantage of simplifying what can grow into a bewildering nest of references in a large model.

DATA OUTPUT

Revit Architecture is very efficient at presenting building or component information in schedules. The data tabulated in schedules can be exported to a spreadsheet or database such as MS Access via ODBC, the Microsoft-developed process for sharing field information between applications. This is a limited and cumbersome process compared to the Data Extract function in AutoCAD, and while it allows for updating an external file based on changes in the Revit Architecture model, the process at present cannot be reduced to a one- or two-click operation, nor automated.

Revit's Export Schedule capability does allow for quick conversion of any scheduled information into *.txt* format for import into other applications, and holds promise of developing into a live link between the Revit Architecture model and external data-parsing applications.

PROGRAMMING INTERFACE

Revit Architecture does not contain an exposed programming interface, as does AutoCAD. One of AutoCAD's strengths from the very beginning was the decision early on to embed a LISP interpreter in the application and make this interpreter available to users. This allowed developers of all stripes to augment basic "vanilla" AutoCAD with routines of sometimes startling ingenuity to aid in creating CAD content, to automate repetitive tasks, and to read values from the drafted geometry. In recent years with the rise of Microsoft Visual Basic as a programming language common to many Windows applications, AutoCAD has developed Visual Basic for AutoCAD as a way for the application to interface with and react to other applications.

As a single-purpose building modeler rather than a general-purpose drafting tool, Revit Architecture does not have the same need for external hooks as AutoCAD, and it has been carefully designed to eliminate many of the repetitive tasks that AutoLISP is often used to speed up. All building model objects are complex, with properties that can placed into and read from schedules. Revit Architecture contains an API (advanced programming interface) that is available for members of the Autodesk Developer Network. There is, at present, no capability for the user to create scripts, macros, or toolbar customization within the program.

WHAT ELSE IS COMING?

Autodesk has made it clear that Revit is the company's platform for architectural building information modeling. The original program has become Revit Architecture. A second application, Revit Structure, with bi-directional links to AutoCAD Architecture, Autodesk Building Systems and external (third-party) analysis applications, appeared in the summer of 2005. Revit MEP (Mechanical, Electrical and Plumbing) that enables the Revit Architecture model to provide the same types of specialized output as DWG-based files, has gone through two iterations as of mid 2007.

PREPARE FOR SUCCESSFUL DESIGN OFFICE IMPLEMENTATION

SHOCK TROOPS

Only the foolishly optimistic would expect to proclaim a new company standard without working to overcome inertia and objections. Introducing new concepts and tools to a work arena is a process, not an event, and takes preparation and education even more than vision and enthusiasm.

Most firms have members who are eager to learn, willing to push the limits of their knowledge, ready to try new things. All firms have members who are most productive when comfortable with routine, not very interested in expanding their skills if risk is potentially involved, and slow to adapt to new circumstances. Quick and ready learners are not necessarily better designers or more useful team members than steady-as-she-goes types (and may not be the company's CAD experts), but can be valuable as "early adopters" who spearhead an effort to learn, utilize and broadcast an important shift in work standards.

- Assemble a team whose members are willing and capable of working with new software and techniques.
- Develop a plan for training.
- Set a schedule for individual and group training.
- Establish a forum for communication: team members to each other, to CAD manager and to project manager.
- Use an initial project as a focus for the training and development of team work practices.

PROJECT BY PROJECT

Decide beforehand on the first project for the spearhead team to work on in Revit Architecture. Considerations for this project may involve building size, client requirements and schedule.

- Understand your company's design process, including your current standards.
- Understand the scope of the project.
- Clearly define and communicate the project goals.
- Establish project and team evaluation criteria for later review.
- Expect and allow for differences in workflow from current company process.

REMEMBER YOUR ROOTS

Revit Architecture is built to take maximum advantage of AutoCAD—"interoperability" can be key. Revit Architecture reads AutoCAD files either via link (equal to external reference) or import, and will retain layer, color and linetype information. It will also export linework from its models to AutoCAD on a by-view basis. Its export DWG files use layer, color and linetype information per default built-in standards, which are fully customizable to your current company standards.

- Team members can import existing site data in DWG, DGN (MicroStation), or DXF format. Use Revit Architecture's topographic tools to build the 3D site surface.

- Link or import existing plans for reference in creating new project content.

- Scan in and trace over hand concept sketches.

- Link or import DWG/DGN/DXF files received from consultants.

- Export DWG/DGN/DXF files to consultants.

- Link or import standard details and symbols from company archives during CD phase.

EXPAND THE TEAM

Once a first project has been completed or nearly completed in Revit Architecture, the company can build on that experience to develop the use of Revit Architecture as useful enterprise software.

During post-project analysis from the first Revit Architecture project, let other members of the company know about the successes of the spearhead team, and be sure to evaluate it honestly and communicate problem areas. Was training adequate? Were expectations clearly expressed? Was support available in a timely way?

Target the next projects, divide the Revit Architecture team between the projects, and plan and schedule training for new team members. In this way, knowledge and experience with Revit Architecture can spread the most effective way—from user to user. People learn most efficiently when actually working with their new knowledge, and when there is a free flow of information—tips, tricks, questions and answers pertaining to work in hand.

STUDENT'S PULL-DOWN MENU

The exercises in this book are designed to give you a thorough grounding in the most powerful and complete building modeler on the market today. The exercises and review questions do not presume that you have any experience in architectural design or drafting. They do presume that you can follow step-by-step instructions explicitly and carefully. Wherever necessary, explanations will be provided as to why certain steps should happen in a certain order, so that you understand how this modeling program works most efficiently.

You should be aware by now that modeling is not the same as drafting, even though many views (plans, elevations, sections) in Revit Architecture may look just like plans, elevations and sections that you may have created before now in other design software applications. Building components are complicated objects with multi-part definitions that control how they appear, behave and interact with all other parts of a project model.

The ease with which designs can be developed and presented using Revit Architecture has changed the former definitions of "basic" and "advanced" skills. You will be presented with tasks such as file importing and setting various element properties along with instructions on creating basic wall layouts, for instance. This mix not only reflects the skills and usages you will need to master in your work life, but is designed to make you comfortable with the interface and controls that make Revit Architecture the useful application that it is.

QUICK TOUR—DESIGN, VIEW AND DOCUMENT A BUILDING

This book contains exercises and lessons that are substantially different from the tutorial lessons provided with the software. First up is a short exercise to give you a very quick look at Revit Architecture's most basic tools and integrated display capabilities.

REVIT ARCHITECTURE COMMANDS AND SKILLS

Walls

View Properties

Doors

Windows

Editing—Copy, Mirror, Element Properties

Dimensions—EQ toggle

Add Level

Filter Select

Floor

Roof

Section

Sheet—Drag and Drop Views

View Scale

Activate View

3D View

Shade View

EXERCISE 1. LAY OUT A SIMPLE BUILDING

1.1 Launch Revit Architecture from the desktop icon. (You can also start Revit Architecture by picking Start> All Programs>> Autodesk> Revit Architecture 2008.)

Revit Architecture will open an empty project.

The View Window (drawing area) will open to a Level 1 plan view, as shown in Figure 1.7.

Four elevation symbols will be visible in the View Window. Your elevation symbols may be round if working in a Metric environment.

Menus on the Menu Bar at the top of the screen and in the tabs of the Design Bar control modeling and other tools.

The Project Browser located between the View Window and Design Bar shows available views in customizable tree form.

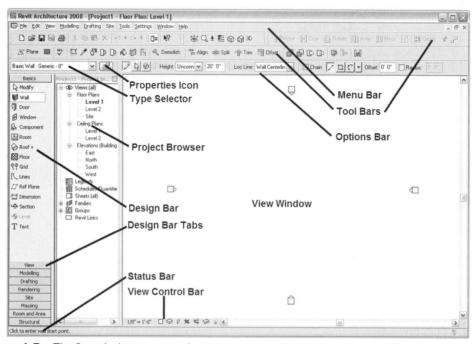

Figure 1.7   *The Revit Architecture interface*

1.2 From the Basics tab on the Design Bar, pick Wall.

The cursor will change appearance to a pencil symbol and the Type Selector drop-down box on the Options Bar will display a generic wall type, as shown in Figure 1.8.

Figure 1.8   *The Wall tool on the Basics tab*

1.3 Click on the down arrow at the right of the Type Selector, as shown in Figure 1.9, to expand the list.

Figure 1.9 *Expand the list of default wall types*

1.4 Scroll up the list and pick (Imperial) Basic Wall: Exterior – Brick and CMU on MTL Stud or (Metric) [Basic Wall: Exterior – Render on Brick on Block], as shown in Figure 1.10.

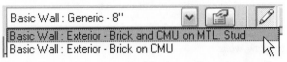

Figure 1.10 *Wall types; these two are compound horizontally and vertically*

 Note: If you are working in a regionalized project template you may not have vertically compound wall types available by default. You can work the exercise using any wall type; elevation and 3D views may not display the same complexity of wall surfaces as the ones listed in Step 1.4.

1.5 Accept the default Height and Location Line values on the Options Bar.

- Check the Chain Option so that your picks will draw walls continually until you stop the command.

- Leave the Straight Line tool depressed.

- Leave the Offset value at **0' 0" [0.0]** (see Figure 1.11).

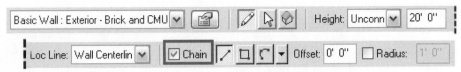

Figure 1.11 *Settings for the walls; select the Chain option*

1.6 Place the cursor in the View Window left of center and above the middle, but inside the elevation markers.

- If you pause the cursor a Tooltip will appear, reading Click to enter wall start point, as shown in Figure 1.12. This message will also appear in the Status Bar at the bottom of the screen.

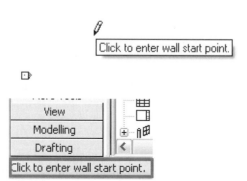

Figure 1.12 *About to draw walls*

1.7 Left-click to start the wall.

- Pull the cursor to the right.

  This will create the wall with its exterior face to the top of the screen. A temporary dimension will appear showing the length of your wall, and an angle value.

  If you pull directly right as shown in Figure 1-13, a dashed green reference line will appear and a Tooltip will verify that the wall is horizontal.

- Pull the cursor until the dimension reads **40'-0" [12000.0]** and left-click to set the wall endpoint.

Figure 1.13 *The first wall, drawn left to right*

1.8 Pull the cursor down to start the second wall.

- If you need to zoom in closer to see your work, right-click to get a context menu with zoom options, and pick Zoom In Region.

- Then pick two points as appropriate, as shown in Figure 1.14.

  You can also use the scroll wheel if your mouse is so equipped.

Figure 1.14 *Right-click to access zoom controls*

As you pull the cursor down, the temporary dimension, angle readout, reference line and Tooltip all appear, as shown in Figure 1.15.

1.9 Make this wall **40' [12000]** long as well, and then left-click.

 Note: The actual dimensions of the walls we are creating in this exercise do not matter. Nor do we really care at this point if the walls are perpendicular to one another. Revit Architecture is designed for quick sketching and contains many tools to establish/revise dimensions and geometric relationships of building elements later in the design process. Ironically, this can be a point of confusion for skilled drafters. These tools will be illustrated extensively in later lessons.

1.10 Pull the cursor to the left **40' [12000]** and left-click.

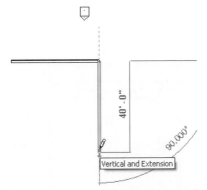

Figure 1.15 *The second wall is perpendicular to the first*

Revit Architecture will pick up alignment points, snap to them, and display them in the Tooltip, as shown in Figure 1.16.

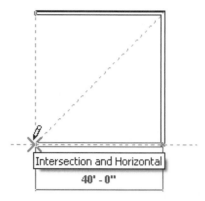

Figure 1.16   *The third wall ends even with the first one*

    1.11 Pull the cursor up and Revit Architecture will snap to your start point.

- Click to finish this simple exterior (see Figure 1.17).

- To exit the Wall tool, hit ESC twice, or pick Modify from the Design Bar, as shown in Figure 1.18.

    This puts the cursor back into select mode.

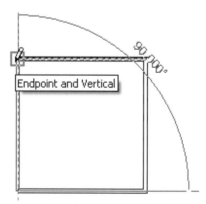

Figure 1.17   *Finish the outside walls*

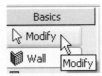

Figure 1.18   *The Modify command is at the top of all the Design Bar tabs. Use it to terminate many command routines.*

1.12 Put the cursor in the View Window and right-click.

The bottom option is View Properties, as shown in Figure 1.19.

- Select (click) View Properties to bring up Revit Architecture's main visibility control, as shown in Figure 1.20.

Figure 1.19   *Select View Properties from the right-click menu*

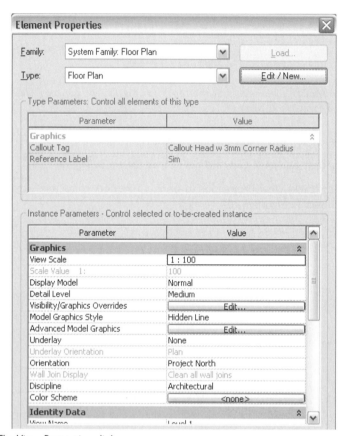

Figure 1.20 *The View Properties dialogue*

1.13 Click in the Value field for Detail Level and change it from Coarse to Medium, as displayed in Figure 1.21. In a metric file, this value may already be set.

Detail Level	Coarse
Visibility/Graphics Overrides	Coarse
Model Graphics Style	Medium
Advanced Model Graphics	Fine

Figure 1.21 *Three levels of view detail*

1.14 Click the Edit button in the Visibility value field to bring up the Visibility/Graphic Overrides dialogue. It has three tabs.

- Click on the Annotation Categories tab and clear Elevations as shown in Figure 1.22. This will turn off the Elevations in this view to simplify it.

- Click OK twice to exit the Visibility and View Properties dialogues to reveal the model.

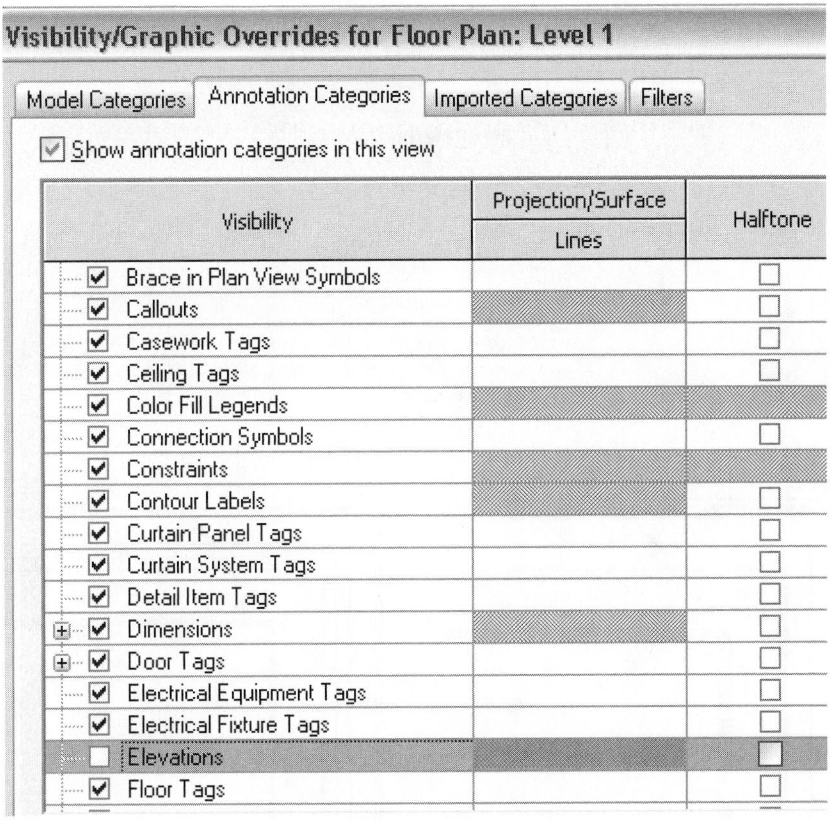

Figure 1.22 *Turning off the elevation symbols in this plan view*

1.15 Right-click the context menu and pick Zoom to Fit, as shown in Figure 1.23.

Figure 1.24 shows a portion of the adjusted view, a complex wall structure with hatching.

Figure 1.23 *The quickest way to zoom to the edges of the model*

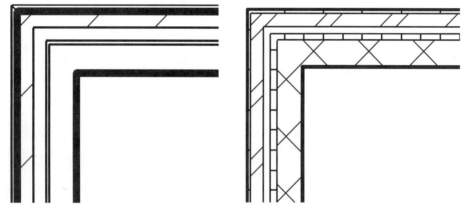

Figure 1.24 *The walls now display complex structure with hatching: Imperial – left, Metric – right*

1.16 From the Menu Bar, choose File>Save and save the project as *quicktour.rvt* in a convenient place or as instructed by your trainer.

File saving options will be explained in detail in later exercises.

EXERCISE 2. ADD DOORS AND WINDOWS

2.1 Pick Door on the Basics tab of the Design Bar.

- Pick Single Flush: 36" x 84" [M_Single-Flush: 0915 x 2134mm] from the available choices in the Type Selector.

- Locate one in the left wall 4' [900mm] down from the upper wall.

Revit Architecture will snap to the walls and read the wall location as you move the cursor. Holding the cursor to the interior or exterior side will set the swing in or out.

- Press the spacebar to toggle the hinge side.

- Set the door so it swings out with the hinge to the top, as shown.

  A door tag will appear with the door.

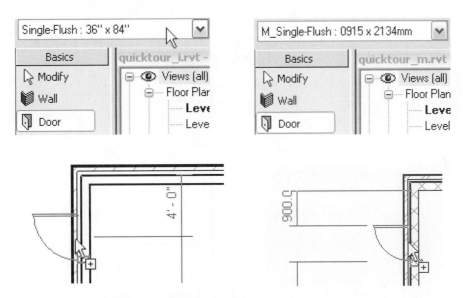

Figure 1.25   *Locate an exit door near the upper wall*

2.2 Set a second door 4' [900mm] up from the bottom wall.

If you need to adjust the swing of a door, control arrow sets appear when it is selected.

- Click the appropriate control arrow to toggle the swing or hand, as shown in Figure 1.26.

  You can do this while in the door placement routine.

  When modifying doors, the cursor changes to the circle/slash "not allowed here" symbol to indicate that clicking will not place a door instance. You can click on the control arrow symbols to activate them.

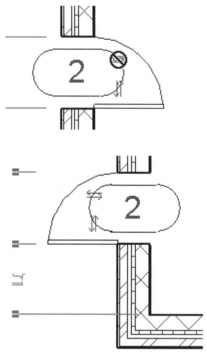

Figure 1.26 *Locate a second door and adjust the door swing if necessary*

The default list of doors does not include a double door.

2.3 While the door tool is active, click the Load button on the Options Bar to bring up a content file browser dialogue, as shown in Figure 1.29.

- Select the *Doors* folder in the *Imperial [Metric] Library* folder, and then pick Open.

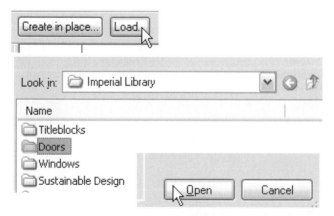

Figure 1.27 *Look for content from the library. In this dialogue you can create your own folders, or browse the web.*

2.4 In the library *Doors* folder, select *Double-Glass 1.rfa [M-Double-Glass1.rfa]*, as shown in Figure 1.28. Click Open.

You can now place different size doors of this type.

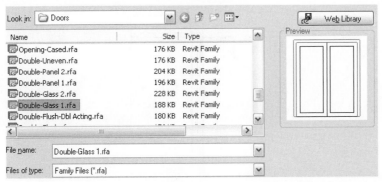

Figure 1.28 *The double glass door type selected. Note the web link above the preview window.*

2.5 Locate an instance of the double glass door midway between the other two doors (Revit Architecture will snap to their centers or edges, depending on settings), as shown in Figure 1.29.

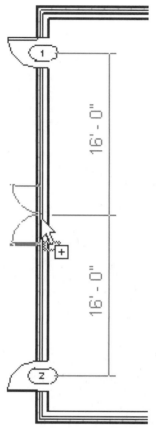

Figure 1.29 *Three doors in the west wall*

2.6 Pick the Window tool button in the Basics tab on the Design Bar.

 Note: You can switch from one tool to another (door placement – window placement) without having to terminate an operation explicitly. To quit placing doors and windows without starting another tool immediately, use esc or click on Modify, at the top of each Design Bar tab.

2.7 Select Fixed: 24" x 48" [M-Fixed: 0610 x 1220mm] from the list of available windows.

2.8 Locate an instance of this window in the west wall, midway between the lower and middle doors, as shown in Figure 1.30.

- Hold the cursor to the exterior side of the wall to set the window with its exterior face at the left (exterior) side.

  A tag will appear with the window.

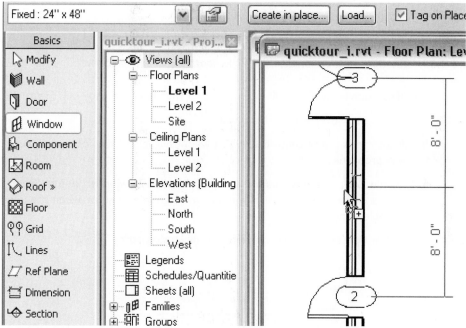

Figure 1.30 *The first window, equidistant from the door opening centers; the metric file will use faces*

2.9 Add a second window midway between the center and upper doors.

2.10 Change the window selection in the Type Selector to **Fixed: 36" x 72" [M_Fixed: 0915 x 1830mm]** and place 5 instances in the lower wall.

Do not be concerned about their location during placement.

- Use the control arrows if necessary to keep the glazing on the exterior side of the wall. See Figure 1.31.

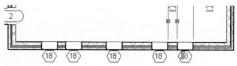

Figure 1.31 *Place windows; you will set their location in the next steps*

EXERCISE 3. EDIT THE MODEL

3.1 Pick Dimension on the Basics tab of the Design Bar, as shown in Figure 1.32.

Figure 1.32 *The Dimension command, used for locating components precisely*

The dimensioning symbol will appear next to the cursor, as shown in Figure 1.33.

3.2 Hold the cursor over the left-hand wall.

It will snap to the wall centerline, which will highlight, and the Tooltip will read out the wall properties.

- Pick the wall, and then hold the cursor over each window in turn from left to right—a center mark for each window will highlight to show when the window is selectable—and click.

- Pick the right wall last to complete the dimension string.

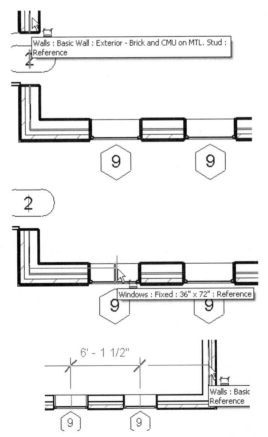

Figure 1.33 *Starting the dimension string—pick the wall, then windows left to right*

3.3 Pull the cursor above the wall, inside the building, and left-click to place the dimension string.

After you pick, the grips for the dimensions will appear, and the letters EQ with a slash through them (see Figure 1.34).

• Pick the EQ symbol—it is a toggle to lock the objects dimensioned in an equidistant relationship, as shown in Figure 1.35.

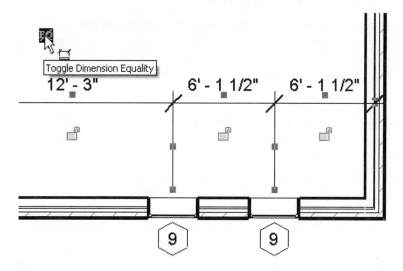

Figure 1.34   *Setting the EQ toggle*

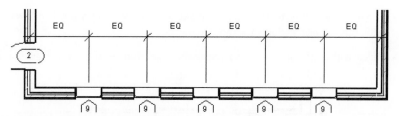

Figure 1.35   *The results of the EQ toggle*

3.4 Pick Modify from the Basics Tab of the Design Bar to terminate the Dimension tool.

3.5 Select the dimension string.

• Press the Delete key.

A Warning dialogue will appear.

• Click OK, as shown in Figure 1.36.

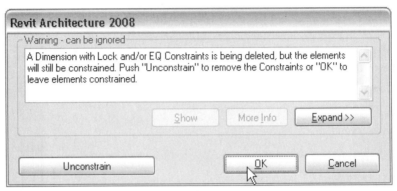

Figure 1.36 *You can delete a dimension and still use its constraint*

In the Select mode, where the cursor appears as an arrow, Revit Architecture allows you to select more than one entity by holding down the CTRL key while picking. This builds a selection set upon which you can perform an action. Selected items will appear in red. To remove an item from a selection set, hold down the SHIFT key and choose it.

Revit Architecture follows the Autodesk standard for selection windows—picking a point in the view window **not** over an entity and holding down the cursor left button starts a selection window with different properties depending on its orientation. Dragging **left to right** starts an enclosing window that selects only entities completely enclosed within its border, which is a solid line on the screen. Dragging from **right to left** starts a crossing window that selects any entity within or touching its border, which is a dashed line.

3.6 Place the cursor below and to the left of the left-hand window.

- Left-click and drag to the right to make a window around the windows in the lower wall, but do not include the entire wall.

- Release the cursor button to make the selection, as shown in Figure 1.37.

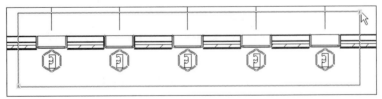

Figure 1.37 *Windows highlighted in the enclosing window*

3.7 When the windows are highlighted, pick the Mirror command on the Toolbar, as shown in Figure 1.38.

Figure 1.38   *The mirror command*

3.8 Pick the center of the center door in the left wall to define the mirror line, as shown in Figure 1.39.

A reference line will appear to assist your pick. Windows will appear in the upper wall.

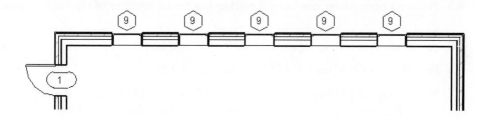

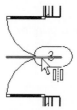

Figure 1.39   *Use the door as the mirror midpoint to create windows in the upper wall*

3.9 Save the file.

EXERCISE 4. ADD ANOTHER LEVEL, UPPER WINDOWS AND A FLOOR

4.1 In the Project Browser, find Elevations (Building Elevations).

- Double-click on West to open up that view.

4.2 Place the cursor in the view window.

- Right-click and pick Zoom in Region.

  The cursor changes to a magnifying glass.

- Click and drag around the left door.

- Note the appearance of the vertically compound wall structure at close range.

4.3 Right-click and pick Previous Scroll/Zoom (or Zoom to Fit) to restore the entire elevation on the screen.

4.4 Pick the Level tool on the Basics tab of the Design Bar, as shown in Figure 1.40.

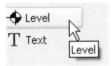

Figure 1.40   *The Level tool*

4.5 Place the cursor over the Level 2 marker line at the left side of the building face.

Revit Architecture will snap to it and read out an offset dimension as you move the cursor up, as shown in Figure 1.41, so long as you keep the cursor over the model.

4.6 Pick to start a new level **8' [2400]** above Level 2.

You can type **8 [2400]** to set the vertical offset at **8' [2400 mm]** if you have any problem getting the temporary dimension to read the desired distance.

- Pull the cursor to the right of the building.

- Left-click and the new Level 3 will appear.

- Note that the Project Browser now shows a floor plan and ceiling plan for Level 3.

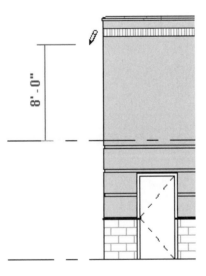

Figure 1.41   *The new level started*

4.7 Hit ESC or pick the Modify button to terminate the command.

- Zoom to Fit in the view.

4.8 Select the dashed line for Level 2.

- Select its left end as shown in Figure 1.42 and drag it closer to the left side of the building model.

  The end of Level 1 will move with it. The padlock icon is a toggle to lock or unlock the alignment of level ends with one another.

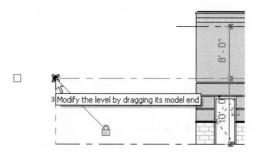

Figure 1.42 *Select the left end of Level 2*

4.9 Repeat this adjustment for the right ends of Levels 1 and 2, and for the left end of Level 3, as shown in Figure 1.43.

Snap lines will appear as components align. This adjustment will make later work with this view easier.

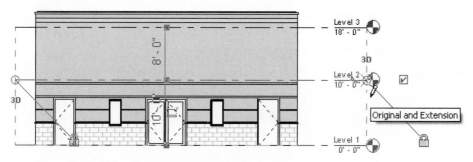

Figure 1.43 *The West Elevation with Level lines brought close to the model; local templates will show different level symbols*

4.10 Use a window pick box as before (pick left to right) to select the windows and middle door, as shown in Figure 1.44.

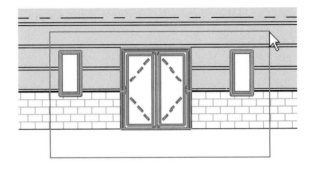

Figure 1.44 *Select windows and a door*

4.11 Choose the Filter Selection icon from the Options Bar.

- In the dialogue that opens, clear Doors and pick OK, as shown in Figure 1.45.

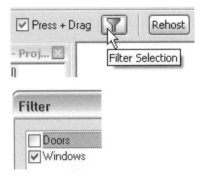

Figure 1.45 *The Filter Selection tool can be much quicker than picking individual objects*

The two windows will remain highlighted.

4.12 Choose Copy from the Toolbar, as shown in Figure 1.46.

Figure 1.46 *Copy the two windows*

Revit Architecture will need a start point and end point to define the relative distance and direction for the copy.

4.13 Select the left end of Level 1, as shown in Figure 1.47.

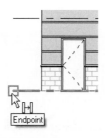

Figure 1.47 *Start the copy by referencing Level 1*

    4.14 Pull the cursor straight up until it snaps to Level 2 and click, as shown in Figure 1.48.

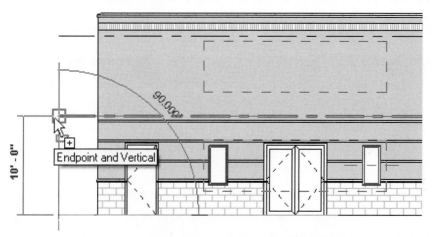

Figure 1.48 *Select Level 2 to make copies of the two windows one level up from the originals*

New windows appear above the originals, as shown in Figure 1.49.

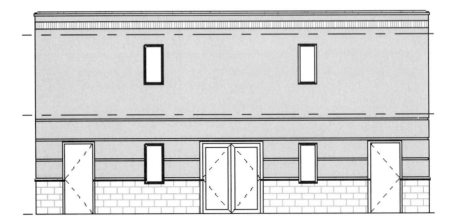

Figure 1.49 *New upper story windows aligned with the ground story windows*

4.15 Double-click Floor Plan Level 2 in the Project Browser to open that view.

- If your local floor naming convention uses Level 1 for the first upper story, double-click that Floor Plan name. See Figure 1.50.

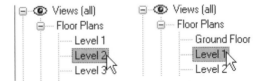

Figure 1.50 *Use the Project Browser to move between views*

Working in different views is an important way to take advantage of Revit's three dimensional modeling. Each level of a multi-story file has its own plan view by default, and the level forms the work plane of the view. Objects created or placed in a plan are based at the level of that view.

4.16 Pick the Floor tool on the Basics tab of the Design Bar, as shown in Figure 1.51.

Figure 1.51 *The Floor tool*

The walls in the View Window will gray out. The Design Bar will go into Sketch mode, with the Pick Walls tool activated. The Status Bar reads Pick walls to create lines.

4.17 Leave the Offset value at **0' 0" [0.0]** and the Extend into wall (to core) option checked on the Options Bar, as shown in Figure 1.52.

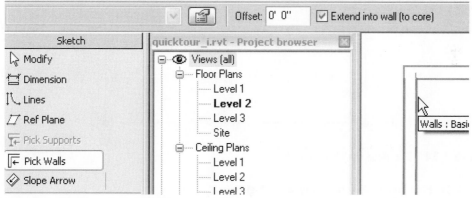

Figure 1.52   *Floor Options*

4.18 Hold the cursor over the interior surface of the left wall—the wall will highlight and the wall properties will appear in a Tooltip and in the status bar, as shown in Figure 1.53.

Figure 1.53   *Ready to pick a wall in Floor Sketch mode*

4.19 Pick the wall and a magenta sketch line will appear on its interior surface.

- Use the control arrows that appear to adjust the sketch line position to the interior side if necessary.

4.20 Pick the remaining three walls to complete a sketch of the floor edges, as shown in Figure 1.54.

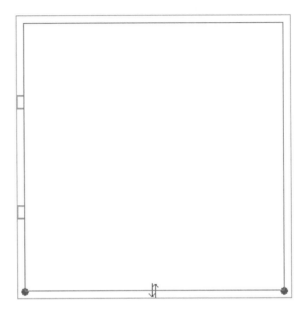

Figure 1.54 *The floor sketch is complete*

    4.21 Pick Finish Sketch, as shown in Figure 1.55.

        • Pick Yes in the question box that appears, so that the floor structure will meet the wall structure and finishes will be trimmed back appropriately in section views.

        When sketch mode completes, the Basics tab will reappear in the Design Bar. The View Screen will not change appearance.

Figure 1.55 *Finish Sketch*

    4.22 Save the file if you intend to take a break.

EXERCISE 5. ADD A ROOF AND VIEW IT IN SECTION

    5.1 Double-click on Level 3 in the Project Browser.

        • Pick the Roof>>Roof by Footprint tool in the Design Bar, as shown in Figure 1.56.

Figure 1.56   *Roof by Footprint*

5.2 Revit Architecture shifts into Sketch mode, as with the Floor tool.

This will be a flat roof below the parapet cap on the exterior walls.

- On the Options Bar, clear Defines Slope and check Extend into wall (to core), as shown in Figure 1.57.

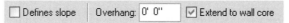

Figure 1.57   *No slope for this roof, which will extend into the wall past the finish layer*

5.3 Pick the interior surface of the walls, as with the Floor command.

The walls will highlight in turn as the cursor passes near them, and reference lines will appear.

- Adjust the position of the sketch lines to interior faces using the control arrows, as before.

5.4 Pick Finish Roof.

- Pick No to the Attach Walls to Roof question box.
- Pick Yes to the join/cut/overlap question box.

The exterior walls will extend past the roof to form a parapet and will not attach to its underside, but will trim their finish layers at the floor structure. The view will not change appearance when the roof is created.

5.5 Double-click on Floor Plans Level 1 [Ground Floor] in the Project Browser to open that view in the View Window.

5.6 Pick the Section tool on the Basics tab of the Design Bar, as shown in Figure 1.58.

Figure 1.58   *The Section tool*

5.7 Pick a start point for the Section line to the left of the upper window in the left wall.

5.8 Pull the cursor to the right horizontally, and pick a spot to the right of the wall to locate the Section, as shown in Figure 1.59.

You can select and move the section line if necessary so that it passes through the window.

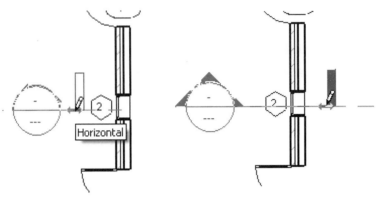

Figure 1.59 *Click two points to create a section line through the wall at the window*

The Project Browser will show a + sign to the left of a new Sections (Building Section) branch.

5.9 Click on the + sign and Section 1 will appear in the Project tree, as shown in Figure 1.60.

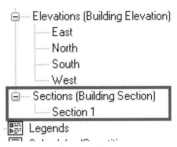

Figure 1.60 *The new Section view in the Project Browser*

5.10 Double-click on Section 1 in the Project Browser to open that view.

5.11 Select the detail level icon on the View Control Bar at the bottom of the view window, to the right of the view scale control, as shown in Figure 1.61.

• Click Medium.

Figure 1.61 *Change the view detail level in two clicks..*

5.12 Pick the Level 2 floor.

- Use the Type Selector drop-down list to change its type from Generic to Steel Bar Joist 14" – VCT on Concrete [Concrete-Commercial 362mm].

5.13 Pick the Roof.

- Use the Type Selector drop-down list to change its type from Generic to Steel Truss – Insulation on Metal Deck – EPDM [Steel Bar Joist – Steel Deck – EPDM Membrane].

- Note the new appearance of these components.

EXERCISE 6. DOCUMENT THE MODEL

So far you have been working on the building model and creating standard views of it. Now you will collate views on a sheet suitable for printing. Revit Architecture can hold as many sheets in a project file as you need to document your project. Annotations that reference sheets (section heads in a plan or elevation view, callouts, etc.) report sheet and detail number automatically.

6.1 Right-click on Sheets in the Project Browser, and pick New Sheet, as shown in Figure 1.62.

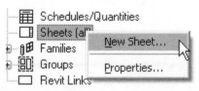

Figure 1.62 *Creating a new sheet for plotting*

6.2 Pick OK to accept the default title block for the new sheet.

The new sheet appears in the View Window.

6.3 Pick Level 1 in the Project Browser (do not double-click), hold down the left mouse button, and drag it into the sheet, as shown in Figure 1.63.

- Release the button and the outline of the view at scale with a title below it will appear.

- Move the cursor and click to place a viewport into the Level 1 floor plan on the sheet (see Figure 1.64).

Figure 1.63 *Starting to drag a view onto the sheet*

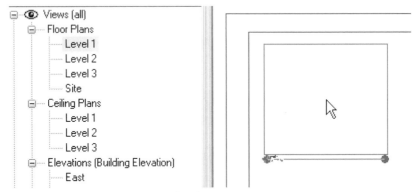

Figure 1.64 *The view outline appears under the cursor*

The viewport is small on the page. You can adjust this quickly.

6.4 Press ESC to terminate the placement.

- Select Floor Plan Level 1 in the Project Browser again, right-click and select Properties.

6.5 In the Properties dialogue for the Floor Plan as shown in Figure 1.65, change the View Scale to **1/4" = 1'-0" [1:50]**.

- Click OK.

Instance Parameters - Control selected or to-be-created instance	
Parameter	**Value**
Graphics	
View Scale	1/8" = 1'-0"
Scale Value 1:	3/8" = 1'-0"
Display Model	1/4" = 1'-0"
Detail Level	3/16" = 1'-0"
Visibility/Graphics Overrides	1/8" = 1'-0"

Figure 1.65 *Changing the view scale without opening the view*

6.6 Repeat this scale change for the West Elevation and Section 1 views in the Project Browser.

6.7 Select Floor Plan Level 1 in the Project Browser and drag it onto the sheet, as before.

- Note the new size indicated in the lower-right corner of the sheet.

6.8 Select and drag views West Elevation and Section 1 from the Project Browser to the sheet and arrange them, as shown in Figure 1.66.

Views will snap to alignment lines to aid in placement.

- Align the Elevation view title with the title of the Plan view, and align Level 2 of the Section with Level 2 of the Elevation view.

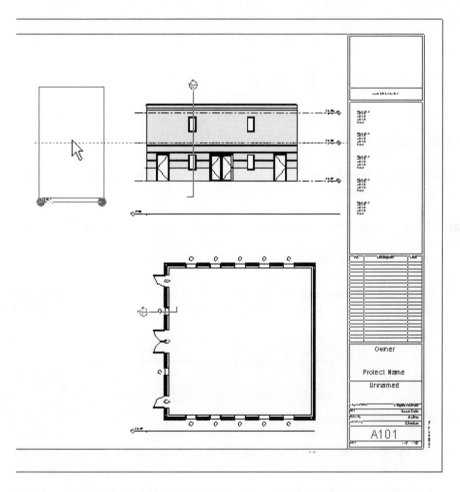

Figure 1.66   *Views will snap to other view titles and to level lines for viewport alignment*

6.9 Zoom in over the plan viewport so you can see the left wall components clearly.

- Place the cursor over the viewport so that its border becomes highlighted. Right-click and pick Activate View, as shown in Figure 1.67.

This will allow you to change the model from the sheet view.

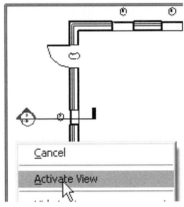

Figure 1.67 *Making the viewport active to work in it*

6.10 Pick one of the windows in the left wall, hold down the CTRL key, and pick the other so both are selected.

 Tip: If you have difficulty picking the windows, you can zoom in closer. You can also use the TAB key on your keyboard to cycle through the available choices until the window you want to modify becomes highlighted and you can select it.

6.11 When both windows are selected, use the Type Selector drop-down list to change their type to Fixed: 16" x 72" [M_Fixed: 0406 x 1830mm], as shown in Figure 1.68.

Figure 1.68 *Changing the window type*

6.12 Pick the Properties icon to bring up the Element Properties dialogue shown in Figure 1.69.

- Change the Sill Height for these two windows to 1' [300 mm] (you can simply type 1 or 300—Revit Architecture's default Imperial unit is the foot and the default metric unit is the millimeter).

- Pick OK.

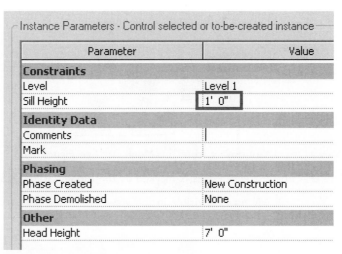

Figure 1.69 *Adjusting the window's sill and head height*

6.13 Right-click and pick Deactivate View.

6.14 Right-click and pick Zoom to Fit.

6.15 Examine the Elevation and Section views to verify that the windows have changed.

6.16 Click the Default 3D View button on the Menu Bar, as shown in Figure 1.70.

The first time you use this tool it will create and open a Southeast isometric view of the model. This view is named {3D}. Once you have created this view, clicking the icon will open it.

Figure 1.70 *The button to create a 3D view*

6.17 Click the Model Graphics Style control on the View Control Bar at the bottom of the view.

This control is to the right of the Detail Level control.

• Pick Shading with Edges, as shown in Figure 1.71.

Figure 1.71 *Change the display to Shading with Edges*

6.18 Pick View>Orient>Southwest from the Menu Bar. See Figure 1.72.

- Note that Shading with Edges is marked as the Model Graphics Style in the menu. The West wall will now be visible in the view.

Figure 1.72 *Changing the view orientation to a standard isometric angle*

6.19 Pick View>Orient>Save View.

- Click OK to accept the default name.

6.20 Double-click Sheets>A101 – Unnamed from the Project Browser to return to the plotting sheet.

- Note that 3D Views now has a + sign to the left of it in the Browser.
- Click on the + sign to reveal the saved view 3D Ortho 1 and the default 3D view, named simply {3D}.

6.21 Pick and drag view 3D Ortho 1 onto sheet A101.

- Release the left mouse button to see the outline of the new viewport.

It will come in small in proportion to the other views.

- Locate it in a convenient place to the left of the View Window.

The new viewport will be highlighted red.

6.22 If not, select it.

- Pick the Properties icon on the Options Bar.
- Change the scale to ¼" = 1'-0" [1:50] as before.
- Click OK.

The viewport will change size.

- Drag the right end of the title to correspond with the new size of the view.

6.23 Move the view to the lower-left corner of the sheet.

Its title will snap into alignment with the title of the Plan view, as shown in Figure 1.73. You can arrange views and view title lines separately for a tidy page.

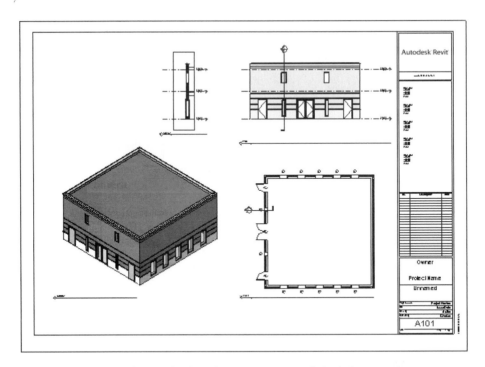

Figure 1.73 *The plotting sheet with plan, elevation, section and shaded isometric views*

6.24 Save the file.

SUMMARY

Congratulations! You have created, edited and documented a building shell in Revit Architecture, using its intelligent components and their innate properties entirely. This sheet is ready to print and you have not drawn any 2D lines, but worked entirely with objects, views and their associated properties.

This model is quite basic, but you have gotten a taste of Revit Architecture's simplicity, ease of use and power. The remaining chapters will show how Revit Architecture's strengths in modeling, representation, documentation, data extraction and change propagation function and prepare you to implement this new architectural design tool in a work environment.

REVIEW QUESTIONS – CHAPTER 1

MULTIPLE CHOICE

1. To make objects visible or not, use controls found in

   a. the Design Bar

   b. the properties of each view

   c. the Project Browser

   d. the Options Bar

2. Hatch lines inside walls in a plan

   a. are always visible if you zoom in close

   b. are only visible in the Level 1 Floor Plan view

   c. are visible when the View Detail level is set to medium or fine

   d. have to be drawn in by hand

3. Zoom controls are found

   a. on the upper Toolbar

   b. in the View menu

   c. in the right-click context menu

   d. all of the above

4. Revit Architecture will allow you to edit the swing and hinge controls of a door

    a. during placement

    b. if it is selected by picking after placement

    c. by using the View properties

    d. a and b, but not c

5. For construction documentation purposes, Revit Architecture provides

    a. Sheets with Views placed on them at scale

    b. Annotations that keep track of their Sheet placement

    c. both a and b

    d. Revit Architecture can't produce documents.

TRUE/FALSE

    6. True _ False _ Sections and Elevations require update procedures to reflect model changes.

    7. True _ False _ When creating floors and roofs, Revit Architecture uses Sketch mode to allow the user to pick walls or draw lines.

    8. True _ False _ Views placed on Sheets do not require update procedures to reflect model changes.

    9. True _ False _ A floor object can be offset from its base Level.

    10. True _ False _ Entities cannot be selected in 3D views.

 Answers will be found on the CD.

CHAPTER 2

Under the Surface: Standards and Settings

INTRODUCTION

This section is of importance to managers, supervisors and team leaders in design firms; university professors or those leading courses of instruction will also find the concepts and exercises useful.

Students and new users will benefit from examining content and structure in template files, importing a CAD file, and exporting a file to CAD.

OBJECTIVES

- Understand the mechanics and value of template files in Revit Architecture
- Experience Revit Architecture's input/output interface with AutoCAD and other applications
- Discuss Revit Architecture's file management systems: import/export, linking, worksets

STUDENT OVERVIEW

In Chapter 1 you worked through a series of exercises in which you created a simple building model using objects that Revit Architecture supplied for you. These predefined objects and the views in which you manipulated them, while complete enough to use from start to finish on a project, are only a portion of the working setups available to you.

In the exercises in this chapter, you will examine template files in some detail. You will adjust various settings that affect the display of imported, drawn and exported items.

CAD MANAGER/INSTRUCTOR SUMMARY

If your responsibilities include deploying the design software your organization uses, keeping designers equipped with software tools, organizing files, and maintaining standards, you have headaches enough without having to adopt (or worse, half-adopt) a new standard. As your company or institution considers using Revit Architecture, what does Revit Architecture have in store for you?

The bad news is that Revit Architecture is different enough from AutoCAD and AutoCAD Architecture that even the people who listen to you and make an effort to follow your directives, instructions, specifications and standards will be confused at times by this new software. The good news is that Revit Architecture is much simpler than AutoCAD and AutoCAD Architecture in significant ways—*not* having to use layers to govern the appearance of objects on screen and paper, for example, does away with an entire, often complicated, set of standards that nobody ever follows completely anyway.

Revit Architecture does contain many of the same mechanisms that you currently use—text and dimensions have styles; reference files, content files and central files all use coherent path setups; drawing sheets work the same way as Paper Space layouts, only more efficiently. Once you understand the strengths and operational underpinnings of this building information modeler, you will be able to serve the needs of your designers, project leaders, clients and students.

TEMPLATE FILES HOLD STANDARDS AND SETTINGS

Revit Architecture uses template files to hold an array of settings, exactly as does AutoCAD. It works with two types of template files: rft files for family templates and rte files for project (building model) files. Families are a major organizing system in Revit Architecture. They do the work both of Styles and Multi-View blocks in AutoCAD-based CAD applications. Families will be treated in later exercises.

When first opening, Revit Architecture uses a template file that corresponds to the *acad.dwt* template file for AutoCAD—a basic setup with minimal predefined content. Just as AutoCAD and AutoCAD Architecture allow the user to specify and modify other template files with enhancements for specific purposes, Revit Architecture provides template files with significantly developed views, content and sheet setups for specific types of building models. These templates are starting points to create company-standard, project-specific, or client-specific templates. As with AutoCAD products, completed Revit Architecture project files can be converted to templates.

When installing Revit Architecture, the information you supply about your country of origin will affect the template files and content libraries that the install routine creates. This text will assume that users working in imperial units will have the template file *default.rte* available to them, and that users working in metric units will have the file *DefaultMetric.rte* available. Detailed examination of localized templates and content is beyond the scope of this text. Working with the template files listed in this chapter will prepare you for future work with other templates. Project template files mentioned in exercise steps are available on the CD.

In this next exercise, you'll examine some of the settings and setups that will make your job, be it designer or manager, easier.

TEMPLATE AND INPUT/OUTPUT INSPECTION

REVIT ARCHITECTURE COMMANDS AND SKILLS

Templates

Families

Views

Sheets

Phases

View Properties

View Phase Filters

View Graphic Overrides

Predefined Schedules

Titleblock

Parameters

Labels

Project Information

Import File

Import Settings – Lineweights

Export File

Export Settings – Layers

EXERCISE 1. EXAMINE TEMPLATES

THE DEFAULT TEMPLATE—THE VERY BASICS

1.1 Launch Revit Architecture.

Revit Architecture will open to a Floor Plan View of Level 1 (the words **Level 1** will be highlighted under the Floor Plans section of the Project Browser tree under Views (all), as shown in Figure 2.1).

• Study the contents of the Project Browser to the left of the View Window.

You can open predefined Floor Plans for Levels 1 and 2, Ceiling Plans for Levels 1 and 2, and four Elevations (graphically indicated on Level 1 Floor Plan by elevation markers). There are no legends, schedules, or sheet views defined.

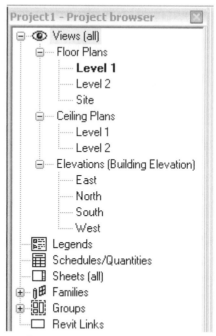

Figure 2.1 *The Project Browser for an empty file using the default template*

1.2 Click the + symbol next to Families in the Project Browser.

- Scroll down the list of Families and expand (click the + symbol next to) Basic Wall under Walls (see Figure 2.2).

Figure 2.2 *Available wall types in the default templates – imperial, left; metric, right*

All Revit content, including model objects, annotations and profiles, are organized by family. The families in a project file are listed in the Project Browser. You can use the Project Browser listing to manage a family or types within a family. You will edit, duplicate and insert family content from the Project Browser in later exercises.

1.3 Place the cursor over the point of the West Elevation symbol in the View Window.

1.4 Right-click and pick Go to Elevation View to open the West Elevation view, as shown in Figure 2.3.

This has the same effect as double-clicking the West View listing in the Project Browser.

Figure 2.3 *Locate, right-click, and pick to open the West Elevation*

1.5 The West Elevation shows the two Levels in this file, at 0'-0" [0] and 10'-0" [4000], as shown in Figure 2.4.

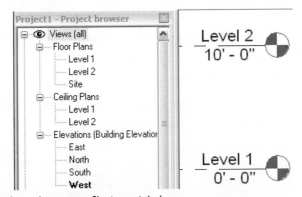

Figure 2.4 *The levels in this empty file; imperial shown*

EXAMINE THE RESIDENTIAL (IMPERIAL) TEMPLATE—PREDEFINED VIEWS AND SHEETS

1.6 From the File menu, pick File>New>Project, as shown in Figure 2.5, to open the New Project dialogue.

Figure 2.5 *Opening a new project*

1.7 In the New Project dialogue, pick the Browse button to choose a template file for the new project, as seen in Figure 2.6.

The path to your Templates folder will be different than what is shown here.

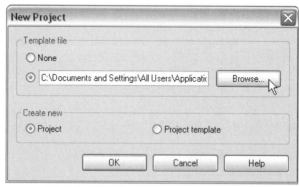

Figure 2.6 *The New Project dialogue*

- In the Choose Template dialogue, pick *Residential-Default.rte*, as shown in Figure 2.7.

- If you are working in metric units, navigate to the folder where you have installed content from the CD enclosed with this book.

- Open the folder *Library/Chapter 2* to find templates you will examine in this exercise.

- Pick Open to return to the New Project dialogue.

- Pick OK to create a new project using this residential template.

Figure 2.7 *Choose the Residential-Default template*

1.8 When the new residential project file opens, study the Project Browser (shown in Figure 2.8).

There are many more Floor Plans than in the default file, including Footing, Framing and Electrical Plans. Levels 1 and 2 have been renamed to First Floor and Second Floor, and a Basement and Roof Level added. There are Schedules predefined as well: Door, Lighting Fixture, Room Finish and Window.

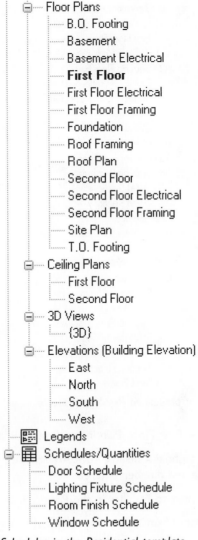

Figure 2.8 *Floor Plans and Schedules in the Residential template*

1.9 Click the + next to Sheets to expand the list of Sheets.

This template includes Plans, Elevations, Detail Sheets, Sections and a Schedule page.

1.10 Click the + next to A0 – Basement Plan and A1 – First Floor Plan, as shown in Figure 2.9.

These expand to show that the Basement Floor Plan and First Floor Plan have already been placed on these two sheets. Similar views have been placed on other sheets: Floor Plans, Reflected Ceiling Plans, Roof Plans and Electrical Plans.

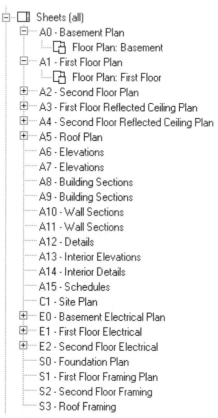

Figure 2.9 *The expanded list of predefined Residential sheets*

1.11 Double-click on A1 – First Floor Plan in the Project Browser to open that view.

It has been defined to hold a default Revit Architecture D size 22" x 34" titleblock, with the sheet name and number entered.

1.12 Click the + symbol next to Families in the Project Browser.

- Scroll down the list of Families and expand Walls (see Figure 2.10). Expand Basic Wall under Walls.

- Study the different content available in this template compared to the default file.

Figure 2.10 *Wall types in the residential template*

1.13 From the Window menu, pick Window>Project 1 – Floor Plan: Level 1 to open up this view.

The Project Browser now shows the views, sheets and families appropriate to this file.

1.14 From the Window menu, pick Window>Project 2 – Floor Plan: First Floor to return to the residential file.

1.15 Double-click on East Elevation in the Project Browser to open that view.

- With the cursor in the View Window, right-click.

- Pick Zoom in Region.

- Pick two points around the Elevation symbols at the right of the screen to see the predefined Levels in this empty file (see Figure 2.11).

There is a Foundation Level 1'-3" below the First Floor level, a Second Floor at 9'-0", and a Roof level at 18'-0".

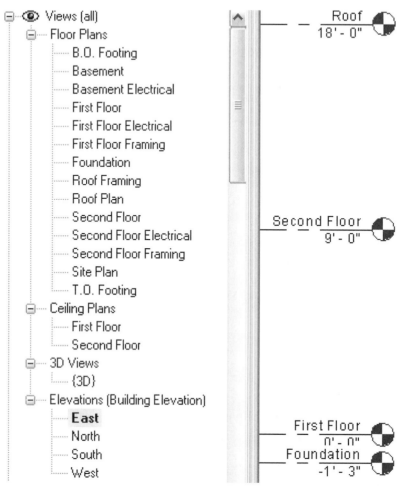

Figure 2.11 *Default Residential levels: Foundation, First Floor, Second Floor and Roof*

EXAMINE THE CONSTRUCTION (IMPERIAL) TEMPLATE – VIEWS WITH PHASES, SCHEDULES

1.16 From the File menu, pick File>New>Project as before.

1.17 In the New Project dialogue, pick Browse.

1.18 In the Choose Template dialogue, pick *Construction-Default.rte*.

- If necessary, navigate to the CD library folder as before. Pick Open.

1.19 In the New Project dialogue, click OK.

1.20 When the new construction project opens, note that the Elevation symbols have been turned off in the Level 1 View window.

- Study the Project Browser: note the Floor Plans located below Level 1 and the list of 3D Views.

1.21 Pick 3D View 02 – Demo in the Project Browser.

- Right-click, and pick Properties, as shown in Figure 2.12.

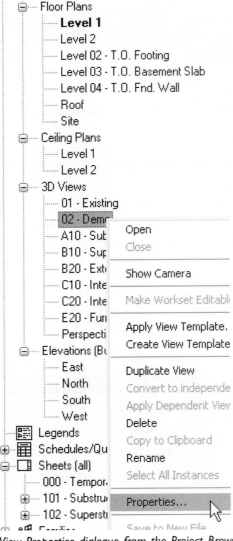

Figure 2.12 *Opening the View Properties dialogue from the Project Browser*

1.22 In the Element Properties dialogue that appears, as in Figure 2.13, scroll down and note the Phase Filter and Phase that have been applied to this view.

This file has been set up to use Phases—building elements will be differentiated according to the time they come into existence or disappear. Phases can be named and filtered. Each Phase can appear differently (or not at all) in each view of the project file.

- Click Cancel to leave the Element Properties dialogue for the current view.

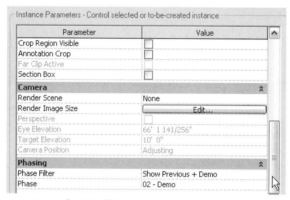

Figure 2.13 *The Phase settings for this 3D view*

1.23 From the Settings menu, pick Settings>Phases, as shown in Figure 2.14.

Figure 2.14 *Opening Phase Settings*

1.24 In the Phasing dialogue that opens, study the first tab, Project Phases, as shown in Figure 2.15.

This file has nine phases defined in it: **Existing**, **Demo**, five **New Construction** phases, **Furnishings** and **Project Completion**.

The other two tabs in the dialogue are Phase Filters and Graphic Overrides.

	Name	Description
1	01 - Existing	Existing
2	02 - Demo	Demo
3	A10 - Substructure	New Construction - Substructure
4	B10 - Superstructure	New Construction - Superstructure
5	B20 - Exterior Enclosure	New Construction - Exterior Enclosure
6	C10 - Interior Constructio	New Construction - Interior Construction
7	C20 - Interior Finishes	New Construction - Interior Finishes
8	E20 - Furnishings	Furnishings
9	Project Completion	Project Completion

Figure 2.15 *Project Phases in this template*

1.25 Pick the Phase Filters tab, as shown in Figure 2.16.

Phase Filters set up view conditions in which phases are displayed according to their categories, displayed with overrides, or not displayed. In this way each element can be viewed in different ways in different views. Revit Architecture supplies the Temporary phase for elements that are created and demolished during the project.

	Filter Name	New	Existing	Demolished	Temporary
1	Show All	By Category	Overridden	Overridden	Overridden
2	Show All	By Category	Overridden	Overridden	Overridden
3	Show Complete	By Category	By Category	Not Displayed	Not Displayed
4	Show Demo	Not Displayed	Not Displayed	Overridden	Not Displayed
5	Show Demo + New	By Category	Not Displayed	Overridden	Overridden
6	Show New	By Category	Not Displayed	Not Displayed	Not Displayed
7	Show Previous + Demo	Not Displayed	Overridden	Overridden	Not Displayed
8	Show Previous + New	By Category	Overridden	Not Displayed	Not Displayed
9	Show Previous Phase	Not Displayed	Overridden	Not Displayed	Not Displayed

Figure 2.16 *Phase Filters*

1.26 Pick the Graphic Overrides tab shown in Figure 2.17.

Here the visual properties of Phases are applied—line weight in plan and section, color, linetype and material for 3D views.

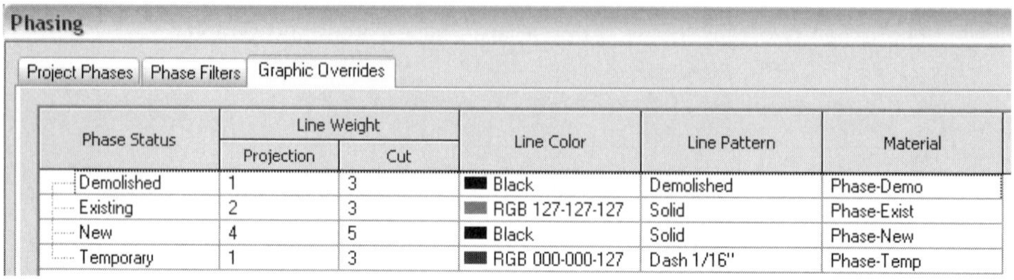

Phase Status	Line Weight		Line Color	Line Pattern	Material
	Projection	Cut			
Demolished	1	3	■ Black	Demolished	Phase-Demo
Existing	2	3	■ RGB 127-127-127	Solid	Phase-Exist
New	4	5	■ Black	Solid	Phase-New
Temporary	1	3	■ RGB 000-000-127	Dash 1/16"	Phase-Temp

Figure 2.17 *Graphic Overrides for Phasing*

1.27 Pick Cancel to exit the Phasing dialogue.

1.28 Pick the + symbol next to Schedules/Quantities to expand the list in the Project Browser, as seen in Figure 2.18.

This template includes examples of nearly all the available standard schedules.

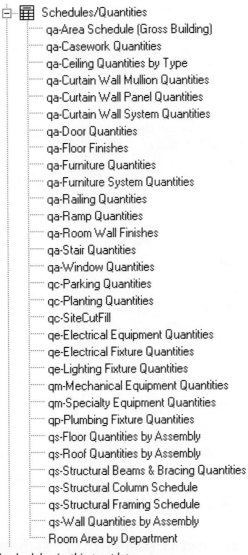

Figure 2.18 *Predefined schedules in this template*

1.29 Expand Elevations in the Project Browser.

- Double-click Elevations>East to open that view.

- Zoom in around the Elevation symbols in the right of the View Window.

- Study the predefined Levels and their heights in this template, as shown in Figure 2.19.

1.30 Click the + symbol next to Families in the Project Browser to expand the list.

- Click the + symbol next to Walls.

1.31 Click the + symbol next to Basic Wall to expand the list in this template.

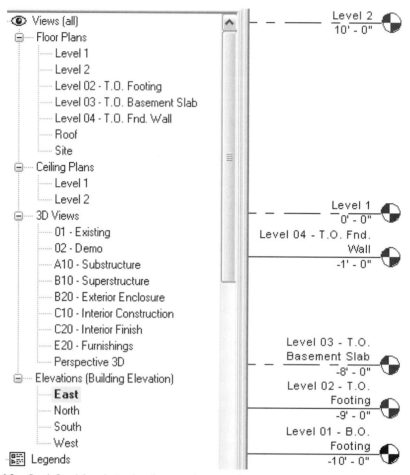

Figure 2.19 *Predefined Levels in the Construction template*

EXAMINE THE COMMERCIAL (IMPERIAL) TEMPLATE

1.32 From the File menu, pick File>New>Project to open the New Project dialogue.

1.33 Pick the Browse button to choose another template.

1.34 In the Choose Template dialogue, pick *Commercial—Default.rte* in the list of templates and click the Open button.

- Navigate, if necessary, to the CD library folder, as before.

- Click OK in the New Project dialogue.

1.35 In the Project Browser, study the list of default Floor Plans.

- Double-click Elevations>East.

- Zoom in around the Elevation symbols in that view (see Figure 2.20).

- Note the location of the predefined levels in this template.

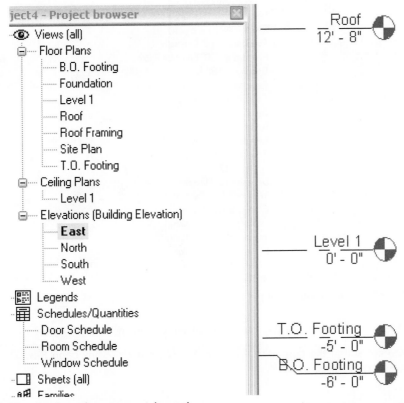

Figure 2.20 *Levels in the commercial template*

1.36 In the Project Browser, expand the list of predefined sheets.

- Click the + symbol next to Families to expand the list of categories.

1.37 Click the + symbol next to Walls>Basic Walls to expand that category (shown in Figure 2.21).

- Study the list of available wall types in this template.

Figure 2.21 *Walls in the commercial file*

1.38 Double-click Level I in the Project Browser to return to that view.

You are still in the commercial project file. Do not close any open Revit Architecture files.

EXERCISE 2. COMPARE SPECIFIC CONTENT IN DIFFERENT TEMPLATES

2.1 From the Window menu, pick Window>Close Hidden Windows, as shown in Figure 2.22.

This will close all the views that you have open except the current one in the current file and the last open view in other open files. Keeping open windows to a minimum will free up display system resources.

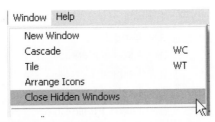

Figure 2.22　*How to close hidden view windows*

2.2 In the Project Browser, click the + symbol next to Families>Profiles to view the list of predefined and named profile shapes available for sweeps or edges in this template.

- If necessary, widen the Browser pane by holding the cursor over its right border until the cursor changes to a two-headed arrow, then click and pull the pane border to the right so you can see the expanded tree view (see Figure 2.23).

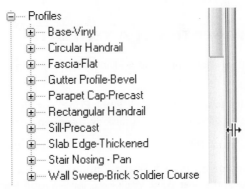

Figure 2.23　*Profiles in the commercial template*

2.3 From the Window menu, pick the available view from *Project 1* (the default template).

- When that file is loaded, click the + symbol next to Families>Profiles to view the list of profile shapes in the default template.

2.4 From the Window menu, pick the available view from *Project 2* (the residential template).

- When that file loads, click the + symbol next to Families>Profiles to view the list of profile shapes in the residential template. See Figure 1.24.

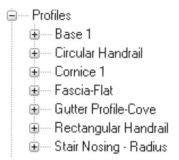

Figure 2.24 *Profiles in the residential template*

2.5 From the Window menu, pick the available view from *Project 3* (the construction template).

- When that file loads, click the + symbol next to Families>Profiles to view the list of profile shapes in the construction template.

EXAMINE VIEW PROPERTIES IN A TEMPLATE

2.6 From the Window menu, pick the available view from *Project 2* (the residential template).

- When that file loads, double-click Floor Plan>First Floor in the Project Browser to make that view active if it is not already the open view.

2.7 On the Basics Tab of the Design Bar, pick Wall to start the Wall tool.

2.8 In the Type Selector on the left of the Options Bar, click the down arrow in the drop-down box and select Basic Wall: Exterior – Wood Shingle on Wood Stud, as shown in Figure 2.25.

- Do not change the Height or Location Line properties.

Figure 2.25 *Exterior wall properties*

2.9 In the sketch lines area of the Options Bar, pick the Rectangle button (see Figure 2.26) so that by picking two points in the View Window you can draw four connected walls with their exterior sides correctly situated.

2.10 Pick a point to the upper left of the View Window and pull the cursor down and to the right to start defining the new walls.

- Use the temporary dimensions that appear to make the new walls approximately 30' wide by 25' long (see Figure 2.26).

The actual wall lengths do not matter.

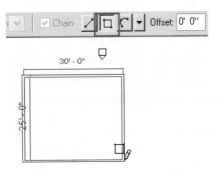

Figure 2.26 *Drawing four connected walls*

2.11 With the Wall tool still active, change the wall type to Basic Wall: Interior – 4 1/2" Partition.

- Change the sketch line type to Line and pick the Chain checkbox to sketch new walls continuously until you exit the command.

2.12 Draw a wall from the midpoint of the left wall (approximately) horizontal to the middle of the interior space, then up vertically to the middle (approximately) of the upper wall (see Figure 2.27).

The actual wall dimensions or locations do not matter.

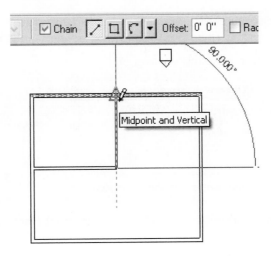

Figure 2.27 *Interior walls in place*

2.13 Pick Component from the Design Bar to terminate the Wall tool and start the Component placement tool.

- Use the drop-down list to select Outlet Duplex : Single, as shown in Figure 2.28.

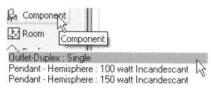

Figure 2.28 *Pick Duplex Outlet in the Component command*

2.14 Place an outlet at the approximate midpoint of the right side interior wall, facing left into the smaller room you have created with the new walls (see Figure 2.29).

The actual placement does not matter.

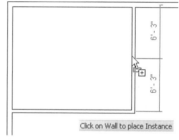

Figure 2.29 *Locating an electrical outlet*

2.15 A Revit Architecture warning, shown in Figure 2.30, will appear to let you know that the settings for this Floor Plan view will not let you see the new electrical fixture.

- Click the X to close this warning.

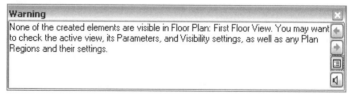

Figure 2.30 *Revit Architecture warns you about this view*

2.16 In the Project Browser, double-click Floor Plans>First Floor Electrical to open that view.

You will see the new electrical outlet correctly placed. Note the light appearance of the walls.

2.17 With the cursor in the View Window but not directly over any part of the model, right-click and pick View Properties (always the bottom choice in this menu).

2.18 In the View Properties dialogue that opens, pick the Edit button in the Value field to the right of the Visibility/Graphics Overrides Parameter field.

2.19 In the Visibility/Graphics Overrides dialogue that opens, study the Model Categories tab (see Figure 2.31).

For most component categories except Curtain Systems, Electrical Components and Electrical Fixtures, the Halftone graphic override has been checked. This accounts for the dim appearance of the walls in this view.

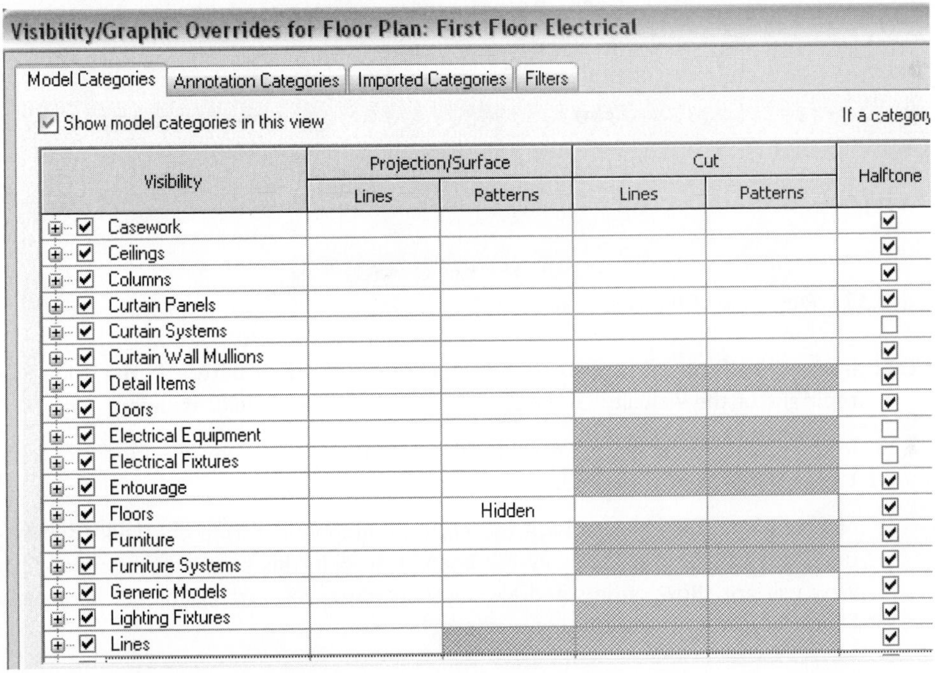

Visibility/Graphic Overrides for Floor Plan: First Floor Electrical

Model Categories | Annotation Categories | Imported Categories | Filters

☑ Show model categories in this view

Visibility	Projection/Surface		Cut		Halftone
	Lines	Patterns	Lines	Patterns	
☑ Casework					☑
☑ Ceilings					☑
☑ Columns					☑
☑ Curtain Panels					☑
☑ Curtain Systems					☐
☑ Curtain Wall Mullions					☑
☑ Detail Items					☑
☑ Doors					☑
☑ Electrical Equipment					☐
☑ Electrical Fixtures					☐
☑ Entourage					☑
☑ Floors		Hidden			☑
☑ Furniture					☑
☑ Furniture Systems					☑
☑ Generic Models					☑
☑ Lighting Fixtures					☑
☑ Lines					☑

Figure 2.31 *Halftone Settings in the Electrical Plan*

2.20 Click Cancel twice to exit the View Graphics and View Properties dialogues.

2.21 Pick Floor Plans>First Floor Plan in the Project Browser to highlight it.

- Right-click and pick Properties, as shown in Figure 2.32, to open the View Properties for the First Floor Plan view without having to open the view itself.

Figure 2.32 *Properties of the First Floor Plan*

2.22 In the View Properties dialogue that opens, pick the Edit button in the Value field to the right of the Visibility/Graphics Overrides Parameter field, as before.

2.23 In the Visibility/Graphics Overrides dialogue that opens, study the Model Categories tab, as shown in Figure 2.33.

No component categories have the Halftone graphic override checked. Visibility of the Electrical Fixtures category has been cleared. In this template, the Floor Plan view will not show outlets and electrical switches; you just examined the separate Electrical Plan created for that purpose.

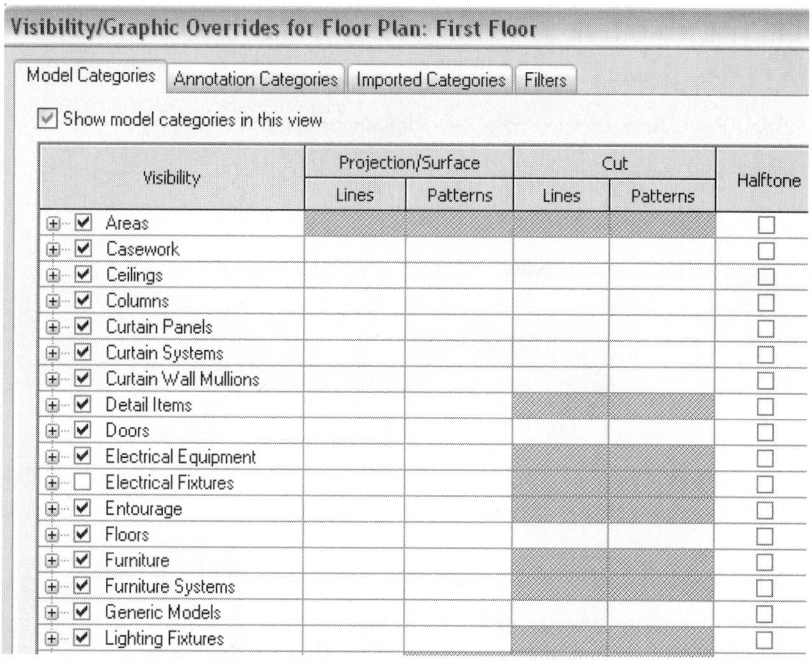

Figure 2.33 *Halftone and visibility controls for the First Floor Plan*

2.24 Pick Cancel twice to exit the Visibility/Graphics and View Properties dialogues.

2.25 From the File menu, pick File>Exit (always the bottom selection on this menu) to exit Revit Architecture.

- Do not save any of the open files.

EXPLORE TEMPLATES ON YOUR OWN—USE WHAT'S ALREADY MADE FOR YOU

Revit Architecture contains other templates—imperial and metric—besides the ones you have just studied. All the templates are worth examining for useful view setups, sheets, families and schedules that have been pre-loaded for users and managers to adopt and adapt. At present there is no mechanism in Revit Architecture—such as the Design Center file/content browser in AutoCAD—to allow drag-and-drop transfer of settings or content from one file to another. Revit Architecture does support standard copy/paste from one file to another, and has a Transfer Project Standards command to move settings, family types, line weights, and rendering materials maps from one open file to another. You'll look at this tool in future exercises.

EXERCISE 3. EXAMINE A STANDARD TITLEBLOCK—EDIT PARAMETERS

3.1 Open Revit Architecture to a new default project.

It will have no sheets defined.

- Select Sheets in the Project Browser.

- Right-click and pick New Sheet, as shown in Figure 2.34.

Figure 2.34 *Adding a sheet to a new file*

3.2 In the Select a Titleblock dialogue that opens, pick the Load button.

3.3 In the Open file browser dialogue, make sure the Look In: value is the *Imperial [Metric] Library* folder and select the folder *Titleblocks*. Click Open.

3.4 Pick each of the default titleblocks in turn to examine its appearance in the preview window, as shown in Figure 35.

- Select Cancel to return to the previous dialogue.

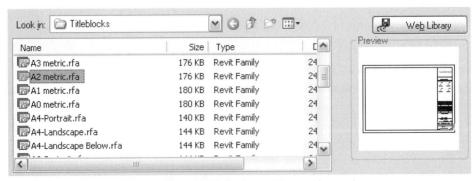

Figure 2.35 *Examine various supplied titleblocks in the library. Metric shown.*

EXAMINE THE DEFAULT SELECTIONS

3.5 Pick OK to create a sheet with the default titleblock.

Revit Architecture will open the new sheet view.

3.6 Click on the new sheet titleblock to select it.

- On the Options Bar, pick the Properties icon.
- Fill in the Parameter Value fields as shown in Figure 2.36.
- Use today's date for the Sheet Issue Date.
- Pick OK.

Parameter	Value
Graphics	
Scale	
Identity Data	
Sheet Name	Site Plan
Sheet Number	A101
Sheet Issue Date	07/02/07
Checked By	Michael Peterson
Designed By	Danielle Browning
Approved By	Lucy Van Pelt
Sheet Width	840.0
Sheet Height	594.0
Other	
Date/Time Stamp	07/02/07
File Path	
Drawn By	Christopher Fox

Figure 2.36   *Titleblock parameter fields*

3.7 With the cursor in the view window, Zoom in around the lower-right corner of the title block.

Some, but not all of the information fields (labels) have updated.

- Select the titleblock, and the linework will turn red to show it has been selected. Labels will turn blue.

3.8 From the Menu Bar, select Settings>Project Information.

This will open the Type Properties Project Information dialogue.

- Fill in the project information as shown in Figure 2.37.
- You will need to pick the Edit button in the Project Address field to open the entry box for that information.
- When you have entered the address information, pick OK.
- Pick OK again to close the Project Information dialogue.

Settings Window Help

Project Information...

Parameter	Value
Energy Analysis	
Energy Data	Edit...
Other	
Project Issue Date	09/09/2007
Project Status	Conceptual Design
Client Name	Smithson, Inc.
Project Address	Edit...
Project Name	Store #09 Deerfield
Project Number	08-556-SC

Edit Text

7839 Main St. East
Williamsville NY 13487

Figure 2.37 *Project Information Settings*

3.9 Study the labels in the titleblock again.

You can edit the information in any label by selecting it.

- Pick the A101 Sheet Number label and hold the cursor over it.

The Tooltip will show Edit Parameter.

- Select the label again and the data entry field will open.

- Change the Sheet Number from **A101** to **A1**, as shown in Figure 2.38; left-click outside the entry field (or hit ENTER) and the label will update.

Figure 2.38 *Editing a titleblock label*

3.10 Click the drop-down arrow next to the Undo arrow on the Revit Architecture Toolbar, under the Drafting menu button.

- Scroll down the list of actions to Sheet.

Your Undo list may differ slightly from the one shown in Figure 2.39.

- Left-click inside the Undo drop-down list to accept the Undo command.

Revit Architecture will delete the titleblock and sheet view.

Figure 2.39 *Undoing a series of actions*

3.11 From the File menu, pick File>Close. Do not save the open file.

INTERFACE WITH CAD TO CAPTURE EXISTING DESIGN DATA

Import, Don't Re-create, CAD

One of the cardinal rules for digital design productivity is to draw as little as possible, and use editing commands to populate your designs. That is, draw an object once and copy or array it many times, or draw half of a symmetrical design and mirror it to complete the pattern. This approach is not only quicker than repetitive drawing, but less error-prone.

Nearly every design firm on the planet has a library or archive of drawn work that designers mine repeatedly for reuse in new projects. Revit Architecture's ability to import from and export to AutoCAD and MicroStation means that firms setting out to use the power of new technology can make effective, leveraged use of their existing files and applications. If you have already drawn designs, components, symbols, or details in CAD that pertain to your new Revit Architecture project, there is at least one effective way to use them in your Revit Architecture project model or documentation.

Revit Architecture does not use layers or levels, the fundamental organizing mechanism of AutoCAD and MicroStation. CAD files use the concept of layers as corresponding to transparent overlays on a manual drafting table. All lines, text and other objects exist on named layers. Users create and manage layer names and colors, and many standards exist. Lines representing walls will, in a well-organized file, be on a layer named A-Wall (using one widely recognized standard). In a badly organized file, wall lines could be on a layer designated for text or roofing, allowing for confusion and error.

Revit Architecture creates objects and uses object category as its organizing principle. Walls are walls, and their component parts can never be designated as text or roofing, to keep to the previous example. Revit Architecture will recognize layers/levels in linked or imported files, so that the user can benefit from the organization in the originating file. Revit

Architecture exports DWG, DXF, or DGN files with properly constructed layers/levels, so that wall objects are translated into lines on wall layers and Revit text objects become CAD text on text layers.

Autodesk Revit Architecture can read and write ACIS® solids. This provides a way to export models from AutoCAD Architecture (formerly Architectural Desktop) and AutoCAD MEP (formerly Building Systems) software and import or link 3D information into Revit Architecture. You can cut sections and perform visual interference detections in the imported file with this method.

The import process creates a single object in the Revit Architecture file, with control over the appearance of its contained elements—exactly like the Insert Block command in AutoCAD. You can explode imported objects with two levels of depth—all the way to simple lines, curves, text and filled regions. Revit Architecture makes particular use of 3D line/polyline information from imported DWG, DXF, or DGN files to create editable Toposurface site objects. Revit Architecture does not recognize attribute definitions, but does recognize attribute values of inserted blocks in an imported DWG file and will explode those values to text. Revit Architecture uses Windows fonts, and contains an editable mapping file for routing AutoCAD shx font information (an older format for lettering contained in AutoCAD from the days before Windows) to Windows fonts.

Revit Architecture will read proxy graphics for AEC Objects from AutoCAD Architecture, such as walls and floors. Proxy graphics will come in as 3D and explode to 2D Model lines in plan views, within limitations.

Export from Revit Architecture into CAD and Other Applications

Revit Architecture writes out DWG, DXF and DGN files to complete the interchange cycle with today's major CAD applications. You can export an individual view or multiple views.

Revit Architecture will export 2D information from 2D views or 3D information from 3D views using mapping files to assign layers or levels to Revit Architecture objects by category. Exported 3D information comes into AutoCAD as 3D polyline meshes that explode to 3D faces.

Revit Architecture's 3D view settings (visibility, hidden line mode) are ignored—a hidden line view opens in 3D Wireframe mode in AutoCAD. The exported model will hide and shade in AutoCAD.

Revit Architecture maps object categories to preconfigured layer names. As with layer import mapping files, you can create your own export mapping files for projects or clients. You will examine this in the next exercise.

Revit Architecture exports schedule field information to delimited files for import into spreadsheet applications (such as Microsoft Excel and Lotus 123). Revit Architecture can export model components to ODBC databases (such as Microsoft Access) using specific table relationships (between type and instance, for example). ODBC export can then be updated throughout the project to the same database file to give progress reports. Revit Architecture will print to PDF writers. If these are installed, you can export views to PDF format as a print process. Revit Architecture exports DWF, Autodesk's proprietary format for Web-based viewing and sharing of design file information across many platforms. Revit Architecture will export various image types directly from any view or selection of views. These images will import into other applications, such as Microsoft Word. You can attach or embed images in email messages or post them to web pages. Revit Architecture has a walk-though creation routine that allows for export to an avi animation file, which will play on any media player. You will treat these output options in future exercises.

These export capabilities make Revit Architecture powerful and useful throughout a building design project. If you need to send DWG or DGN information to an outside consultant using their layer standards, Revit Architecture makes this operation straightforward. If you need number-crunching information, Revit Architecture will supply this from the very beginning of conceptual design work, and gives you a consistent way to update the figures. If you need images, you can generate them at any time in more than one format. Once you have set up your various export routines, the process is simple enough so that non-experts can readily generate supplementary files from a Revit Architecture model. This will save designer time for designing.

EXERCISE 4. IMPORT FROM AUTOCAD

EXAMINE IMPORTED FILES

4.1 Launch Revit Architecture.

- From the File menu, pick File>Import/Link>CAD Formats.

- Navigate to the folder where you stored the files from the CD for Chapter 2.

- Select *3D import imperial.dwg [3D import metric.dwg]* (see Figure 2.40).

4.2 Verify that the Layer/Level Colors option is set to Invert colors.

Most CAD files are created on black screens; yellow and cyan lines, for instance, will not display well on Revit's white screen.

You can change Revit's screen background to black in the Settings>Options dialogue, Graphics tab. There is a check option to invert the screen color.

If you wish, you can emulate a CAD screen with black background and different colors for walls, doors, windows, etc. This is really the long way around. The original purpose of colored lines in computer drafting was to provide a visual demarcation for what lines were supposed to represent, and to key printed line weights.

Revit's black on white views always display exactly what is going to print, thus removing unnecessary complexity.

- Verify also that the options for Automatically Place and Center to Center are active.

- Click Open.

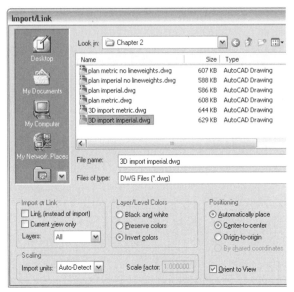

Figure 2.40 *Import an AutoCAD filer*

4.3 Open the default 3D view.

The icon for this is on the View Toolbar.

4.4 Place the cursor over the house that appears in the view.

The view window will show a rectangular outline and the Tooltip will identify it as an Import Symbol, as shown in Figure 2.41.

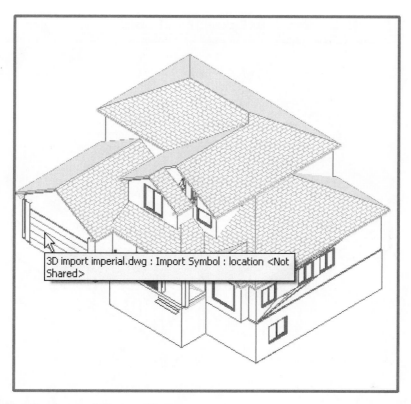

Figure 2.41   *The imported file in a 3D view*

4.5 Click on the import to select it.

The entire model will turn red, the Type Selector will identify it as Import Symbol: 3D import, and the Options Bar will activate.

• Click Query on the Options Bar, as shown in Figure 2.42.

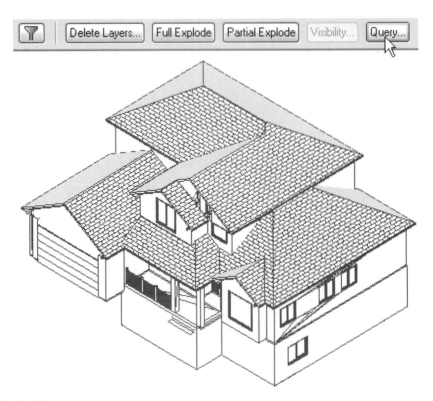

Figure 2.42 *Query the import*

4.6 Place the cursor over the edge of a roof.

- Click to select.

  The Import Instance Query dialogue will identify the layer as A-Roof-Slab. You have options to hide the layer in the view or delete the layer from the import file.

- Click OK to close the dialogue without taking action.

4.7 Click Dimension on the Basics Tab of the Design Bar.

- Click the ends of a foundation wall, drag the cursor away from the building, and click to place a dimension, as shown in Figure 2.43.

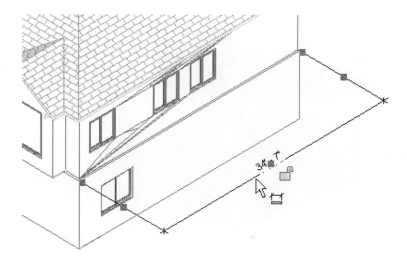

Figure 2.43 *You can dimension to edges of the imported object*

4.8 On the View Control Bar, click on Model Graphics Style.

- Click Shading with Edges.

Faces of the import will shade.

4.9 Hold down the SHIFT key and the scroll wheel on your mouse at the same time.

This activates the spin/3D orbit mode. You can combine spin with zoom and pan.

- Practice with this until you can examine the import file from underneath (see Figure 2.44).

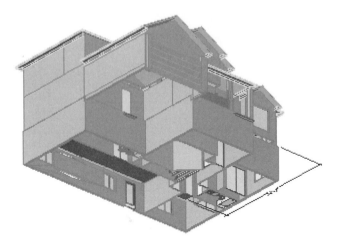

Figure 2.44 *Look at the 3D import from all angles*

4.10 From the Window menu, select Project 1: Floor Plan: Level 1 to return to that view, as shown in Figure 2.45.

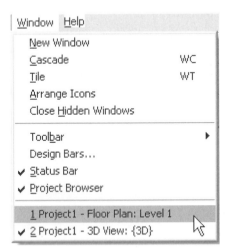

Figure 2.45 *Preparing the export file names and location*

4.11 Select the import.

- Click Query from the Options Bar.

- Click on the Roof.

- Click Hide in View. See Figure 2.46.

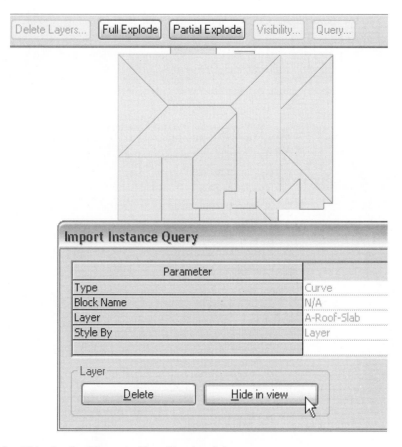

Figure 2.46 *Hide the Roof Layer in Floor Plan Level 1*

4.12 Click the New File icon (see Figure 2.47).

- Do not close the open file.

Figure 2.47 *Open a new file using the default template*

4.13 From the File menu, pick File>Import/Link>CAD Formats.

- Navigate to the folder where you stored the files from the CD for Chapter 2.

- Select *plan imperial.dwg [plan metric.dwg]*.

- Set the Layer/Level Colors option to Black and white, as shown in Figure 2.48. Click Open.

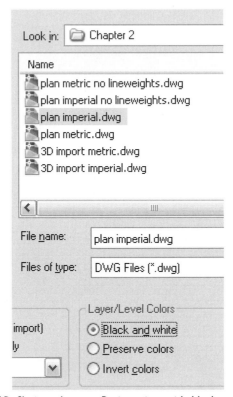

Figure 2.48 *Import a CAD file into the new Revit project with black and white layers*

4.14 Zoom in to examine the linework in the imported file.

Lines representing walls, cabinetry, appliances and stairs appear with different weights. These weights were set in the original file.

4.15 Select the import.

- On the Options Bar, click Query.

- Click on a wall.

The Import Instance Query dialogue will show that its layer is A-Wall. Click OK.

4.16 On the Options Bar, click Full Explode.

The regeneration operation will take about a minute. A warning box will appear.

- Click anywhere in the view window to dismiss the warning.

4.17 Click on a line representing a wall.

The Type Selector reads A-Wall.

- Click the Properties icon.

  The Properties dialogue shows that the CAD file has exploded into Model Lines. The Line Style for this line is A-Wall.

- Click Cancel to exit the Properties dialogue.

Revit places most of its management dialogues under the Settings menu. We will examine these as we move through exercises.

Revit contains default styles for lines and objects, which control the appearance of items in all views. You can modify existing styles and create your own as necessary. You can also change the appearance of lines or objects in particular views. This prevents drawing or drafting one thing many times.

When you explode an imported CAD file, Revit places its lines in new styles with names derived from the layers in the import. You can change imported line styles or move existing lines from one style to another.

4.18 From the Settings menu, click Line Styles.

- Expand the Lines category.

  Revit has created Line Styles from the CAD file (see Figure 2.49).

- Click Cancel.

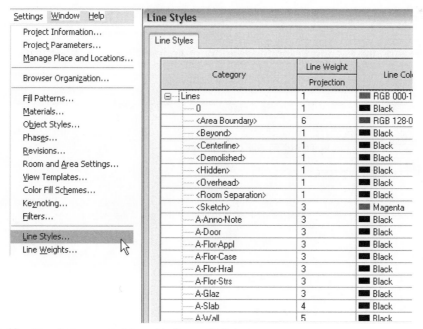

Figure 2.49 *Line Styles created by exploding a CAD import*

4.19 Open a new project file, as in Step 4.12.

- Do not close the open files.

4.20 From the File menu, pick File>Import/Link>CAD Formats.

- Navigate to the folder where you stored the files from the CD for Chapter 2.
- Select *plan imperial no lineweights.dwg [plan metric no lineweights.dwg]*.
- Set the Layer/Level Colors option to Black and white, as before.
- Click Open.

4.21 Zoom in to examine the lines in this import.

This file was set in AutoCAD so that all layers have the same default lineweight. The thickness of printed lines is controlled by color.

4.22 Click the Undo arrow on the Toolbar to undo the CAD import.

The view window will revert to the original view.

4.23 From the File menu, pick File>Import/Export Settings>Import Lineweights DWG/DXF.

- In the Import Lineweights dialogue, set the value for color 4 (cyan) to **5.** See Figure 2.50.

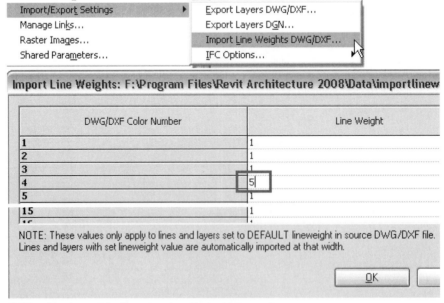

Figure 2.50 *Set the weight for color 4 to Revit thickness 5*

4.24 In the dialogue, click Save As.

You will create a file to hold your new layer-to-lineweight mapping.

- Enter the name *Chapter 2.*

- Click Save.

4.25 Repeat the CAD import from Step 4.20.

4.26 Zoom in, and you will see that walls now have thick lines.

4.27 Close all open files.

- Use the File menu: File>Close.

Since Revit allows you to keep many views from many files open at once, the most efficient way to close a file is by the menu. Clicking on the X symbol at the upper right of the view screen closes only the current view, unless that is the only view open in a file.

Revit will ask about saving a file if you have changed the model or the properties of a view. Zooming or panning in a view, or opening views, is not a change and will not prompt save questions.

- Do not save any files unless you wish to keep them for reference.

EXERCISE 5. EXPORT 2D AND 3D TO AUTOCAD

5.1 From the File menu, pick File>Open.

- Navigate to the *Library/Chapter 2* folder where you installed files from the CD.

- Select *Chapter 2 residence.rvt [Chapter 2 residence metric.dwg].*

- Click Open.

  The file will open to a 3D view.

5.2 From the File menu, select File>Import/Export Settings>Export Layers DWG/DXF. See Figure 2.51.

Figure 2.51 *Open the Export Layers control dialogue*

This dialogue references an external mapping file. Revit makes a distinction between projected lines (as those seen in an elevated wall) and cut lines (the same wall seen in plan).

5.3 Scroll down the Category list until you see Walls.

- Note the layers for Projected and Cut wall lines.

- In the dialogue, click Standard.

5.4 In the Export Layers Standard dialogue, pick the AIA Option (if you are in a metric file) or the ISO13567 Option (if you are in an imperial file).

- Click OK.

5.5 Scroll down the Category list to Walls, as before.

- Note the different standard that has been applied.

    You can create your own standards by filling in appropriate fields and using the Save As button.

- Click Cancel.

5.6 From the File menu, select File>Export>CAD Formats.

- In the Export dialogue, as shown in Figure 2.52, select an appropriate AutoCAD format for your system.
- Select Automatic>Short for File Naming.
- Check Export each view or sheet as a single file.
- Choose Selected Views/Sheets under Range.
- Click Select.

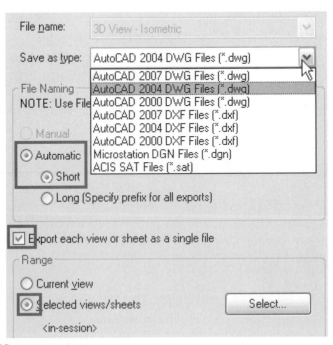

Figure 2.52 *CAD export options*

5.7 In the View/Sheet Set dialogue, check views as shown in Figure 2.53: 3D View: Isometric, Elevation: South, Floor Plan: 1st Flr. const.

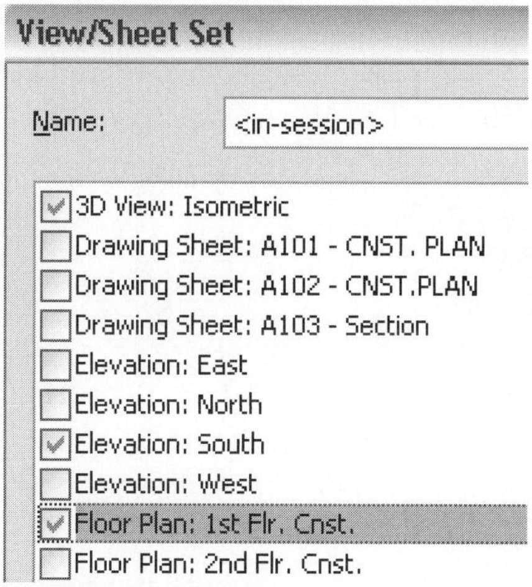

Figure 2.53 *Make a selection of views to export*

5.8 Click OK.

- Click No in the question box about saving your selection.

- Click Save to create three separate export files.

   The Export will take about a minute.

- If a Warning dialogue appears, click Hidden Line Removal to eliminate duplicated lines.

5.9 Launch AutoCAD if it is installed on your system.

- If you do not have access to AutoCAD, skip the remaining exercise steps.

5.10 Open the new files Revit Architecture created in turn to examine their contents and layer structure.

5.11 Zoom Extents in Model Space in the *3D View - Isometric* drawing.

- Change the view from Shaded (Realistic) to Hidden Line to 2D Wireframe and view the results each time.

- In the 2D Wireframe view, pick a railing section on the interior stairwell and note its layer in the Object Properties Toolbar.

- Right-click, pick Properties from the context menu, and note that the railing section is a block.

- Pick a few more entities and note what they consist of: blocks and polyface meshes.

5.12 Open the Layer Properties Manager dialogue box for this file and note the layers Revit Architecture has created (see Figure 2.54).

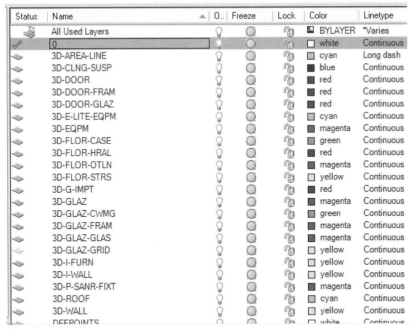

Figure 2.54 *Layers in the 3D file*

5.13 Close the Layer Properties Manager.

- Close this file without saving and examine the *Floor Plan – 1st Flr – Cnst* file.

5.14 Use a crossing window to pick a section of wall including a window and window tag.

- Type **LI**, and then hit ENTER to list the contents of your selection set: it consists of lines and blocks.

5.15 Open the Layer Properties Manager for this file.

- Note the layers that Revit Architecture has created (see Figure 2.55).

  Two layers have a Dash or Long Dash linetype. All layers except layer 0, Defpoints, and a dimension layer have lineweights applied. The drawing has been set up to use color-based plot styles.

Status	Name	▲	On	Freeze	Lock	Color		Linetype	Lineweight
🏛	All Used Layers		💡	○	🔒	■ BYLAYER		*Varies	*Varies
✓	0		💡	○	🔒	☐ white		Continuous	——— Default
↔	A-ANNO-DIMS		💡	○	🔒	■ red		Continuous	——— Default
↔	A-ANNO-DIMS-48		💡	○	🔒	■ red		Continuous	——— 0.05 mm
↔	A-ANNO-TEXT		💡	○	🔒	☐ yellow		Continuous	——— 0.09 mm
↔	A-AREA-LINE		💡	○	🔒	☐ cyan		Long dash	——— 0.09 mm
↔	A-DOOR		💡	○	🔒	■ red		Continuous	——— 0.18 mm
↔	A-DOOR-GLAZ		💡	○	🔒	■ red		Continuous	▬▬▬ 0.40 mm
↔	A-FLOR-HRAL		💡	○	🔒	■ red		Continuous	——— 0.09 mm
↔	A-FLOR-OTLN		💡	○	🔒	■ magenta		Continuous	——— 0.18 mm
↔	A-FLOR-STRS		💡	○	🔒	☐ yellow		Continuous	——— 0.09 mm
↔	A-GLAZ		💡	○	🔒	■ magenta		Continuous	——— 0.09 mm
↔	A-GLAZ-CWMG		💡	○	🔒	■ green		Continuous	——— 0.09 mm
↔	A-SECT-MBND		💡	○	🔒	■ blue		Dash	——— 0.09 mm
↔	A-WALL		💡	○	🔒	☐ yellow		Continuous	——— 0.18 mm
↔	Defpoints		💡	○	🔒	☐ white		Continuous	——— Default
↔	S-GRID		💡	○	🔒	■ red		Continuous	——— 0.05 mm

Figure 2.55 *Layers in the Floor Plan*

5.16 Close the Layer Properties Manager.

- Close the file without saving and open the *Elevation - South* file.

5.17 Pick an area of brick facing and list it—a hatch on layer A-WALL-PATT, an AIA standard layer name.

5.18 Open the Layer Properties Manager and note the names and properties of layers in this file.

- Pick Cancel to close the Layer Properties Manager.

5.19 Close the file without saving and exit AutoCAD.

CONCLUSIONS TO DRAW: STANDARD SETTINGS SAVE STEPS

Proper understanding of template files and program settings benefit all users of complex applications such as Revit Architecture. Whether you are a beginning student or a seasoned user, you can save yourself work by utilizing what has already been created for you in even the simplest file, and the more you know about the underlying framework the more quickly you can be productive.

The exercises in this chapter touched on import/export settings to illustrate the fact that very few building designs are created without having a relationship with previous work. In today's hi-tech environment, the ability to bring in data efficiently from sources outside of the current project file, and to send coherent, smart information from the project file out to other applications reliably, is considered part of a basic skill set.

Revit Architecture Works with CAD

CAD managers or others responsible for maintaining the flow of CAD information through a multi-platform office should take heart from this demonstration. Revit Architecture readily accepts AutoCAD content, and just as readily creates output to AutoCAD that captures in a legitimate and completely adjustable form the content of the Revit Architecture model. Firms with experienced AutoCAD drafters or detailers, or a valuable inventory of standard detail files, can proceed with implementation of Revit Architecture, knowing that Revit Architecture's modeling capabilities will not bypass the strengths of their workforce or archives.

Advanced Teamwork Tools: Links and Worksets

One of the early criticisms of Revit Architecture was of its single building model structure. Most users of AutoCAD are familiar with the xref system, whereby an unlimited number of files can be linked to compose a complex building model or a campus, with very little strain on system resources. External references also allow team members to work independently on parts of an assembled whole.

You have seen that Revit Architecture will import AutoCAD and other files. An import is not a live connection between files—in that way it corresponds to a block insertion in AutoCAD. Revit Architecture uses a system of live links that correspond exactly to external references in AutoCAD. Revit Architecture will create and maintain links with DWG, DXF, DGN and RVT (Revit Architecture) files. Links with other Revit Architecture files are intended to be links between entire building models, as in a campus or site plan. Revit Architecture warns specifically that linking a vertical assembly of floors together to make a single building is not supported. This is the opposite of the way firms use AutoCAD and AutoCAD Architecture to create assembled building models, so be aware that designers moving from AutoCAD-based design to Revit Architecture will have to learn some new methods.

Revit links are tied now to Groups, a mechanism for combining many elements into a named entity that can be placed many times. Intelligent use of groups and links makes it possible for design teams to work on portions of a design efficiently without having to develop Worksets.

Revit Architecture has developed the Workset system for design teams working on a single building model. Worksets—divisions of the model using all sorts of criteria—provide a mechanism for individuals to check out parts of the model from a central storage location and return these parts to update the central file. This is also an opposing concept to the usual xref setup, and one that CAD managers will have to understand to make effective use of Revit Architecture and prevent loss of time or work on a project.

AutoCAD xrefs usually consist of many separate files, often on completely separated company systems and servers. The individual part files are collated in one or more files to create an assembly model or plotting setup, and often the designers or detailers who work on the parts and pieces never see the entire assembly. Revit Architecture's Worksets work the other way around. There is always a central file; worksets are subsets of the central file. Users editing worksets update them periodically in the central file without gaps or lapses in order for information to flow correctly to other members of the team. This system is not a return to the days of mainframes; Revit Architecture has mechanisms to allow work to take place remotely.

You shall look specifically at links and worksets in future exercises.

REVIT ARCHITECTURE IS A CHANGE ENGINE

So far in this chapter you have examined topics that concern CAD managers, who have to maintain a filing system and set of company standards that exist quite apart from what may or may not happen with Revit Architecture in the organization. CAD managers, expert though they are in CAD, may not get the chance to become familiar with Revit Architecture during the period when their company is implementing its use. These managers must make sure that those who do use Revit Architecture can communicate effectively with the network, other departments or offices, and those in the company who will continue to use CAD.

There is an individual in most firms implementing Revit Architecture who needs to be aware of its effect on work habits and methods—the Project Manager. A Project Manager's primary responsibility is to assemble and shepherd the design team through its paces. No matter what your previous training and experience with Revit Architecture, there are bound to be a few surprises when people new to the program go live on a project. As with any foray into the unexpected, preparation and planning are key. You can't emphasize enough the value of organized training. Reasonable expectations, given the size, complexity, and deadlines of the trial project, are also critical for team success. Team makeup will play an important role in reaching project goals and developing the efficiency benefits that Revit Architecture makes possible.

Training

Managers are often responsible for scheduling training or enhancement instruction to keep designers aware of progress with CAD software. People who become relatively efficient with a particular set of tools or methods often become hide-bound and not interested in learning new practices and techniques, which results in loss of efficiency over time. Any manager who has tried to convince drafters who never mastered AutoCAD's Paper Space of the superior efficiency of multiple Layouts, will understand the difficulty of keeping workers up-to-date in this era of constant change.

The interface and tools make Revit Architecture considerably easier to pick up and start working with than nearly any CAD drafting program. It is quite possible for individuals to teach themselves enough in a short time to create the basics of designs and visualizations that are far more advanced than what they could have created in CAD in the same amount of time, even after training. However, Revit Architecture is a powerful, sophisticated apparatus that requires more than a rudimentary amount of knowledge to work effectively and efficiently. More and more firms are realizing that training is essential. The amount of training for experienced designers to become efficient in Revit Architecture is measured in hours or days, not weeks, but a training budget should be considered part of any implementation strategy.

Along with the necessary training, managers can help quite a bit by offering designers the encouragement that these brave souls deserve for tackling the job of learning an entire new software system in the stressful context of a live project.

Firms implementing Revit Architecture are reporting with regularity their conclusions that on-the-job training is not as effective as 16 to 32 hours of dedicated training time with Revit Architecture's online lessons and tutoring from users with previous experience.

Manage Expectations

Successful implementers also report that keeping the first project or two to an appropriate size for new users to handle goes a significant way to ensuring success. Revit Architecture is a complex, comprehensive design package. Nobody ever became proficient with AutoCAD, MicroStation, or AutoCAD Architecture overnight. Despite Revit Architecture's relative ease in creating complex models quickly, project and company management must allow the inevitable learning curve to run its course. Smart building models take a little more time for new users to create than line drawings, but provide downstream benefits.

Resistance and Skepticism

The willingness of designers and drafters to take on new challenges will also play a part in their ability to deal with the tensions of producing a project with new tools. Early adopters who relish the chance to learn and expand their skills will provide a different atmosphere than skeptics or those who work best entirely within their comfort zones. Change always meets some resistance, and new methods have to prove effective under fire, so to speak. It's the project manager's challenge to keep designers focused on solving problems, not on the perceived difficulties of working with unfamiliar tools.

Versatility Is Valuable

Revit Architecture does not remove the need for designers to understand drafting. It does eliminate the primacy of drafting skills in today's building model design process, and will hasten the disappearance of the CAD jockey whose sole function is as a scribe for designers. A Revit Architecture model and documentation package requires far fewer repetitive tasks—designers spend more time designing, and the software takes care of providing views, updating annotations, extracting numbers, and the like. Designers with versatility rather than specialists in CAD skills will play larger roles in Revit Architecture projects.

Changing Roles

What is a manager to expect from Revit Architecture? While each project will be different, and results will vary from firm to firm, certain trends are already clear.

Revit Architecture's ability to produce renderings within the application, and the relative ease of exporting models to other rendering programs, mean that design illustrations and visualizations will come into play earlier than before. This will mean earlier client response to conceptual design options, hardly ever a bad thing. Revit Architecture includes a suite of site development tools that allow building designers to specify site design more accurately and completely than before, without having to export plans to other applications.

Revit Architecture relies heavily on families, which function as combinations of AutoCAD Styles and 3D Blocks. Families manage functional aspects such as titleblocks and building components such as walls or furniture. Revit Architecture provides a number of family templates for the creation of custom content. Skill at creating families for company use will prove as valuable as was skill at composing AutoCAD library material. Creating and managing family content is a subject large enough to merit its own book.

Revit Architecture provides tools for data handling, including schedules for nearly every type of component, with customizable parameters for holding information that can be exported from schedules or shared between projects. Once schedules are set up in a project, updating and handling the information they contain becomes more of an administration task than a job requiring CAD or design skills to initiate. Revit Architecture's space-planning tools also lend themselves to use by administrative personnel.

The designed effect of Revit Architecture's modeling, documentation and data handling structure is to move nearly all project tasks down the food chain, to eliminate repetitive, low-value, but specialist work, and to return control of design projects to designers. This will eventually mean changes in company workforce structure. For managers implementing Revit Architecture in a first design project or two, you can expect to be pleasantly surprised at the ease with which designers will produce complex building models and documentation. The longer lasting value of those complex models will only become apparent over time.

SUMMARY

This chapter has taken a look at some of the more technical aspects of Revit Architecture—templates and titleblocks are important parts of any firm's design standards. Proper understanding and setup of templates will result in significant benefits. You examined Revit Architecture's titleblock structure to illustrate labels, data fields and parameters. You explored how Revit Architecture will accept and work with vector information from other CAD formats. Lastly, you saw how Revit Architecture will export model information to CAD.

REVIEW QUESTIONS - CHAPTER 2

MULTIPLE CHOICE

1. Template files can contain

   a. customized Levels

   b. customized views and sheets

   c. customized family content

   d. all of the above

2. Template files cannot contain

   a. custom menus

   b. custom Design Bar tabs

   c. custom lineweights

   d. a and b, but not c

3. Revit Architecture will export

   a. to specified layers

   b. the current view or selected views/sheets

   c. 2D and 3D information

   d. all of the above

4. An imported CAD file comes in as

   a. individual line, dimension and text entities

   b. a single object that can be exploded to individual entities

   c. a single object that can't be modified

   d. an image

TRUE/FALSE

5. **True _ False _** Revit Architecture allows you to delete Layers from imported AutoCAD files.

6. **True _ False _** Once you save a template file, you can't edit it later.

7. **True _ False _** Revit Architecture does not recognize text in imported files that have been exploded.

8. **True _ False _** Project information automatically appears in certain titleblock label fields.

Answers will be found on the CD.

CHAPTER 3

Modeling with Building and Site Elements

INTRODUCTION

This section begins with a series of related exercises using basic Revit Architecture techniques to create a project file. The exercises show you how to bring existing AutoCAD files into Revit Architecture, and use those files as references to create a complex site object and complete building shell.

OBJECTIVES

- Import CAD files as references for design project setup
- Model a building shell and apply Phases
- Create and subdivide a site object
- Start detailing the building shell using modeling, editing and sketching techniques

REVIT ARCHITECTURE COMMANDS AND SKILLS

Settings

File import

Import Query

View Properties: Scale, View Range, Phase, Phase Filter, Crop Region

Visibility/Graphics Overrides

Model Graphics Style

Walls

Type Editing

Elevations

Levels

Reference Planes

Toposurface: Use Imported

Pad

Subregion

Split Surface

Materials

Roofs

Phases

Demolish

Windows

Load From Library

Sketching: Line, Arc, Tape Measure

Editing commands: Move, Copy, Align, Trim, Split, Array

Grids

Group

Wall Properties: Constraints, Layers, Split Layer, Sweep

Profile Editing

EXERCISE OVERVIEW

The exercises in the rest of this book will follow a hypothetical design project. The program of this project is as follows:

- Project setup tasks: Work with an existing site containing one multistory building, an access road, and a parking lot. The building holds classrooms and offices. Model the existing building shell and site in Revit Architecture. The reference material consists of AutoCAD files—a site plan in 3D, and 2D exterior elevations of the existing building.

- Project Phase 1 tasks: Design a second multistory building near the first one. The new building is to hold offices, classrooms, labs and a kitchen/cafeteria facility. Transform one wing of the existing building into an elevated passage to the new building. Move the parking lot underground; turn the space between the two buildings into a pedestrian courtyard; create a pedestrian plaza over the new parking garage. The reference materials consist of scanned sketches.

- Project Phase 2 tasks: Design a second building to contain offices, a dining facility, and a residence tower. Move the occupants of the offices in buildings 1 and 2 to building 3. Convert the office space in building 2 to classrooms. Convert building 1 into a residence hall. There is no reference material for this phase.

- Project Phase 3 tasks: Design a performing arts building to sit between buildings 2 and 3. This building is to have offices; rehearsal halls; an art gallery; and theater space with a stage, wing, overhead space, a projection booth and raked seating. The building will feature a central atrium with skylights, cafe with seating and a ceremonial staircase. There is no reference material for this phase.

The exercises will be designed to cover all the basic Revit Architecture skills necessary to complete a design and documentation project using multiple files. The exercises will develop skills and techniques as they would be used in the course of a typical design project; for this reason the exercises are best done in order.

Teamwork is an important part of every design project not produced by a single individual. Autodesk long ago developed wblock insertions and the external reference system for AutoCAD so that more than one person could work on a project and share information using a couple of different modes. Sheet Set Manager, Drawing Manager and the Project Navigator are current mechanisms for project setup and management. Revit Architecture contains a similar import/link system for CAD files, and a single building model can include links to numerous other Revit Architecture files, so that designers can coordinate models in a multibuilding or campus development. Revit Architecture contains Worksets and Phases, mechanisms for dividing and coordinating design tasks.

A project of the size and scope we have outlined takes many hours of work to complete, and a large portion of that work is inevitably repetitious, even with a design application as complete as Revit Architecture. Our exercises will touch on each task phase of the hypothetical project, but we will not carry these phases through to completion, once the appropriate concepts and techniques have been illustrated.

 INSTRUCTOR'S NOTE The scope of this hypothetical project will provide opportunities for auxiliary exercises for projects or extra credit. Students can explore alternate design ideas or flesh out areas of the model (classroom, office or residence layouts).

For those individuals working these exercises on their own and not as part of a group, we admire your gumption and encourage you to work through the grouping, file linking and Workset exercises in later chapters so that you will understand how files and projects are subdivided and linked together. Even if you will work entirely by yourself for the foreseeable future, you will certainly find a use for a multifile project, or Worksets.

To begin, in this chapter, you will open a blank file and import a series of AutoCAD files—a floor plan, site plan and elevations for the existing building on the site. You'll use those files as background to make a Revit Architecture model of the existing building, and use that model to export an elevation of one face of the building that incorporates new design information out into AutoCAD.

EXERCISE 1. CREATE A PROJECT—ESTABLISH SETTINGS

1.1 Launch Revit Architecture. Revit Architecture will open to a Floor Plan view of Level 1 in an empty project file using the default template.

Four Elevations appear in the View window.

- Place the cursor anywhere in the View window (but not over an Elevation or Elevation View control arrow), and right-click.

    This will open the right-click context menu (see Figure 3.1). View Properties is always the bottom command.

1.2 Select View Properties to open the Element Properties Dialogue for Level 1 (see Figure 3.2).

- Study the properties of this view, noting the default View Scale (1:96, or 1/8" = 1'-0" in an imperial file, 1:100 metric) and Phase information.

    You will have to scroll down to see the Phase field.

- Click Cancel.

Figure 3.1 *You can reach View Properties with a right-click.*

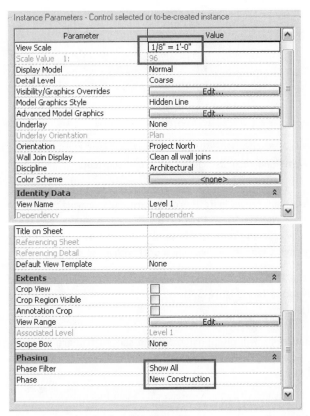

Figure 3.2 *The View Properties for the Level 1 Floor Plan in an empty Imperial file*

1.3 Double-click on Site in the Project Browser to open that view.

This view also shows the four Elevations.

- Type **VG** at the keyboard to open the Visibility/Graphics dialogue for the current view.

VV opens the same dialogue.

 Tip: Revit Architecture contains a number of two-letter keystroke shortcuts for common commands and the user can create more. Shortcuts appear on drop-down menus (see Figure 3.3). We will introduce a number of them in the exercises in this and future chapters.

Figure 3.3 *Menu items and their associated keystroke shortcuts*

 Note: Views are crucial in Revit Architecture, and managing them is an important skill to develop early. The **View Properties** dialogue contains overall controls for the current view plus the **Visibility/Graphics** dialogue, which contains visibility settings for every object type in the building model within the current view.

1.4 Choose the Annotation Categories tab of the Visibility dialogue for the Site view and clear Elevations (see Figure 3.4).

- Click OK to exit this dialogue.

  The Elevations will disappear from the screen. You are removing them in this view to make future work less cluttered.

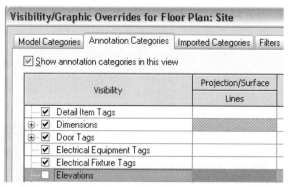

Figure 3.4 *Making the Elevations not visible in the Site View*

1.5 From the Settings menu, pick Settings>Options (the bottom selection).

- In the Options dialogue that opens, on the General tab, verify that the Save Reminder interval value in the Notifications section is set to 30 minutes.

  Revit Architecture does **not** auto-save (see Figure 3.5).

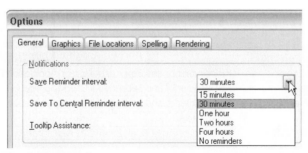

Figure 3.5 *Setting the Save interval*

1.6 On the File Locations tab, click Browse to set the Default path for user files as directed by your instructor if you're in a class (see Figure 3.6).

- Otherwise, set the path to your work project directory.

  This can be a network location. Your path will be different from the one pictured.

- Click OK in the Browse for Folder dialogue to return to the Options dialogue.

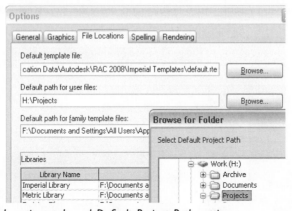

Figure 3.6 *The File Locations tab and Default Project Path setting*

1.7 In the Libraries area of the File Locations tab, click the icon shown in Figure 3.7 to create a shortcut to a file location.

The new library shortcut will be named New Library 1.

- Click the Library Name to rename it.

- Enter **Revit Book CD Files** or another name that makes sense to you.

- Click in the Library Path field.

- Navigate to the folder where you stored files from the CD included with the book.

- Click OK twice.

Library Name	Library Path
Imperial Library	F:\Documents and Settings\All Users\Application Data\A
Metric Library	F:\Documents and Settings\All Users\Application Data\A
Training Files	F:\Documents and Settings\All Users\Application Data\A
Work	H:\Projects
Revit book CD files	H:\Projects\Revit Book 2008\CD Files\Library\

Figure 3.7 *Create a new library shortcut by clicking this icon.*

1.8 From the File menu, select Save As.

- In the Save As dialogue, pick the Options button.

- In the File Save Options dialogue that opens, set the number of backup files to save to 1, or as specified by your instructor (see Figure 3.8).

  The default setting is for 3 backups, identified by a 00x addition to the file name. Revit Architecture will save up to 50 copies of Workset-enabled projects. This option is *not* a global Revit Architecture setting, and *must be adjusted for each new file*.

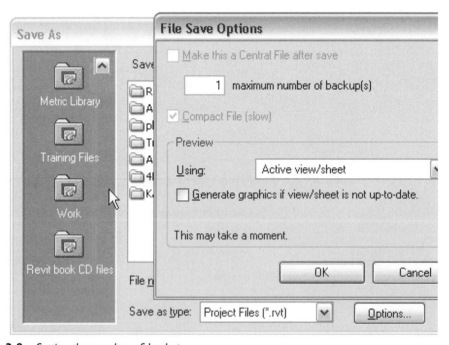

Figure 3.8 *Setting the number of backups*

Note: The new library you defined in Step 1.7 is visible in the shortcut pane on the left side of the File Save dialogue. This shortcut, and any others you create in the Options dialogue, appears whenever you navigate to open or save files.

1.9 Click OK to close the File Save Options dialogue.

- In the Save As dialogue, give this new project the name **Chapter 3 Phase 1**, and save it as type *Project Files (*.rvt)* in the location specified by the class instructor (Figure 3.9).

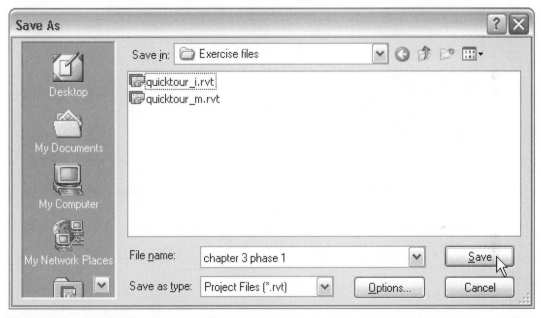

Figure 3.9 *Give the file a name and location.*

EXERCISE 2. START A SITE PLAN USING IMPORTED CONTENT

2.1 Open the file *Chapter 3 Phase 1.rvt* from the previous exercise if you closed it after Step 1.8, or continue with the open file.

- From the File menu, pick File>Import/Link>CAD Formats.

- In the Import/Link dialogue that opens, use the Look in: drop-down list to navigate to the folder that contains the file *site plan chapter 3 imperial.dwg/site plan chapter 3 metric.dwg* (from the CD).

- Pick that file so that the name highlights in the folder contents window and the name appears in the File name: field, as shown in Figure 3.10.

  A preview image will appear.

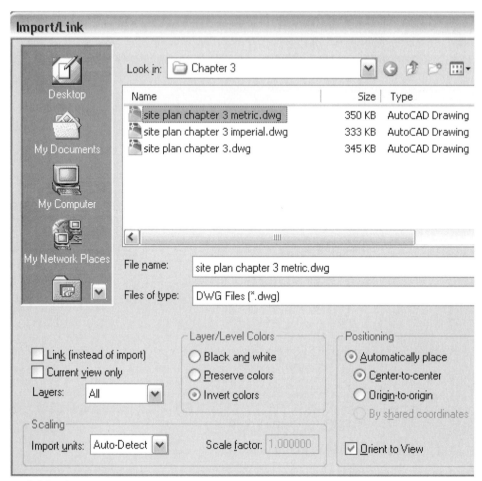

Figure 3.10 *Import settings*

2.2 Accept the default options for Import or Link, Scaling, Layer/Level Colors and Positioning (see Figure 3.10).

- Click Open to start the import.

2.3 When the screen shows linework, type **ZF**, the keyboard shortcut for Zoom to Fit.

2.4 When the cursor is placed anywhere over the imported object, a Tooltip will appear that identifies it as an Import Symbol, and a border will appear around the import.

- Click to select the imported symbol.

   The cursor will change to a double-headed arrow, to signify that you can now drag the symbol to another position. The contour lines will change from black to red, Revit's color for selected items. The Type Properties drop-down list on the Options

Bar will show the type category of the import (Import Symbol: site plan chapter 3 .dwg [Import Symbol: site plan chapter 3 metric.dwg]), the Properties icon will become active, and other action buttons become visible on the Options Bar.

- With the cursor over the selected symbol, right-click and study the right-click context menu.

With an object selected, commands available on the Options Bar are also available by using right-click. In this case, Delete Selected Layers and Explode options appear in the context menu (see Figure 3.11).

Figure 3.11 *Available commands for an Imported Instance*

2.5 Click anywhere in the view window off the context menu to dismiss it.

- Click if necessary to select the Import object.

- Pick Query from the Options Bar (see Figure 3.12).

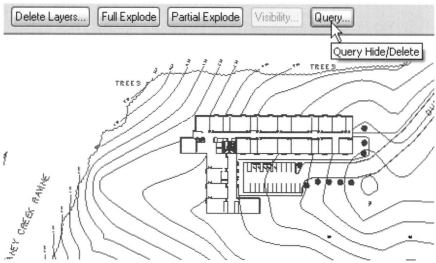

Figure 3.12 *Query the imported file*

At this point, you begin working with walls, which are obviously important in architectural design. In Revit you can place walls by drawing lines or by picking lines. When you have a sketch or plan to trace, a very common condition, you can create many walls quickly by using pick options. Later exercises in later chapters will provide opportunity to practice drawing walls.

CREATE WALLS USING THE IMPORTED FILE

2.6 Zoom in using the scroll wheel so you can clearly see the outline of the building. Pick one of the blue lines.

- Note the name of the Layer in the dialogue, as shown in Figure 3.13. Click OK.

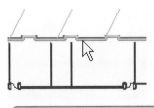

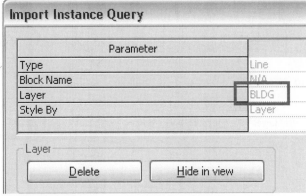

Figure 3.13 *Use Query to find the Layer of the building outline.*

2.7 Type **VG** to open the Visibility/Graphics dialogue for the view.

- Clear checkboxes under the name of the imported drawing except for BLDG, as shown in Figure 3.14.

- Click OK.

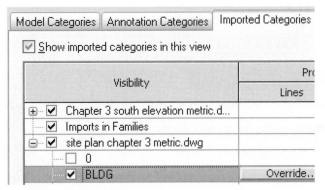

Figure 3.14 *Isolating the building outline*

2.8 Zoom in to the lower-left corner of the building outline so you can clearly see the wall lines.

- From the Basics Tab of the Design Bar, click Wall.

- In the Type Selector, choose Basic Wall: Generic: 12"[Basic Wall: Generic: 300mm].

2.9 Click the Pick (arrow) option on the Options Bar.

- Set the Location Line Option to Finish Face Exterior.

2.10 Hold the cursor over a line representing an exterior wall face.

Revit will display a green dashed line between the blue wall lines to indicate where a wall will be placed if you click.

2.11 Hit TAB to cycle the pick options until both edges of all visible walls highlight.

The Tooltip and Status Bar (lower left) will read Chain of Walls or Lines, as shown in Figure 3.15.

- Click to place walls all around the building outline.

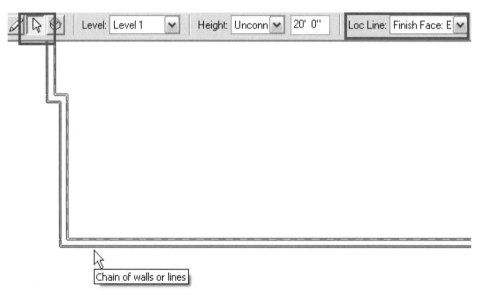

Figure 3.15 *Place walls all around the building outline with one click.*

2.12 Click the Default 3D view icon on the View Toolbar to open that view.

2.13 Click Modify to terminate wall placement.

- Select the CAD import.

- Right-click and select Hide in View>Elements (see Figure 3.16).

- Zoom to Fit.

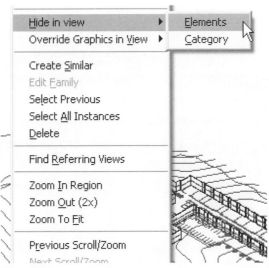

Figure 3.16 *Hide the CAD file in the 3D view.*

2.14 Double-click Floor Plan: Level 1 to return to that view.

- Type **VG** to bring up the Visibility dialogue.

- On the Imported Categories tab, clear the check box next to the name of the import file.

- Click OK.

    This will turn off all CAD elements in the view.

ADD FOUNDATION WALLS

2.15 Click Wall on the Basics Tab of the Design Bar.

- In the Type Selector, choose Basic Wall: Foundation – 12" Concrete [Basic Wall Foundation – 300mm Concrete].

- Click the Properties icon next to the Type Selector.

    You are about to create a new wall type that does not exist in this project file by duplicating and changing an existing type. This is a basic skill in Revit.

2.16 In the Element Properties dialogue, click Edit/New to open the Type Properties dialogue.

- Click Duplicate.

- In the Name dialogue type the name **Foundation – 16" Concrete [Foundation – 450mm Concrete]**. Click OK. See Figure 3.17.

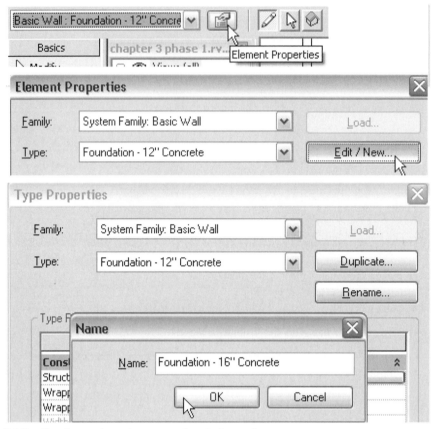

Figure 3.17 *Creating and naming a new foundation wall type*

    2.17 In the Type Properties dialogue for the new wall type, click the Preview button in the lower-left corner.

        • Click Edit in the Value Field for Structure, as shown in Figure 3.18.

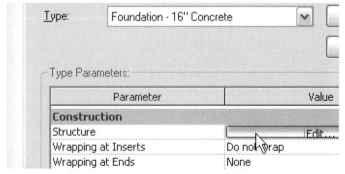

Figure 3.18 *Edit the structure of the new wall type.*

2.18 In the Edit Assembly dialogue, click inside the Thickness field for wall layer 2 (Structure).

- Edit the thickness value to **1'4 [450]** (see Figure 3.19).

Revit's default imperial unit is the foot, and default metric unit is the millimeter. You can type foot/inch values out completely using single and double quote marks to indicate feet and inches, or you can indicate the foot value by hitting the Spacebar. You need not add a suffix for millimeters when inputting metric distances.

The preview will update to show the new thickness value. The Total Thickness reading will also update.

- Click OK to exit all dialogues.

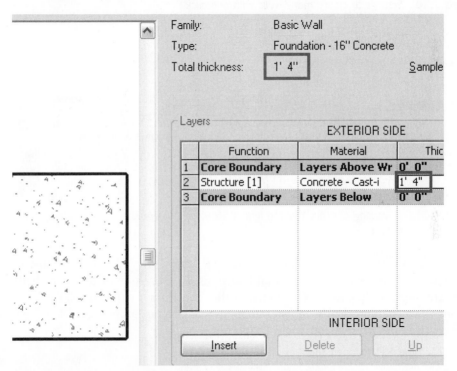

Figure 3.19 *You change the thickness of floor elements in this dialogue. Floors can be complex objects.*

2.19 Select the Pick option on the Options Bar, and set the Location Line to Wall Centerline.

Note that foundation walls are created from the top down, so the Options Bar indicates a Depth value rather than Height.

2.20 Place your cursor over a wall and tab to cycle until the entire chain of walls highlights, as before (see Figure 3.20).

- Click to place walls.

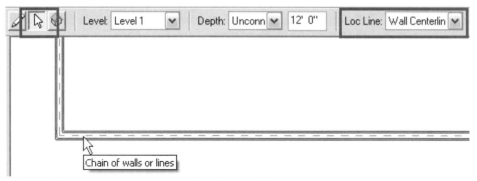

Figure 3.20 *Settings to create many walls with one click*

A warning box will appear.

2.21 Click on the X to dismiss it.

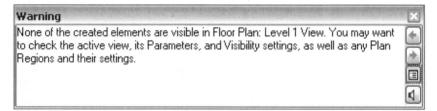

Figure 3.21 *A warning appears. You can ignore warnings, but errors force you to stop and try something else.*

All views in Revit Architecture except Drafting Views are 3D. Orthographic ("straight-on") views such as plans, elevations and sections do not show distance or depth, but in Revit Architecture they have depth settings that control what they display. Elevations and sections have graphic depth controls that also appear as properties. Plan View Depth settings are found only in the View Properties dialogue.

2.22 Type **VP** to open the View Properties dialogue.

- Scroll down to the Extents section and click Edit next to View Range.

- In the View Range dialogue, set the View Depth value to Unlimited (see Figure 3.22).

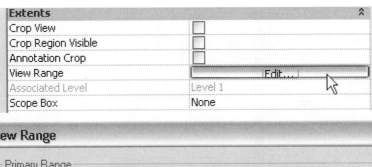

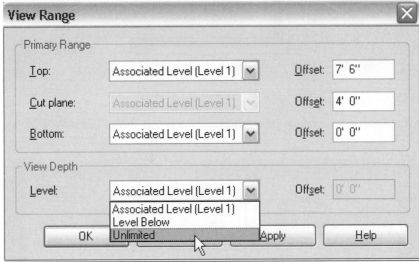

Figure 3.22 *Set the View Depth to reveal the foundation wall. Plan Views are 3D.*

2.23 Click OK twice to exit the dialogues.

Foundation walls will appear. They are below the cut plane of the view, so they display with lighter lines than the exterior walls, as shown in Figure 3.23.

Figure 3.23 *Greater View Depth shows walls below Level 1*

2.24 Open the default 3D view.

External walls and foundation walls will appear, as shown in Figure 3.24.

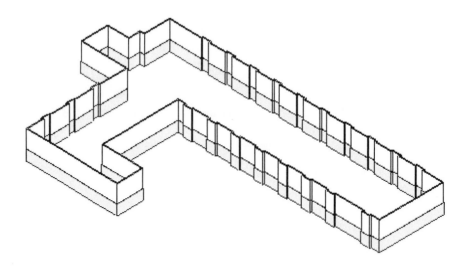

Figure 3.24 *The building outline; walls up and down from Level 1*

You have created the outline of a complex building shape in 3D, delineating two different types of walls, with only a few mouse clicks, using information from another file format. You have not needed to draw a line.

- Save the file.

EXERCISE 3. ADJUST AND ADD FLOOR LEVELS

3.1 Open the file from the previous exercise or continue in the open project.

3.2 Open Floor Plan Level 1 view.

- Zoom to Fit.

ADJUST ELEVATION POSITION, CREATE A NEW ELEVATION

You inserted the CAD file used to create the building outline without regard to its location in the Revit project. Since you created walls according to a location on an imported object, and not in the center of the project file, the elevations do not point at the walls. Since you will use the CAD file again to create a site object known as a toposurface, it will be easier to move the elevations.

Elevations in Revit are complex objects. They consist of the elevation object, the symbol (tag) for which can be varied, and the elevation views associated with the Elevation (up to 4—see Figure 3.25). You can activate or delete elevation views by checking or clearing the boxes that appear when you select the elevation symbol. You can rotate the elevation by clicking on the rotation control and dragging it.

You can place one interior elevation and activate four views from it. The views will seek walls to set their limits.

ID fields in the elevation and views will fill in automatically when views are placed on sheets. You will see this behavior in later exercises.

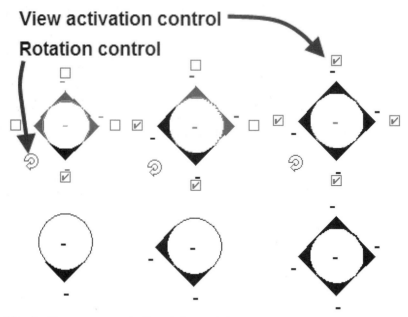

View activation control

Rotation control

Figure 3.25 *An Elevation tag with (from left to right) one, two and four elevation views activated. Above, the elevation controls activate when it is selected; below, the normal state. Round (metric) tag shown.*

To modify an elevation, you should select both the elevation and its associated views. In this case, you want to move more than one elevation at a time.

To build a selection set in Revit, hold down CTRL while clicking on items to select them. Use SHIFT + select to remove an item from a selection set.

You can also use a window selection, which will often be quicker than selecting items one at a time. Window selection in Revit works the same as in AutoCAD. Left-click in open space, hold the mouse button down, and drag a window. Moving the cursor left to right creates an enclosing window—anything you wish to select must be completely within the window, and the window is shown by a solid line. Moving the cursor right to left creates a crossing window—anything inside or touched by the window is selected, and the window shows with a dashed line.

3.3 Drag a window around the North and South elevations, as shown in Figure 3.26.

- Place the cursor over an elevation so that it changes to the double-headed arrow.

- Hold the left button down and drag the elevations so they face the upper and lower horizontal walls.

- Release the button and click anywhere to terminate placement.

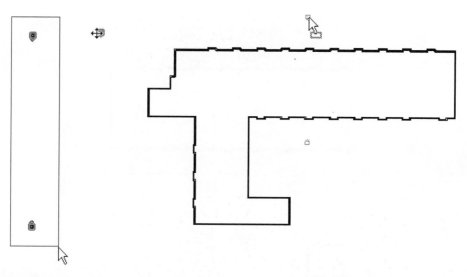

Figure 3.26 *Select elevations (left), drag (center) and relocate (right) the North and South elevations*

3.4 Repeat the select and drag operation on the East and West elevations so they face the building walls.

3.5 Select the South elevation and view.

- From the Edit Toolbar, click Copy.

  Copy is a two-step process.

- Click anywhere to establish a start point, then drag the cursor straight down approximately 100' [30000mm] and click again.

  The exact location is not important, so long as it is below the building. Revit will create a new Elevation named South1 in the Project Browser. See Figure 3.27.

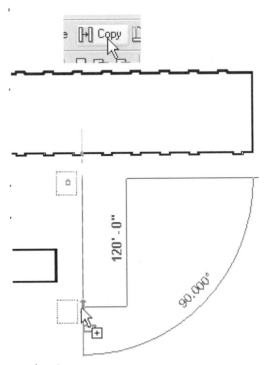

Figure 3.27 *Create a new elevation*

3.6 Place the cursor over the elevation arrow for Elevation: South.

- Right-click. In the context menu, select Go to Elevation View, as shown in Figure 3.28.

  You can also double-click on an elevation arrow to open the indicated view.

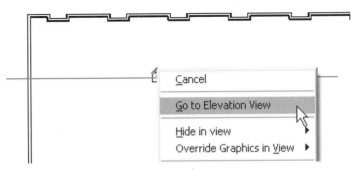

Figure 3.28 *Use the right-click menu to open an elevation*

3.7 In the elevation view, the imported CAD file is visible.

You imported it into all views so that you can use its 3D volume to create a site object, but you do not need to see it in elevations.

- Click on the import, right-click, and select Hide in View> Hide Category. See Figure 3.29.

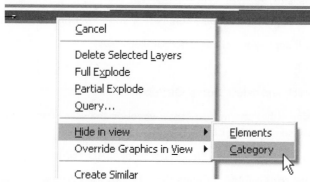

Figure 3.29　*Hide this CAD file in the view*

IMPORT A CAD FILE TO SET FLOOR LEVELS

3.8 Zoom to Fit.

You will see walls and two Levels with symbols on their right ends. Select either level line. Controls on it will be come active. You can turn the symbols at either end on or off, change the name and elevation, and change the properties of the line.

- Use the drag control to pull the left end of the line to the left of the building model, as shown in Figure 3.30.

The other level will adjust as well. Level ends are locked together by default. You can use the padlock icon to toggle the lock on and off if you need to stretch level lines individually.

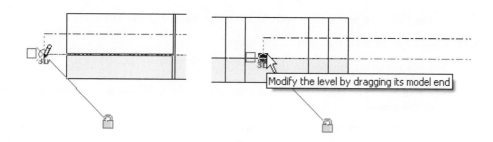

Figure 3.30　*Drag the level lines past the building*

3.9 Zoom in so you can see the level symbols clearly.

- Drag the right ends of the levels closer to the building.

3.10 Click Propagate Extents from the Options Bar.

- In the dialogue, check Elevation: North and Elevation: South 1. Click OK.

  This will apply your changes to parallel views.

3.11 Zoom to Fit.

- Right-click and select Zoom Out (2x).

- Pan using either the mouse scroll wheel or the scroll bar at the bottom of the view window so the model is on the left side of the view.

3.12 From the File menu, select File>Import/Link/CAD Formats.

- Navigate to the folder holding the CD files for Chapter 3.

- Select *Chapter 3 elevation south.dwg* [*Chapter 3 elevation south metric.dwg*].

- Set the import options as shown in Figure 3.31: Check current view only, Black and White, Manually place, Cursor at origin.

- Click Open.

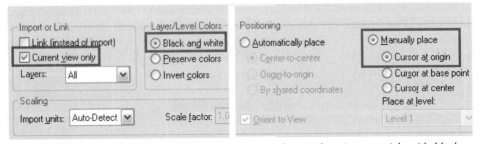

Figure 3.31   *Bring this CAD file in to the current view only, at a location you pick, with black and white lines*

The CAD file will appear with its lower-left corner under the cursor.

3.13 Click to place it to the right of the model.

  You will set exact location in the next step.

- Zoom in so you can see the level indicators in the CAD file and the level line heads of the model.

Revit provides a 3D Align tool that allows you to snap objects to faces of other objects. The objects do not have to be parallel or close together. You can lock the alignment once it has been created; thus, if the reference item moves, then items aligned with it will also move. The effects of this in design workflow are tremendous. You can place elements without specifying their initial position exactly, and align them precisely at a later stage.

3.14 Select the Align tool from the Toolbar.

This tool works with two picks: the first is the reference object, and the second object picked will move into alignment with the first.

- Pick the level line for Level 1.

  A dashed green reference line will appear.

- Pick the level line identified as 1ST FLOOR in the imported file.

  The import will align so its 1ST FLOOR line is even with Level 1, and a lock toggle (unlocked) will appear. Do not select the lock.

  The Align tool is still active.

- Select the line identified as 2ND FLOOR in the import to make it a reference.

- Select the level line for Level 2. See Figure 3.32

  Level 2 will move up, its elevation field will read **13' – 4" [4064],** and the lock option appears. Do not select the lock.

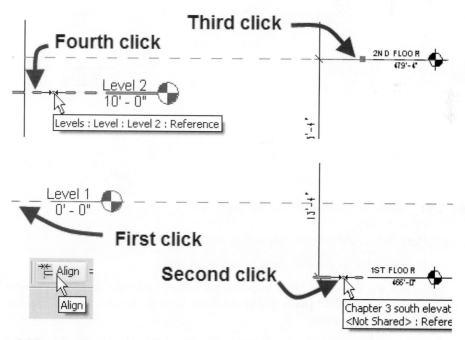

Figure 3.32   *Align the import with Level 1, then align Level 2 with the import*

3.15 Zoom to Fit.

- From the Basics Tab of the Design Bar, select Level to add more Levels to the project, to match the CAD file.

- Move the cursor to the left of the View window above the Level 2 level line.

  Revit Architecture will read from that line and show a temporary dimension up from the level line.

- Type **13 4 [4064]** at the keyboard to match the dimension in the import object.

  Revit Architecture will apply that dimension and show the level's Elevation bubble (see Figure 3.32). When typing imperial units you can use the Spacebar in place of the foot (') symbol.

- Move the cursor to the right and Revit Architecture will snap to an alignment point above the existing bubbles.

- Left-click to accept that alignment point.

  A new Floor Plan and Ceiling Plan named Level 3 appear in the Project Browser. Revit Architecture will start another Level.

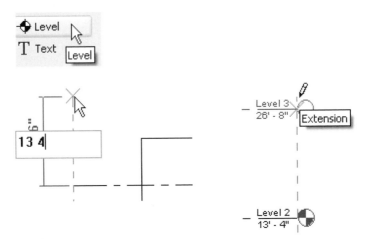

Figure 3.33   *Specify the Level offset from the one below*

3.16 Place the cursor over the left end of existing levels.

You may have to pan and scroll to see them clearly.

- Type the dimension offset between the Level 3 and the new level, as shown in the import file (**14'8" [4470]**), and either hit ENTER or left-click outside the new level's elevation value field.

- Align the tag as before, and click to place (see Figure 3.34).

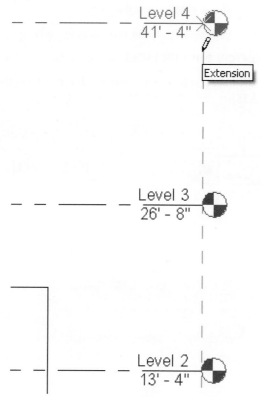

Figure 3.34 *Bubbles will snap to align*

3.17 Pick Modify from the top of the Basics Tab in the Design Bar to exit the Level tool.

Tip: You can also hit esc twice to exit the Level command completely. Many commands in Revit Architecture are multi-level routines, which allow the user to cancel a certain part of the procedure while still remaining in the command set. AutoCAD users should be well familiar with the esc-esc reflex.

3.18 Select the name field for Level 1.

The Level and associated fields will highlight.

- Select the name field again to open the value for editing.

- Type **1ST FLOOR MILLER HALL** and click outside the field or hit ENTER.

- Select Yes in the question box that appears about renaming the plan views of this level (see Figure 3.35).

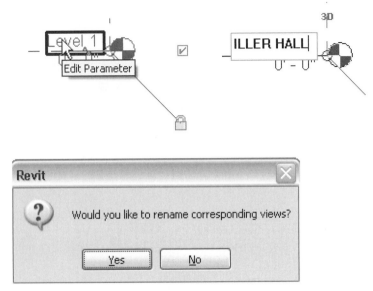

Figure 3.35 *Edit the name for Level 1 and rename its views*

3.19 Edit the names for Levels 2, 3 and 4 to match, as shown in Figure 3.35.

- Accept the rename question each time.

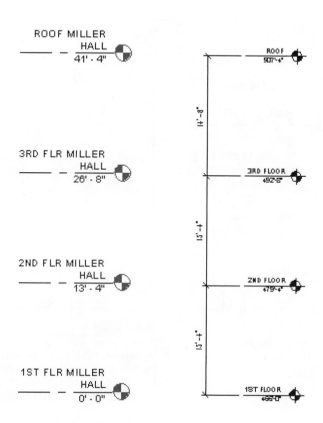

Figure 3.36 *Rename all the levels*

EXTEND WALLS TO THE ROOF LEVEL

3.20 Select any of the walls above the first floor.

- Right-click and pick Select All Instances from the context menu.

All the exterior walls will turn red.

- Click the Properties icon from the Options Bar.

3.21 Set the Top Constraint value to Up to Level: ROOF MILLER HALL (see Figure 3.37).

- Click OK.

The walls will update to the new height. If you change the elevation of that Level, the walls will adjust to follow it.

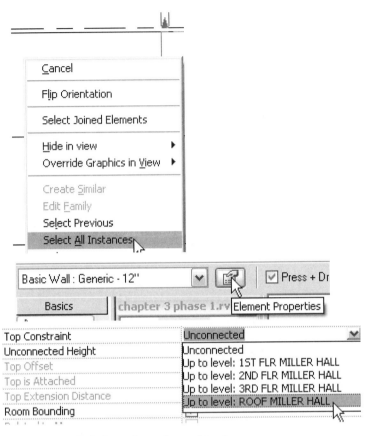

Figure 3.37 *Set the tops of all the walls to the roof level*

3.22 Open the default 3D view.

- If Revit Architecture has not yet reminded you to save your file, save it now.

EXERCISE 4. CREATE AND MODIFY A TOPOSURFACE

Revit contains tools for creating Site and Landscaping objects. These objects are parametric, not simply drafted, and include parking components, property lines, landscaping and site-located objects such as streetlights or bicycle racks. The Site menu and Site Tab of the Design Bar contain the tools.

The one you will work with probably the most is the Toposurface, which represents surface and depth of earth around your building. A toposurface is a 3D object; you control its surface by placing and editing points. You can subdivide toposurfaces or split them, and create what Revit names Graded Regions to show and calculate cut and fill. Toposurfaces can have materials applied, so they can represent earth, sand, water, pavement, etc. You can edit the display of toposurfaces so they show standard patterns in section, and you can set the display of contour lines to match your requirements. You can create Pads, or vertical cuts, so section display of earth and walls is displayed correctly by 3D objects, without drafting.

One major time-saving feature of a toposurface is the ability to read 3D points from an imported CAD file. Using a properly set-up site plan to create an accurate 3D landform is very quick, as you will see.

4.1 Open the file from the previous exercise or continue with the open file *Chapter 3 Phase 1.rvt.*

- Double-click on Site in the Project Browser to return to that view.

- Take a moment to examine topographic information in the CAD file.

At the building, the contours are at 466' [141m], which we know (from the CAD file we imported into the South elevation view) is the level of the ground floor. The lowest contour elevations are at 447' [136m].

This CAD file has been created with its lowest contours at 0 elevation. It needs to be lowered 19' [5m] so its contours are correct against the building model.

4.2 Select the CAD file.

- Click Properties.

- Set the Base Offset value to **-19 [-5000]**.

Revit will supply the units. See Figure 3.38.

- Click OK.

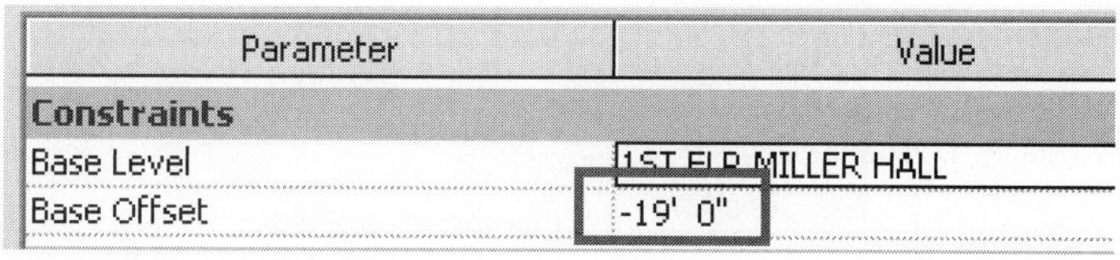

Parameter	Value
Constraints	
Base Level	1ST FLR MILLER HALL
Base Offset	-19' 0"

Figure 3.38 *Lower the CAD file in 3D space*

4.3 Click Query from the Options Bar.

- Click a cyan topo line.

The dialogue that appears will identify the layer (see Figure 3.39).

- Click OK.

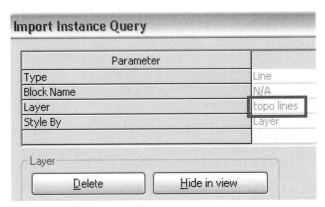

Figure 3.39 *Identify the layer for the topo lines*

4.4 Place your cursor over the Design Bar.

- Right-click to bring up a list of available Tabs.

- Click Site, as shown in Figure 3.40.

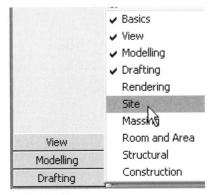

Figure 3.40 *Making the Site tab available*

The Site Tab will activate.

4.5 Click Toposurface.

The Design Bar will change to Toposurface mode, as shown in Figure 3.41.

- Click Use Imported>Import Instance.

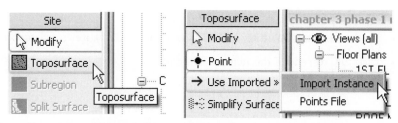

Figure 3.41   *Create a toposurface from the imported file*

4.6 Click on the CAD file.

- In the Add Points from Selected Layers dialogue, clear all layers except topo lines (see Figure 3.42).

- Click OK.

Revit will place data points along the polylines in the CAD file.

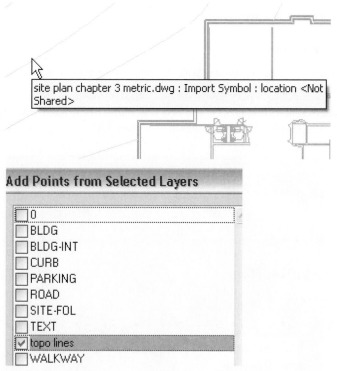

Figure 3.42   *Select the file and the layer to search for contour points*

If a very large number of points is created, this can affect performance.

4.7 On the Toposurface Tab, click Simplify Surface.

- Set the Accuracy value to **3 [1000]** (see Figure 3.40). Revit will supply the units.

- Click OK to reduce the number of points.

Figure 3.43 *Simplify the new toposurface*

4.8 Click Finish Surface.

- Click the Model Graphics Style control on the View Control Bar. Click Shading with Edges.

Revit has created the surface without wall information from your model, so it penetrates the building model, as shown in Figure 3.44.

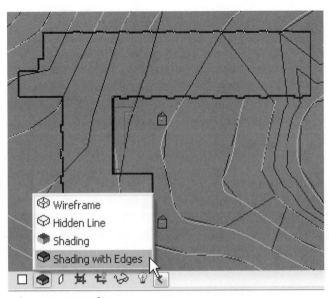

Figure 3.44 *View the new toposurface*

PLACE A PAD TO FIT THE WALLS

Revit's Pad tool allows you to place vertical offsets in a toposurface object so it will display correctly in section views and the landform will not appear to penetrate building walls. You can do this by picking walls or drawing lines, without having to edit individual points in the toposurface.

First you will adjust the view to make your work easier.

4.9 Click the Model Graphics Style control, as you did in Step 4.8.

- Click Wireframe.

- Zoom into the building outline so the foundation wall lines are clearly visible.

4.10 Click on one of the internal wall lines to select the CAD file.

- From the View Control Bar, click the Temporary Hide/Isolate control. Click Hide Element (see Figure 3.45).

The CAD file will disappear, and the view window will show a cyan border that also appears under the Hide/Isolate icon.

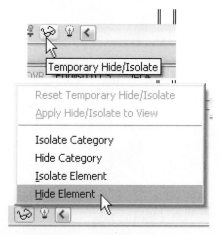

Figure 3.45   *Temporarily hide the CAD file in the Site view*

4.11 Click Pad from the Site Tab of the Design Bar.

The Design Bar will change to Sketch mode. The Pick Walls option is active.

- Place your cursor over the exterior surface of a foundation wall.

The Tooltip and Status Bar will identify the type of wall.

Tip: You can use the tab key with the cursor placed over a wall to cycle through the selection choices (interior face, center line, exterior face). This will make your selection precise without having to zoom all the way in.

- TAB until the exterior surface of the foundation walls highlight and the Tooltip and Status Bar read Chain of walls or lines.

- Click to create the pad sketch.

- If the magenta sketch lines appear on the interior face of the wall, click on the double-headed arrow Flip control, as shown in Figure 3.46.

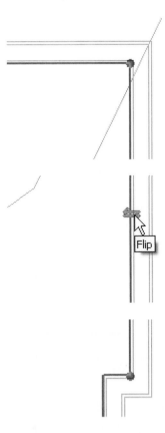

Figure 3.46   *You can flip a sketch from interior to exterior, and vice versa, just like a wall, door or window*

4.12 Click Pad Properties.

- Set the Height Offset from Level value to **-12 [-4000]** to set the top of the pad at the base of the foundation walls.

- Click OK. Click Finish Sketch.

- Zoom to Fit.

The toposurface no longer passes through the building.

WORK WITH VIEWS AND LEVELS

4.13 Click the Temporary Hide/Isolate control on the View Control Bar.

- Click Reset Temporary Hide/Isolate.

4.14 Click the Show Crop Region control on the View Control Bar.

- Zoom to Fit to show the new view border.

- Pick the crop region.

It will highlight red and controls will appear, as shown in Figure 3.47.

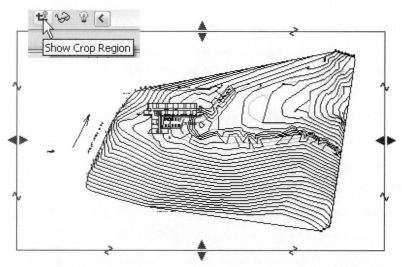

Figure 3.47   *Show the adjustable border for the view. All model views have crop regions you can apply.*

4.15 Pick the double-arrow controls on each face of the crop and drag them closer to the model, so that the Elevation and driveway lines are visible just inside the border (see Figure 3.48).

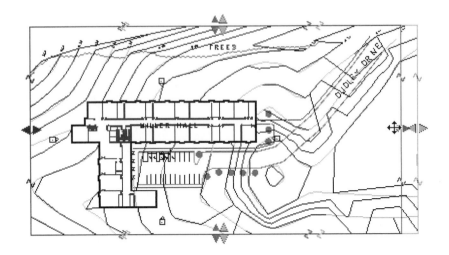

Figure 3.48 *Adjust the crop of the view closer to the building*

4.16 Zoom to Fit.

- Click the Hide Crop Region toggle on the View Control Bar to turn off visibility of the border.

This will eliminate the possibility of picking it by mistake while working.

In order to perform the next operations on the Toposurface, it will be better to have real-world height values for the building levels.

4.17 Open Elevation view South.

- Zoom in around the levels at the right side of the building model.

- Click on the Toposurface object, right-click, and pick Hide in View>Elements.

You will see the pad at the base of the foundation walls.

4.18 Select the level line for 1ST FLOOR MILLER HALL.

- From the Options Bar, click Properties.

- In the Element Properties dialogue, click Edit/New.

- In the Type Properties dialogue, click Duplicate.

Levels, like all other Revit objects, come in types that you can modify and duplicate. Revit's Levels can read their elevation values from within the project, the default setting, or they can report Shared values that the user creates or imports.

- In the Name dialogue, enter **Shared**.

- Click OK.

- In the Constraints field, set the Value to Shared. See Figure 3.49.

- Click OK twice to set the new property.

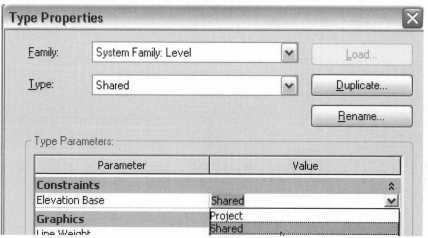

Figure 3.49 *Make the level for IST FLOOR read Shared values. You will apply this change in the next step.*

4.19 From the Tools menu, pick Project Position/Orientation/Relocate This Project. The cursor changes to the Move symbol.

- Click anywhere in the view window.

- Pull the cursor straight up and enter **466 [14100]**. See Figure 3.50.

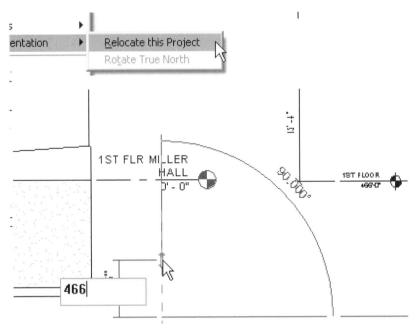

Figure 3.50 *Relocate the project up to the elevation listed in the CAD file*

4.20 Zoom to Fit.

- Zoom in so you can see the levels.

1ST FLOOR MILLER HALL now reads the same as the CAD file, while upper elevations read their offset from the lowest level.

Now you will divide the Toposurface into separate elements and edit their points. First, you will create the driveway, using the CAD site plan file for reference.

DIVIDE THE AND MODIFY THE LANDFORM

4.21 Open the Site view.

- Click on the CAD file.

- Click Query from the Options Bar.

4.22 Pick a magenta line at the edge of the parking lot. This represents the curb.

- Verify the layer and click Hide in View, as shown in Figure 3.51.

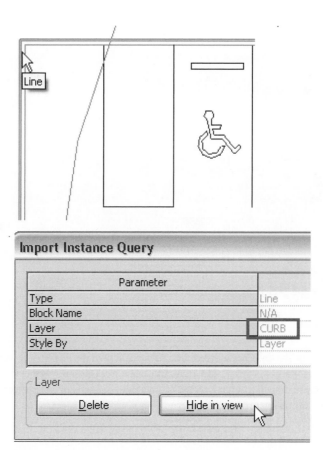

Figure 3.51   *Turn off the curb layer to make later work easier*

4.23 From the Site Tab of the Design Bar, pick Split Surface.

- Click anywhere outside the building to select the Toposurface.

The Design Bar changes to Sketch mode.

4.24 Select the Pick icon on the Options Bar.

- Place the cursor over an orange line indicating edge of pavement and TAB until the Tooltip and Status Bar read Chain of walls or lines.

- Click to select.

4.25 Click the Draw option and draw a line from endpoint to endpoint across the open right end of the sketch, as shown in Figure 3.52. Be careful with your picks.

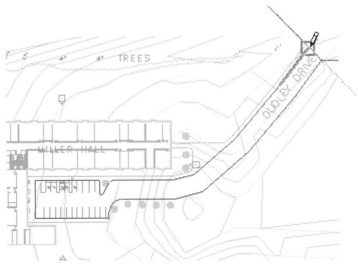

Figure 3.52 *Close the sketch by drawing a line*

4.26 Click Finish Sketch.

Note: If you get an error message, you have not drawn the final line correctly to close the profile. Revit will attempt to show you open areas or intersecting lines. Select the final line you drew if it is not correct. You can drag the ends of a too-short or too-long line to snap points on other lines, or delete it and draw it again.

Simplify the view.

4.27 Click on the CAD file.

- Right-click and pick Hide in View>Category.

- Click on the Model Graphics Style control on the View Control Bar.

- Select Hidden Line.

4.28 Click on the new toposurface.

- Click Properties. In the Element Properties dialogue, click in the value field for Material.

This currently reads <By Category>. An icon appears to the right of the field.

- Click the icon. This opens the Materials dialogue.

- In the Materials pane at the left side of the dialogue, use the list to find and Select Site – Asphalt [Site – Tarmacadam] (see Figure 3.53).

- Click OK twice.

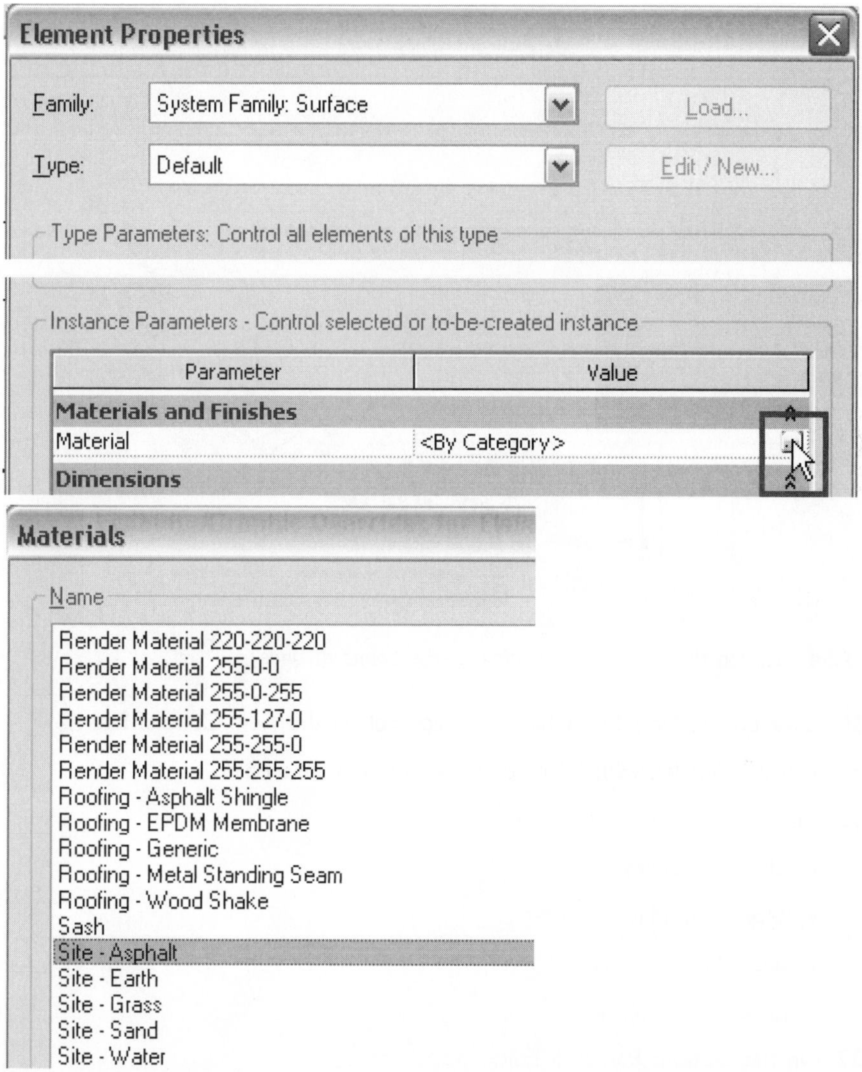

Figure 3.53 *Set the material of the new toposurface*

4.29 On the Options Bar, click Edit. The toposurface sketch opens.

- Window-select all the points that make up the surface.

- On the Options Bar, in the Elevation field, enter **0** (see Figure 3.54).

- Click Finish Sketch.

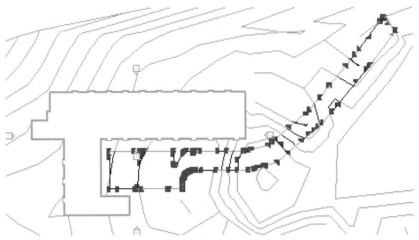

Figure 3.54 *Flatten the driveway by editing all the points at once*

4.30 Click on the Model Graphics Style control on the View Control Bar.

- Select Shaded with Edges to show the new material.

4.31 Click on the original toposurface.

- Click Properties.

- Repeat the Materials edit.

- Select Site – Grass.

- Click OK twice.

4.32 On the Options Bar, click Edit.

- CTRL + select the points along the driveway edge.

- Set their elevation to **0**, as you did for the other surface in Step 4.29.

The surface will change around the driveway, as shown in Figure 3.55.

- Click Finish Surface.

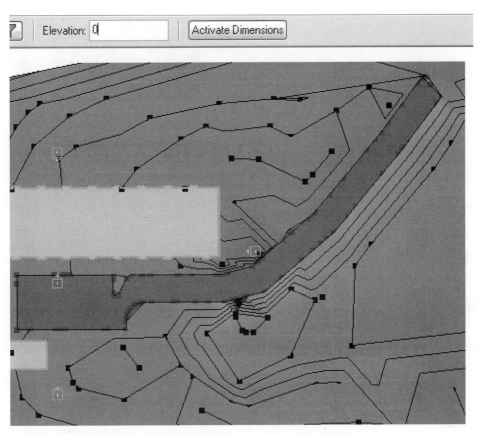

Figure 3.55 *Edit points in the main ground form to meet the driveway*

4.33 Expand the 3D views list in the Project Browser by clicking the + symbol next to that category.

- Double-click view {3D} to open that view, which is the default 3D view.

4.34 Window-select right to left so that the cursor catches everything visible.

- Click Properties.

- Set the Phase Created value to Existing (see Figure 3.56).

- Click OK.

- Click anywhere off the model to clear the selection set.

Note the changed appearance.

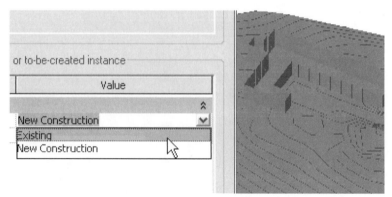

Figure 3.56 *Change the phase of everything you have created. You are modeling existing conditions.*

Revit's views work with phases and phase filters to manage model elements and their appearance. All projects have Existing and New Construction phases set automatically. Views in a default file (such as we are working in) are set to New Construction phase, which is why we changed the building walls and toposurfaces manually. If you set the phase of a view to Existing, the Phase Created property of all items you subsequently create in that view is set to Existing.

You can set up as many phases in a project as you wish. Phases and Phase Filters will be examined in detail in later exercises.

4.35 Save the file.

EXERCISE 5. COMPLETE THE EXTERIOR WALLS

You have sketched a building shell and located it in relation to a landform. Now it's time to develop the building exterior for completeness and accuracy. To do this, you will work with alignment tools, edit the properties of walls, split, relocate and make copies of walls.

ALIGNMENT USING REFERENCE PLANES

5.1 Open the file from the previous exercise or continue working in the open project file. Open Elevation view South.

- The CAD file you imported first served to set building levels. You will now use it to locate walls and a roof.

5.2 Place the cursor over the CAD import.

- Click and drag the import so it is above the building model.

- Right-click and pick Zoom to Fit.

- Select the Align tool from the Toolbar.

- Click the left side of the building for the first reference, and then carefully select the left edge of the wall in the imported elevation.

- Click the padlock icon to toggle on the alignment constraint (see Figure 3.57).

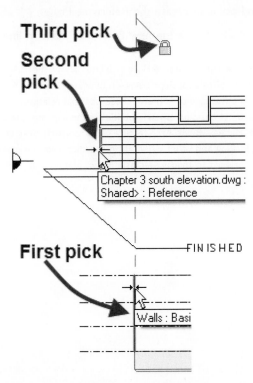

Figure 3.57 *Align the import to the model, then lock the placement*

This will prevent accidental movement of the CAD file during your next actions. You can always unlock something you have constrained in this way.

You can also Pin objects to a specific location. This will be covered in later exercises.

5.3 Select Ref Plane from the Basics Tab of the Design Bar.

The cursor will change to a pencil symbol; the Status Bar and Tooltip will display information relative to the active command.

5.4 Zoom into the sectional part of the elevation file (between column lines N and L).

- Pick the corner, as shown in Figure 3.58, to the left of column line N to start the Reference Plane.

- Pull the cursor straight up, so that the angle vector reads 90° and the Tooltip shows Vertical.

- Pick a second point to create the Reference Plane.

- Click the padlock to lock the new Reference Plane in line with the line you used.

Note: Revit Architecture's parametric capabilities allow the designer to establish a nearly unlimited set of relationships between building elements. You will explore ways to take maximum advantage of some of these possible relationships in future lessons. One of the first, most basic parametric utilities is the alignment lock. When combined with Reference Planes—Revit Architecture's 3D construction lines—in a carefully planned design, locking and other orientation specifications can make even late design changes straightforward to manage—closer to wholesale than retail.

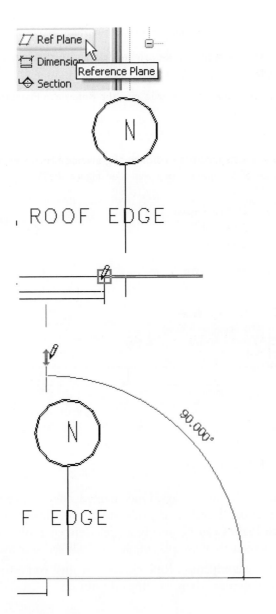

Figure 3.58 *Create a Reference Plane at the left side of the section*

5.5 Pick Modify from the Basics Tab on the Design Bar.

- Click the new Reference Plane, right-click, and pick Element Properties from the context menu.

- In the Element Properties dialogue, give the Reference Plane the Name **WEST SIDE OF WALKWAY**.

- Click OK.

  The name will display at the end of the Reference Plane. This plane will represent the exterior face of an upper story wall (see Figure 3.59).

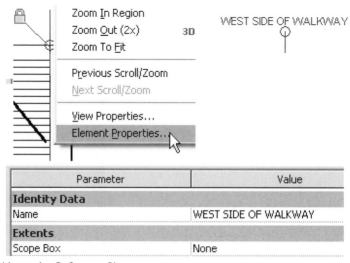

Figure 3.59 *Name the Reference Plane*

Reference Planes are useful in many ways. Once named, they can become work planes for object placement or creation. Levels and grids are automatically horizontal and vertical work planes; Reference Planes can be at any angle. Reference planes are visible in other views than the one in which they are placed, as you will see shortly.

Reference Planes and their companion Reference Lines give you a framework to tie views of different orientation together when creating 3D objects.

5.6 Repeat Step 5.5 to create a second vertical Reference Plane at the right edge of the section next to column line L (see Figure 3.60).

- Name this plane **EAST SIDE OF WALKWAY**.

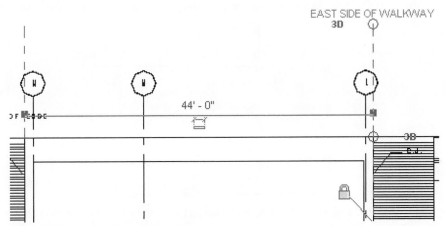

Figure 3.60   *Reference Planes at the Section*

5.7 Zoom to Fit.

- Study the building profile of the section—the first story is wider than the upper stories.

5.8 Select each Reference Plane in turn and drag its lower end to a point below the model (see Figure 3.61.)

A drag control will appear as on a level.

The Reference Planes will be visible in each plan view once they are extended past the level for that plan view. You are preparing to create the upper story walls that the section depicts.

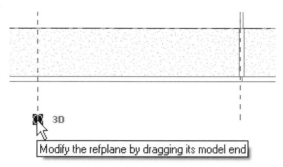

Figure 3.61 *Drag the Reference Planes so they intersect the model below the ground floor level*

5.9 Double-click on Floor Plans>2ND FLOOR PLAN MILLER HALL in the Project Browser to open that view.

- Click on the CAD file.

- Type **VH** at the keyboard.

    This is a shortcut for Hide in View—the same as picking Hide in View>Elements as the right-click menu or clearing the category checkbox in the Visibility/Graphics Overrides dialogue.

CHANGE THE BASE AND HEIGHT OF WALLS

You will now make changes to walls.

5.10 Zoom out as necessary to see the North–South wing of the building.

- Select all the walls, as shown in Figure 3.62, using a left-to-right window.

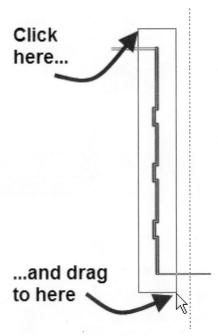

Figure 3.62 *Select the walls with an enclosing window*

5.11 With the walls selected, click the Copy command from the Toolbar.

The cursor changes to the copy/move icon.

- Verify that Constrain and Copy are checked on the Options Bar.

  Constrain restricts object movement to vertical or horizontal only, and will prevent you from moving the walls up or down accidentally.

- Zoom in if necessary so you can see the inside corner, as shown in Figure 3.63.

- Click to establish a start point.

- Pull the cursor to the right until it snaps to the WEST SIDE OF WALKWAY Reference Plane.

- Click to create the copies.

  Revit Architecture will snap to the Reference Plane to make the pick easier.

- Pick Modify from the Basics Tab of the Design Bar to terminate the command (see Figure 3.57).

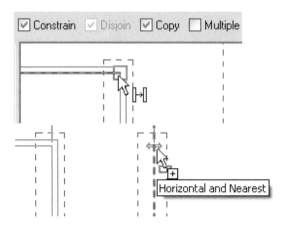

Figure 3.63 *Copy all the walls from their original position right to the Reference Plane*

The walls you just placed are still highlighted.

5.12 Pick Properties from the Options Bar.

- Set the Phase Created to Existing.

  You are working in a view set to New Construction phase, so you need to set the phase of objects you create if they are not new.

- Set the Base Constraint value to 2ND FLOOR MILLER HALL.

- Set the Base Offset value to **-2' [-600]** (see Figure 3.64).

  You will place a roof at the 2ND FLOOR level in a later step; you are extending the walls down past the roof.

- Click OK.

Parameter	Value
Constraints	
Location Line	Finish Face: Exterior
Base Constraint	2ND FLR MILLER HALL
Base Offset	-2'
Phasing	
Phase Created	Existing

Figure 3.64 *Set the Phase and Base Offset for these walls*

Because you copied walls, there is now an area where walls do not meet.

5.13 Extend the horizontal wall, as shown in Figure 3.65.

- Click Trim/Extend from the Toolbar.

  Trim/Extend has three modes: Corner, Single Element to T, Multiple Elements.

- Set the Trim option to Corner (the left icon, and the default).

- Click the horizontal wall first.

- Click the vertical wall second.

  Revit will preview the new wall arrangement when you place the cursor over the second wall.

  Where walls cross, click on the portion of the wall you wish to keep, not the part you wish to eliminate. This will take some practice to become automatic.

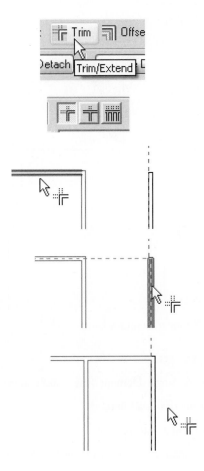

Figure 3.65   *Extend walls to meet*

The segment of wall you just extended will have the same base condition at the walls to its right—it will run from the 2ND FLOOR level to the ROOF level. Since you created it by extending another wall and not by drawing it as a separate object, you will need to separate it.

5.14 Select the Split tool from the Toolbar.

The cursor will change to a scalpel symbol.

- Click on the new horizontal wall segment between the two vertical walls.

The exact location of the split is not critical.

5.15 Select the Trim tool.

- Make a corner on the left, as shown as shown in Figure 3.66.

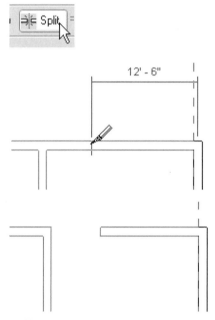

Figure 3.66 *Split the wall and Trim to create a corner.*

The Trim tool is still active.

5.16 Set the Trim option to Single Element (the middle icon).

- Click on the left vertical wall first.

- Click on the horizontal wall second.

This wall will terminate at the face of the left side wall, as shown in Figure 3.67.

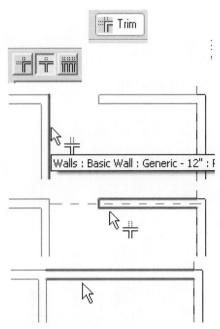

Figure 3.67 *Trim one wall to the face of another*

5.17 Click on the wall segment you just extended.

- Click Properties.

- Set the Base Constraint and Base Offset as you did in Step 5.12: **2ND FLOOR, offset −2' [600]**.

5.18 Repeat Steps 5.14 to 5.17 on the wall at the lower left.

- Split the wall.

- Trim to a corner.

- Trim to a face. (see Figure 3.68)

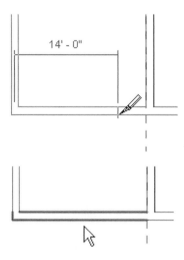

Figure 3.68 *Split and trim this wall, as you did before*

5.19 Use a right-to-left crossing window to select walls, as shown in Figure 3.69.

- Set the Top Constraint to Up to: **2ND FLOOR MILLER HALL**.

  The Walls will change appearance, as they are no longer cut in this view.

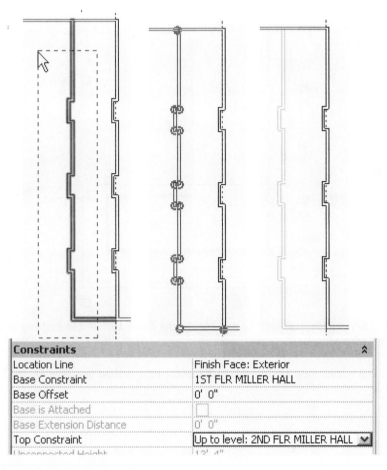

Figure 3.69 *Select and shorten walls*

DESIGNATE WALLS AS DEMOLISHED

The walls you shortened are still selected. These walls, plus the ones that form the L-shaped end of the building, will be removed in Phase 1 construction, when a new building is constructed adjacent to this one.

5.20 Hold down the CTRL key and window-select walls, as shown in Figure 3.70.

- Click on Properties.

- In the Element Properties dialogue, scroll down to the Phasing section.

- Set Phase Created to Existing.

- Set Phase Demolished to New Construction.

- Click OK.

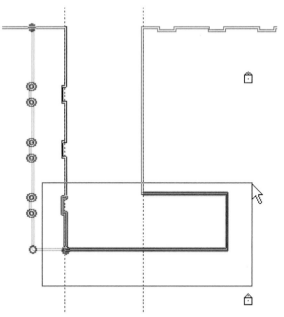

Figure 3.70   *Select additional walls. You will change the Phase Demolished property of all the selected walls.*

The appearance of the walls will update to a dashed line (see Figure 3.71).

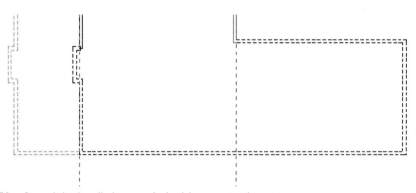

Figure 3.71   *Demolished walls have a dashed linetype in this view*

5.21 Save the file.

You have finished work on the exterior walls of the building model for the time being. Now you will model a number of roofs.

You place roof objects in Revit Architecture in three ways. In plan or 3D views, you create a Footprint sketch of the outline of the roof edge. In elevation or section views, you create a profile line to extrude sideways. If you have created Massing objects by 3D modeling or from imported CAD/SketchUp files, you can place roofs on appropriate Faces on the Mass objects.

With a footprint roof, you declare sloping surfaces; Revit will calculate the ridge line. You can have a combination of roof pitches in one roof object. For conditions where a roof does not slope parallel to an edge, you can place a Slope Arrow on the roof to direct slope. For nominally flat slab roofs that actually slope for drainage, you can indicate points below the roof edge to provide a complex drainage plan. This complex sloping will be in the 3D model, not sketched on the face, and will display correctly in section views.

A roof can be a multiplex object, with layers in it to represent physical components as they are placed or constructed. Each layer has a function, material and thickness. Roof layers can penetrate wall layers so that structural layers meet and represent correctly in section views. You can edit layers at any time.

Each roof is based at a level, with an offset (up or down) that you can apply. Roofs will host openings and roof-based components such as skylights. Walls can attach to roofs, and fill in the gap between the nominal top of the wall and a gable end, for instance.

You can place fascias and gutters on roofs by applying a sweep to the roof edges. You do this most easily in a 3D view. You can create roof soffits between walls and the roof edge. You sketch a soffit by picking walls and roofs in a plan or 3D view, and the resultant object will appear in sections. See Figure 3.72 for an example of a section view at a roof-wall join.

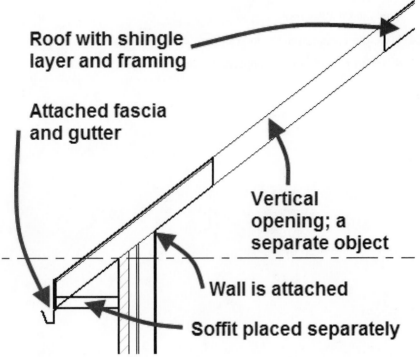

Figure 3.72   *A roof-wall intersection. All the objects are model components—no drafted lines in this view.*

You can combine roofs in various ways to create valleys, dormers and double-pitch (Mansard or Gambrel) roofs. There is a Join/Unjoin Roofs tool specifically for combining or separating roof objects to create complex conditions.

In the next exercise, you will model the existing roofs on the project model; you will create flat roofs and a simple gable, and attach walls to the underside of the gable.

EXERCISE 6. PLACE ROOFS ON THE BUILDING

CREATE A FLAT ROOF AT THE 2ND LEVEL

6.1 Open the file from the previous exercise or continue working in the open project file.

6.2 Zoom or pan so you can see the single-story walls to the left of the vertical wing.

- Pick Roof>>Roof by Footprint from the Basics Tab of the Design Bar.

  The Design Bar will switch to Sketch mode with drawing commands. The Pick Walls tool will be active (depressed) by default.

6.3 On the Options Bar, clear Defines Slope.

This will be a flat roof.

6.4 Place the cursor over the interior face of a one-story wall (left side).

- TAB so that the outline appears, as shown in Figure 3.73.

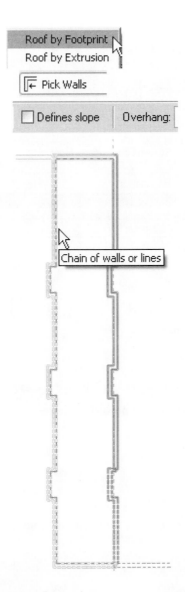

Figure 3.73 *Settings and pick for the flat roof*

- Click to create a roof sketch.
- If the sketch lines are on the wrong side of the walls, use the Flip arrow control.

6.5 Click Roof Properties on the Sketch Tab of the Design Bar.

- Set the Roof Type to **Generic - 9" [Generic - 125mm]**.
- Set the Base Offset from Level to **-1'-0" [-300]**.

This will set the top of the roof down slightly from the top of the walls, a typical condition (see Figure 3.74).

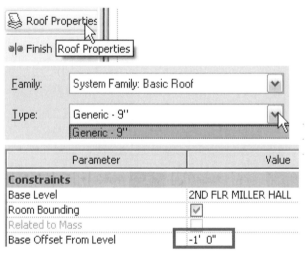

Figure 3.74 *Properties of this roof*

- Click OK.
- Click Finish Roof on the Sketch Tab of the Design Bar.

  Since the roof material has no surface pattern, the new roof will not be immediately visible.

SET UP 3D VIEWS

6.6 Open the default 3D view.

- From the View menu, select View>Orient>Southwest, as shown in Figure 3.75.

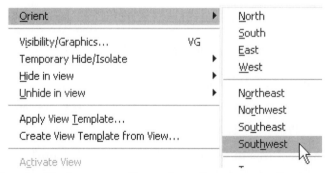

Figure 3.75 *Orient the 3D view so you can see the walls and new roof*

6.7 Pick Show Crop Region from the View Control Bar.

- Click the crop border to activate controls.

- Drag the sides close to the model.

- Zoom to Fit.

6.8 From the View menu, pick View>Orient>Save View.

- In the dialogue that opens, enter **SW - Demo**.

- Click OK.

  The new view name appears in the Project Browser. {3D} remains as a separate view.

  The new roof is shown in deep gray. Its phase is New Construction.

6.9 Pick the new flat roof.

- Open its Properties dialogue.

- In the Element Properties dialogue, scroll down to the Phasing section.

- Set Phase Created to Existing.

- Set Phase Demolished to New Construction.

- Click OK.

6.10 In the Project Browser, right-click on 3D Views: SW – Demo.

- Pick Duplicate View>Duplicate, as shown in Figure 3.76.

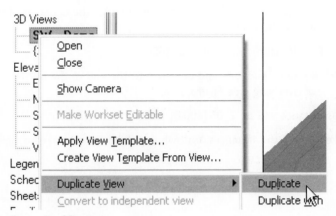

Figure 3.76 *Duplicate the 3D view*

A new view named Copy of SW – Demo will become the active view.

6.11 Right-click the name of the new view; pick Rename.

6.12 In the Rename View dialogue, Enter **SW – Existing**.

- Click OK.

6.13 With the cursor anywhere in the view window, right-click.

- In the context menu, pick View Properties.

- In the Element Properties dialogue, set the Phase to Existing.

- Click OK.

    The view changes appearance. Demolished items now appear the same as others, because in this phase they have not been demolished yet.

GET MORE INFORMATION

6.14 Double-click Elevations>South in the Project Browser to return to that view.

- Zoom to Fit.

- Zoom in to the gable roof at the left side of the imported elevation drawing.

- Select the Tape Measure tool on the Toolbar.

    This will read the distance and angle between any two points.

6.15 Pick the intersection of the horizontal line identified as METAL ROOF EDGE and the vertical line representing the METAL PANEL wall under the right side of the roof for the first point.

- Pull the cursor up vertically to the roof, and pick the endpoint of that line at the line representing the roof fascia.

    The Tape Measure will read 2'-2" [650].

- Click the same endpoint and pull right to the end of the horizontal line, to measure the roof overhang (see Figure 3.77).

    This distance will be 2'-0" [600.0].

- Take note of the distances. You will use them soon to create walls and a roof.

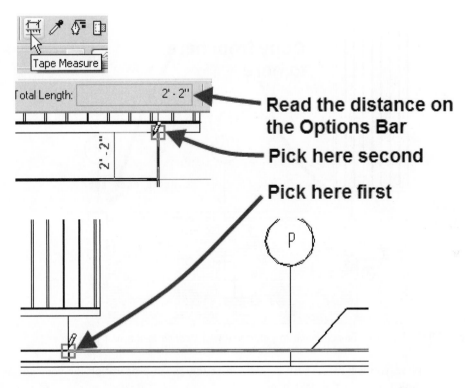

Figure 3.77 *The tape measure tool in use*

You will now use the location of the wall under the roof in the CAD file that you just measured to locate walls in the model.

6.16 Click on Reference Plane WEST SIDE OF WALKWAY.

- Pick Copy from the Toolbar.

- For the first click, pick the intersection of the Reference Plane and the line marked METAL ROOF EDGE.

- Pull to the left and click at the first point you used for the Tape Measure.

  This will place an unnamed Reference Plane to mark the location of this wall in other views (see Figure 3.78).

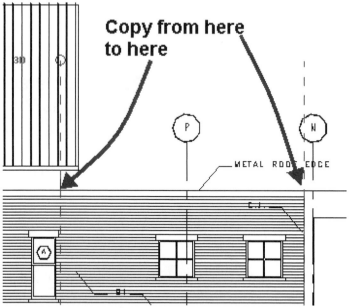

Figure 3.78 *Copy a Reference Plane to determine wall location in the roof plan view*

Next, you will match wall demolition conditions in preparation for your conceptual design work in the next exercise.

6.17 Zoom out until you can see the model.

- Zoom or pan until you can see the vertical wall at the right side of the walkway.

6.18 Pick Split from the Toolbar.

- Place the cursor over the right side wall between levels 1ST FLR and 2ND FLR.

- Click to split the wall; the exact location does not matter.

6.19 Click Align from the Toolbar.

- For the first pick, click the base of the vertical wall at the left side of the walkway.

- Click the split.

  It will align with the base of the first wall.

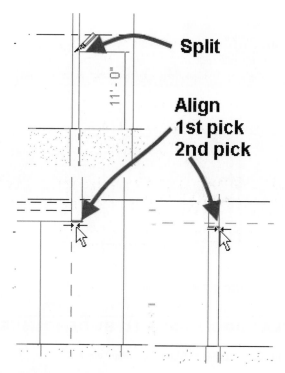

Figure 3.79 *Split the wall and align the split*

6.20 Select the Demolish tool from the Toolbar.

The cursor changes to a hammer symbol, and the Status Bar reads Click on elements to mark them as demolished in the current phase.

- Click on the lower portion of the wall you just split.

This sets the Phase Demolished property to the current phase, which is New Construction for this view. The wall display changes to match other demolished components. See Figure 3.80.

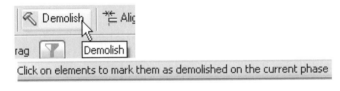

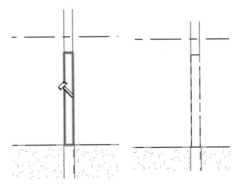

Figure 3.80 *Demolish a section of the split wall*

CREATE TWO FLAT ROOFS—ONE TO BE DEMOLISHED

6.21 Open the ROOF MILLER HALL Floor Plan view.

- Zoom to Fit. Pick Roof>>Roof by Footprint from the Basics Tab of the Design Bar. The Design Bar will change to Sketch mode with the Pick Walls tool active.

- On the Options Bar, clear Extend into wall (into core).

- This will be a flat roof, so leave Defines slope cleared.

6.22 Place the cursor over the interior face of any wall.

- TAB until the entire wall outline highlights and the Tooltip reads Chain of walls or lines.
- Click to create the roof sketch.
- Check the location of the lines and use the flip arrows if necessary (see Figure 3.81).

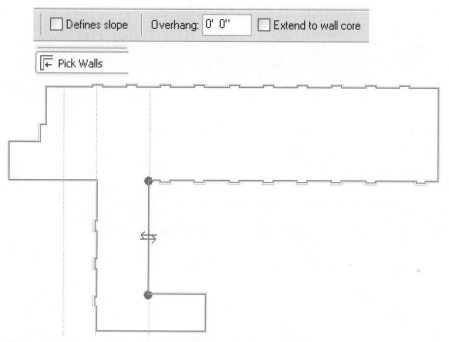

Figure 3.81   *Pick walls to start the roof sketch.*

6.23 Click Lines on the Sketch Tab of the Design Bar.

- Draw a line from the intersection of the outline sketch and the Reference Plane you copied in the elevation view (see Figure 3.82).
- Draw a line approximately 24' [8000].

    The actual distance does not matter.

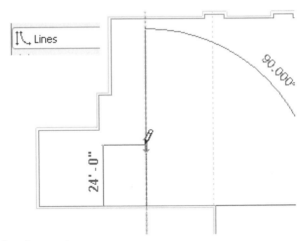

Figure 3.82 *Sketch a line as shown.*

   6.24 Click Trim from the Toolbar.

- Use the Corner option (the default).
- Trim two areas to make corners, as shown in Figure 3.83.

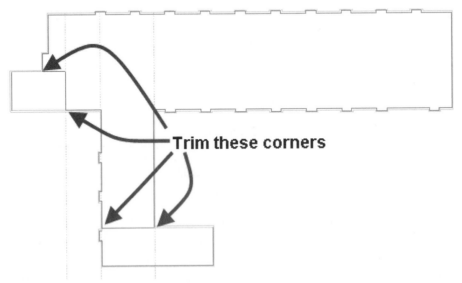

Figure 3.83 *Trim the outline as shown.*

   6.25 CTRL + select lines outside the new border.

- Click Delete from the Toolbar.

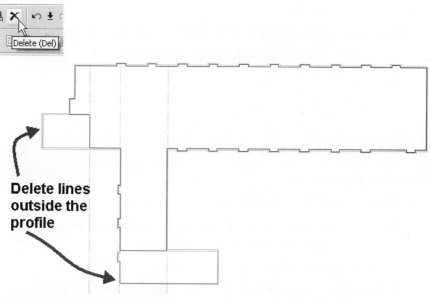

Figure 3.84 *Delete extra lines*

6.26 Select Roof Properties from the Sketch Tab of the Design Bar.

- Set the Base Offset from Level to **-1'-0" [-300]**.
- Click OK.
- Click Finish Roof.

6.27 Click Roof>Roof by Footprint again.

- Zoom into the lower section of the building.
- Pick individual walls, as shown in Figure 3.85, including the short offset walls at the left.
- Click Trim and create the final corner, as shown in Figure 3.85.

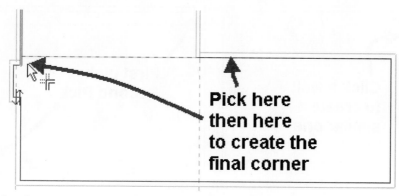

Figure 3.85 *Create the outline by picking walls and trimming lines*

- Click Roof Properties.
- Verify that the Base Offset from level matches the previous roof.
- Set the Phase Demolished to New Construction.
- Click OK.
- Click Finish Roof.

CREATE WALLS AND GABLE ROOF

6.28 Zoom to the upper left.

- Click on a wall. Pick Create Similar from the Toolbar.

  This is the icon at the far right. If your screen resolution is limited, you may have to click the >> (expand) symbol to access that tool.

- Select the Rectangle Option on the Options Bar.

- Click two points upper left and lower right, as shown in Figure 3.86, to create four walls. This will locate the exterior faces on the outside.

  Three walls will sit on top of previously created walls. The fourth will align with the Reference Plane.

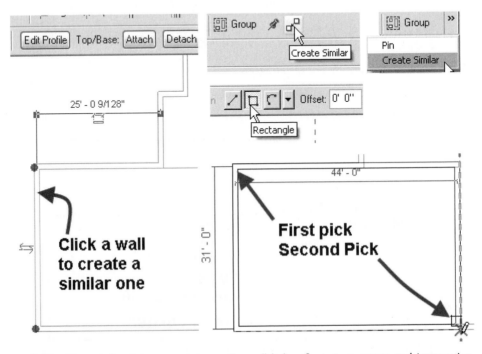

Figure 3.86 *Place walls of the same type as the wall below. Snap to a corner and intersection.*

6.29 Pick Split from the Toolbar.

- Split the upper wall at the edge of the wall under it, as shown in Figure 3.87.

- Click Modify.

- Click the right part of the wall you just split and the right vertical wall.

- Click Properties.

- Set the Base Offset to **-2' 0" [-600]**, so these walls extend down past the roof beside them.

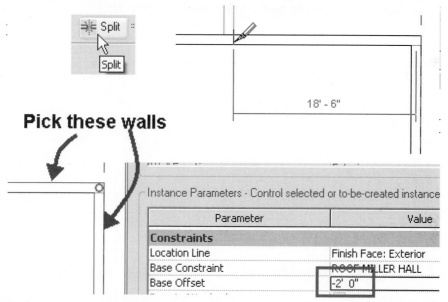

Figure 3.87   *Split and lower walls*

6.30 Pick Roof>Roof by Footprint from the Basics Tab of the Design Bar.

- Set the Overhang value on the Options Bar to **2'-0" [600]**. You measured this dimension earlier.

- Leave Defines Slope cleared and pick the left and right walls you just created so that the sketch line appears to the outside of each wall.

- If necessary, use the Flip control to adjust the line position.

6.31 Check Defines Slope.

- Pick the lower wall and one upper wall.

- Pick Trim from the Toolbar.

- Click twice to close the gap in the outline.

6.32 Click Modify.

- Pick the upper line.

  A slope value will appear in blue, meaning you can click on it to edit. The imperial slope value is in inches of rise per foot of run; the metric value is in degrees of slope.

- Click on the slope value field and change it to **12 [45]**.

  Revit will supply the "or ° annotation for inches of rise per foot [degrees]. This value is given here for you rather than take the steps to import another CAD elevation and measure it.

- Repeat the slope edit on the lower line (see Figure 3.88).

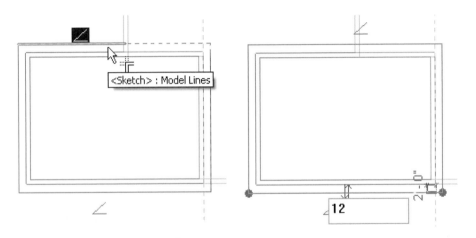

Figure 3.88 *Walls above the third floor roof align with existing walls and the Reference Plane*

6.33 Pick Roof Properties on the Sketch Tab.

- Set the Base Offset From Level value to **2' 2" [650]** (the distance you measured earlier).

- Click OK.

- Click Finish Roof. In the question box that opens, click Yes.

  This will cause the walls to attach to the roof underside.

  The view will update to show the roof. The View Range properties truncate the display, so the ridge is not visible.

6.34 Open 3D view SW – Existing.

- From the Window menu, pick Window>Close Hidden Windows.

- Open 3D view SW – Demo.

- From the Window menu, pick Window>Tile (see Figure 3.89).

6.35 Save the file.

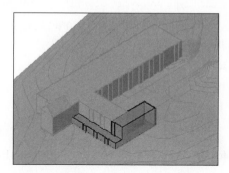 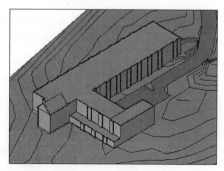

Figure 3.89 *The model outline is complete, and you can display it two different ways*

For the purposes of our hypothetical project, you now need to put wall, door and window information on certain walls of the building so it is an accurate model of parts of the building represented in the CAD file you have been studying.

The first stage of the design project will involve transforming the North–South wing of this building. The ground level will be removed for a pedestrian concourse, and the upper two stories will become walkways between the existing East–West wing and a new building to be located at the south end of the new concourse.

The next chapter's exercises will locate the shell of this new building and those of later phases in the ambitious master plan of this little college.

Once the walls have been accurately represented on the parts that will remain, you will design the faces of support pillars for the concourse.

EXERCISE 7. ADD WINDOWS

LOAD FROM LIBRARY

7.1 Open or continue working in the file from the previous exercise. Make South Elevation the current view.

7.2 Zoom in to the CAD file between Gridlines K - P.

7.3 Click on the Import.

- Click Query.

- Pick a horizontal line representing brick or block in the wall.

- In the dialogue, click Hide in View.

- Repeat twice more until the walls do not show any patterns around the windows.

  There are three window types, identified as A, C and E.

7.4 Click on the Tape Measure tool on the Toolbar.

- Click on the corners of a window of each type and write down the measurements.

  Type A will be 2' 8"w x 6' 8"h [820 x 2040]. Type C is 8' 4"w x 7' 0"h [2540 x 2140]. Type E will be 4' 0" x 4'0" [1200 x 1200].

  The sill heights are as follows: Type A: 3' 0" [914]; type B: 5 4 1/2" [1640]; type E 3' 0" [914]. You can verify the sill heights by placing dimensions to level lines.

7.5 Select Window from the Basics Tab of the Design Bar.

- Pick Load from the Options Bar.

- Navigate to the folder holding files for Chapter 3 from the CD. Select *Type E.rfa* [*Type E metric.rfa*]. Click Open.

7.6 Place the window, as shown in Figure 3.90, roughly in line with Gridline P and above Level 2ND FLR.

Revit Architecture will attempt to locate the windows at the correct sill height for you, and will show temporary dimensions indicating the sill height and distance from the window to other building elements. If it's necessary to adjust the sill height of the first instance, pick the dimension for the sill height and adjust it, as shown in Figure 3.90.

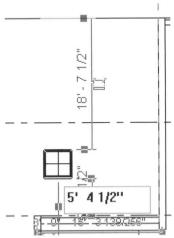

Figure 3.90 *Locate the window approximately and adjust its sill height value if necessary*

EDITING TOOLS – ALIGN, COPY, GROUP, ARRAY

Efficient drafting and design take advantage of repetition and symmetry whenever possible. Editing tools are quicker and more precise than manual placement. In particular, Revit's easy to use Align tool promotes a work pattern where you place an object approximately, then shift and constrain it as necessary. You can apply any degree of precision you wish.

7.7 Click Align from the Toolbar.

- Pick the side of a Type E window on the CAD file.

- Click the corresponding side of the new window instance in the model.

- Click Modify.

7.8 Select the new window.

- Click Copy from the Toolbar.

- Pick the left end of level line for 2ND FLR as the first pick.

- Click the left end of 3RD Floor level to make a copy one story up.

  The new window will be highlighted red.

7.9 Hold down CTRL and click the first window.

- Pick Copy from the Toolbar.

- Zoom in so you can see the Type E windows in the CAD file clearly.

- Pick a corner of one of the left windows as the start point reference.

- Pull to the right and click the corresponding corner of one of the two right windows.

  There will now be four windows in the model.

7.10 Select Window from the Basics Tab of the Design Bar.

- Click Load. Navigate as before to the CD library files.

- Select *Type A.rfa* [*Type A metric.rfa*].

- Click Open.

7.11 Place one instance of the Type A window above 1ST FLR level to the left of the Reference Plane extending down from the roof, as shown in Figure 3.91.

- Adjust sill height if necessary.

- Align with the CAD file windows as before.

- Click Modify.

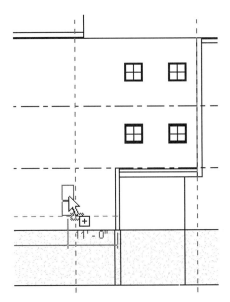

Figure 3.91 *Place the first Type A window*

7.12 Select the new window.

- Click Copy on the Toolbar.

- Check Multiple on the Options Bar.

- Click the left end of 1ST FLR level.

- Click the left of 2ND FLR level and 3RD FLR level to make two copies.

- Click Modify.

Revit's Group function provides a way to make named collections of objects on the fly. These collections can then be copied; editing one instance will update all instances. Groups can be exported, linked and imported, providing a way to share content between projects. We will examine this in depth in later exercises. In the next steps, you will create a group to make the job of copying many windows more efficient and less tedious.

7.13 Window-select left to right over the new Type A windows so all three highlight, but nothing else.

- Pick Group from the Toolbar.

- In the dialogue, enter **3 Type A Windows** for the group name.

- Click OK. See Figure 3.93.

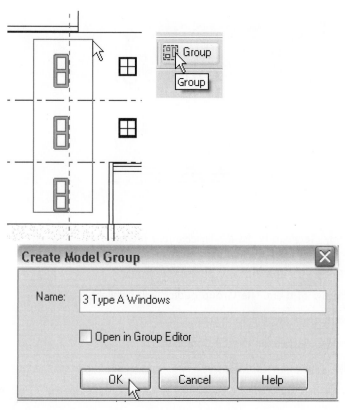

Figure 3.92 *Create a Group from three windows The Group is highlighted.*

Revit's Array tool lets you make multiple copies of items in linear or radial patterns. You define the number of copies and the distance to cover—either between items or overall. You can group and associate an array, which makes it parametric; you can later adjust the number of instances in the array.

7.14 Click Array from the Toolbar. On the Options Bar, Linear will be the default option. Clear Group and Associate. Enter **6** for Number. Check Constrain (see Figure 3.93).

7.15 Zoom in as necessary to see the CAD file clearly to the left of Gridline P.

- Click a point on one of the Type A windows below the roof in the CAD file.

- Pull to the left and click the corresponding point in the next window.

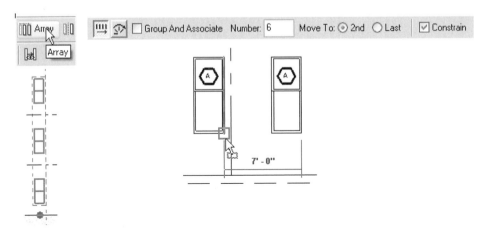

Figure 3.93 *Array the group across the building face precisely*

7.16 Zoom out so you can see the new windows.

- Pick any of them. The Group outline will highlight.

- Pick Copy from the Toolbar.

- Click Constrain on the Options Bar.

- Click Multiple on the Options Bar.

- As before, pick the corner of a window outline in the CAD file for the start reference.

- Zoom or pan to the right until you can see Gridline K.

- Click on the corresponding corner of a window outline to the left of the gridline to place a copy below it.

7.17 The new instance of the group is highlighted.

- Click the corresponding point on a window to the right of the gridline to place a second copy of the group (see Figure 3.94).

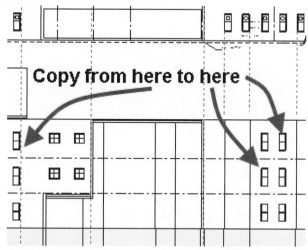

Figure 3.94   *Copy a window group using the CAD elevation for location*

- Click Modify.

PLACE GRIDS TO AID WINDOW PLACEMENT

Grids are basic layout tools. You can place them in plan, elevation or section views and they will appear in other views. You can create linear or radial grids. Structural columns snap to grid intersections, and structural framing snaps to gridlines. You can lock other elements such as walls and floor edges to grids, so that changing a gridline location can move an entire building section. Grids auto number as they appear, and you can change the numbering system at any time. You can modify the linetype and symbol of grids, and adjust their visibility per view.

Quite often commercial or industrial building projects will start with a grid layout even before model components are placed. In this case, you will place grids based on the CAD file grids, to make it easy and precise to place many instances of windows in a long façade.

7.18 Zoom in to the upper-right corner of the CAD file, so you can clearly see gridline A.

- On the Basics Tab of the Design Bar, click Grid.

- On the Options Bar, click the pick option. Click on the line below the bubble for grid A.

- In the grid that appears, click on the blue number I to edit the value.

- Enter **A** (see Figure 3.95).

- Click off the grid bubble to write in the value.

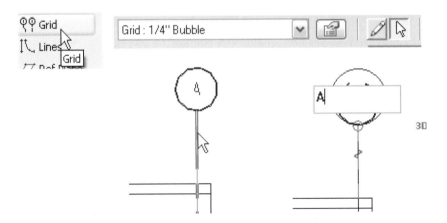

Figure 3.95 *Place grids by picking lines that are already present, to save time*

7.19 Continue placing grids by clicking on gridlines in the CAD file from right to left.

- At Grid Line J, click in the bubble as before.

- Edit the value to J.

- Click off the bubble.

- Place a grid at Grid K.

7.20 Zoom out so you can see the lower end of the grids you have placed.

- There is a circular grip control at the bottom of the last gridline.

- Click on it and drag it below the model.

  All the gridlines will move together (see Figure 3.96).

- Repeat for the upper end.

- Zoom in close enough to pick the drag grip without touching the elbow control that places flex points in the gridline.

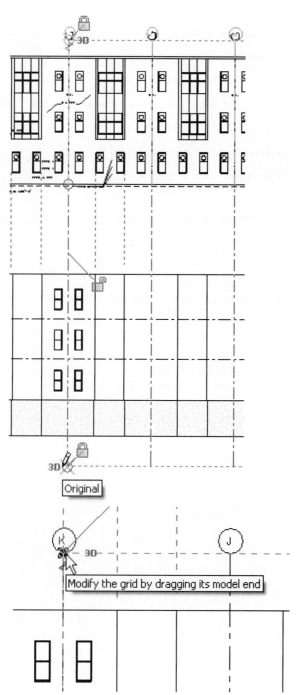

Figure 3.96 *Drag the grid ends down to the model; lower end first (shown above), then the upper ends*

7.21 Click Modify.

- Zoom out to see the model.

- CTRL + select the two window groups at Grid K.

- Click Copy from the Toolbar.

- Pick the end of Grid K for the first reference, then click Grids J through B.

- Click Modify.

7.22 Select the window group to the left of Grid B.

- Copy it to the same place at Grid A.

ADD, GROUP AND COPY THE REMAINING WINDOWS

7.23 Zoom or pan to Grid B.

- Select Window from the Basics Tab of the Design Bar.

- Click Load. Navigate as before to the CD library files.

- Select *Type C.rfa* [*Type C metric.rfa*].

- Click Open.

7.24 Place an instance of the Type C window, as shown in Figure 3.97. It will snap weakly to the midpoint of the wall, and to the default sill height.

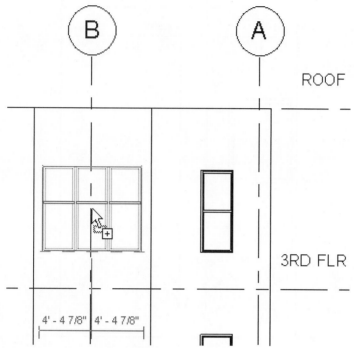

Figure 3.97 *Place a copy of the triple window*

7.25 Place an instance of the window at the 2ND FLR level.

7.26 From the Type selector, pick Type A: 2'8 x 6'8" [Type A: 820 x 2040].

- Place it at the 1ST FLR level.

- Click Modify.

- Click on the new window.

- Use the temporary dimensions that appear to align it at sill height 3' 0" [914] and offset from the wall edge **1' 0" [300]**, as shown in Figure 3.98.

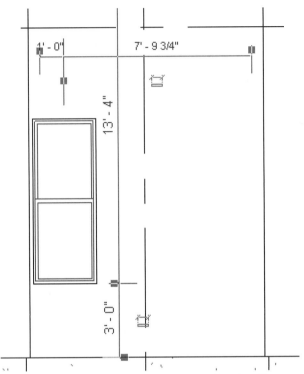

Figure 3.98 *Place the window and check its sill height and left-right offset*

7.27 Place a window to the right of the gridline with the same offset dimensions.

7.28 Zoom out so you can see the four windows you just placed.

- CTRL + select the four windows.

- Click Group on the Toolbar.

- Name the new group **2 Type A and 2 Type 3 windows**. Click OK.

7.29 Click Copy.

- Place the cursor over the top of the wall at Grid B so that the midpoint snap displays, as shown in Figure 3.99.

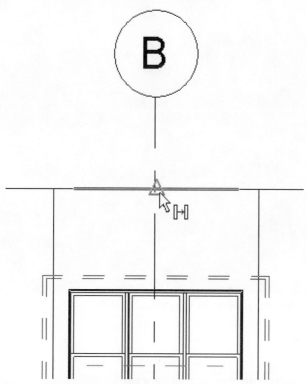

Figure 3.99 *Copy from wall midpoint to wall midpoint*

- Click to establish the start reference.

- Click the midpoint of each of the narrow wall segments to the left.

- Check yourself against the CAD file if necessary.

- You will place 7 copies of the group. Click Modify to terminate the copy.

7.30 Open view SW – Existing to check your work (see Figure 3.100).

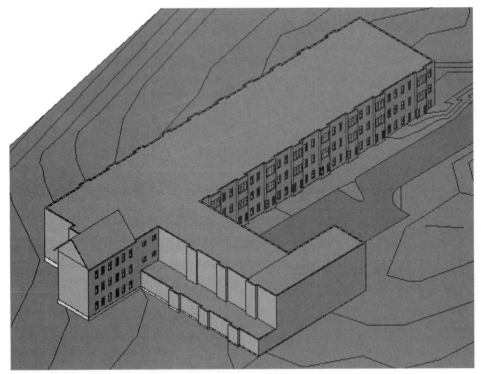

Figure 3.100 *Windows along the south wall, viewed in 3D*

7.31 Save the file.

EXERCISE 8. WORK WITH WALL TYPES.

You have successfully created the exterior shell of a complex building outline. You have added windows to the walls. Now you will adjust the generic walls to show the different finishes indicated in the CAD file.

Basic walls in Revit can be complex both horizontally and vertically. One wall object can contain and represent layers from exterior finish through structural components to interior finish. You apply materials and thickness values to each layer, and set priorities that control how walls clean up at intersections. You can add sweeps and reveals to walls, and split layers so they can hold more than one material. Revit also has Stacked Walls, where you can combine wall types vertically. You can embed one wall type in another.

Curtain wall types are walls based on a grid pattern. You specify the vertical and horizontal layout for the gridlines, either manually or per a formula. You apply mullion components to gridlines and specify material for the panels between grids. You can add doors and windows to curtain walls. You will work with curtain walls in a later chapter.

Walls are as System Families; this means that they do not exist as separate rfa files. You do not load walls in from libraries, but always work with them in projects. You can copy wall types from one project to another, and will almost certainly put your most frequently used wall types in template files. In the next exercise, you will create a number of wall types in the current project.

CREATE A NEW WALL TYPE FROM AN EXISTING ONE

8.1 Open or continue with the file from the previous exercise.

- Open Elevation View South.

- Enter **VV** at the keyboard to open the Visibility/Graphics Overrides dialogue.

- On the Imported Categories tab, click + next to Chapter 3 south elevation.dwg to expand the list of layers.

- Check all A-PATT layers that you hid previously, as shown in Figure 3.101. Click OK.

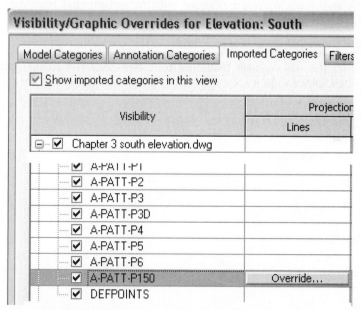

Figure 3.101   *Restore visibility of layers in the CAD file*

8.2 Zoom in to the CAD file.

- Pan across it while studying the notations about wall materials.

  From the ground up, there is a Type 1 facing with three bands of Type 2 and a band of Type 3 at the 2ND FLR level. Above this level are B1 facing and MP, alternating across the façade.

  Type 1 is block, also called Concrete Masonry Unit; Type 2 and 3 are stone ledges of different thickness. B1 is brick, and MP is a metal panel system.

- Use the Tape Measure or place dimensions to note the height above the 1ST FLR level of the three bands of Type 2.

  This will be easiest at the left side.

  Going up, they are at 3' 0" [813], 5' 4" [1524] and 7' 8" [2235]. You will use this information shortly.

Revit provides three main families of walls: Basic, Curtain and Stacked. In any of these walls, Revit lets you work horizontally and vertically. Walls can run from level to level or extend past multiple stories. In the case of the walls you have just examined, the facing changes completely at a certain point. You can create a basic wall type to run to the 2ND FLR level to represent the block, and wall types to show brick and metal panel above that level, or you can create wall types that change materials as they extend up. This is what we shall do in this exercise. Walls created this way are the same thickness all the way up and down, except for sweeps and reveals. Revit also provides a Stacked Wall family that allows you to combine wall types so you can show walls in section that have different thickness at different elevations.

8.3 Zoom to the left of the building model, under the gable roof.

- Select the wall segment that runs from level 2ND FLR to the roof, with the four square windows in it.

- Click Properties.

- Click Edit/New.

- Click Duplicate.

- In the Name box, enter **Brick**. Click OK.

- In the Type Properties dialogue, click Edit next to Structure (see Figure 3.102).

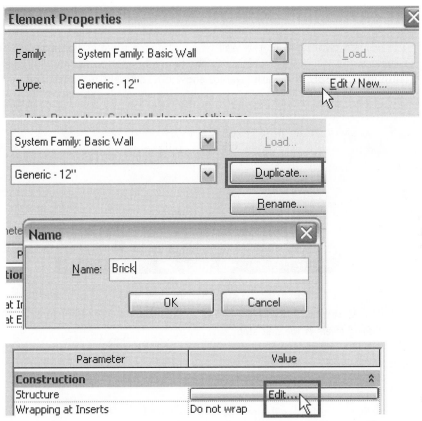

Figure 3.102 *Create a new wall type from an existing one, and prepare to change its structure*

The Edit Assembly dialogue is the place to create complex wall, roof and floor structures. You create and manage layers that will run inside to outside (walls) or bottom to top (roofs and floors). Each layer has a function, material and thickness. The function determines how components intersect so that they represent actual construction conditions without additional drafting. The material property for each layer, which opens another dialogue, is the control point for display of elements in plan, elevation, sections, 3D views and rendered images.

Two layers named Core Boundary are Revit functions, and cannot be deleted. You use the Core Boundary to define the border between structural and finish components of a complex wall, roof or floor. You can use the Core Boundary as the location line when placing walls in plan. This will affect how the wall shifts if you flip its inside-outside orientation.

ADD WALL LAYERS

You will add two layers to this wall type and define them as finish, air gap and structure. You will assign materials to the layers. In later steps, you will change the finish material, split the finish layer to show two different materials, and add sweeps.

8.4 Click Insert twice to add two layers in the wall assembly (see Figure 3.102).

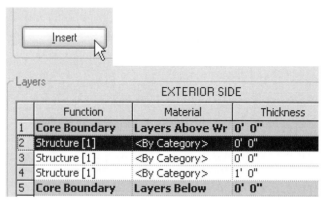

Figure 3.103 *Adding layers is the first step*

8.5 Click in Layer 2.

- Click the Up button.

    The Structure field will move up and become Layer 1.

- Click in Layer 3.

- Click Up.

8.6 Click in the Function field for Layer 1.

- From the list that appears, select Finish 1 [4].

- Click in the Thickness field.

- Enter **4" [100]**.

8.7 Click in the Function field for Layer 3.

- From the list, select Thermal/Air Layer.

- Click in the Thickness field.

- Enter **2" [50]**.

8.8 Click in the Thickness field for Layer 4. Enter **6″ [150]**.

The results should appear as in Figure 3.104. Revit tracks and displays the total wall thickness.

You can adjust the width of the fields in this dialogue.

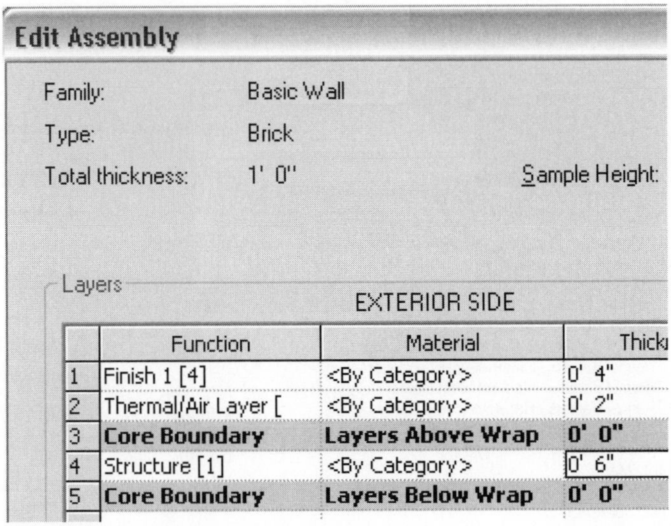

Figure 3.104 *Wall layers with function and thickness*

DEFINE MATERIALS

8.9 Click in the Material field for Layer 1.

- Click the icon that appears, indicating a dialogue is available.

- In the Materials dialogue, scroll down the list in the left pane to Masonry – Brick. Click on that choice.

- The right side pane will fill in with shading, rendering and pattern information (see Figure 3.105).

- Click OK.

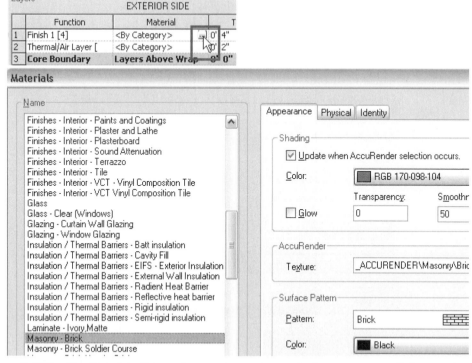

Figure 3.105 *The Materials dialogue is extensive, and controls the appearance of elements in all views*

8.10 Click in the Material field for Layer 4, as in the previous step.

- In the Materials dialogue, click on Metal – Stud Layer.

- Click OK until all dialogue boxes close.

  The Wall will change appearance. If you zoom in, you can see the brick pattern. Revit adjusts the display of model patterns to the zoom.

  Model patterns are sized in real-world dimensions, and their scale does not change. Drafting patterns scale with the view scale. You will see the difference in later exercises.

CREATE A VERTICALLY COMPOUND WALL TYPE

8.11 Click the wall to the left of the one you just redefined.

- Click in the Type Selector and pick Basic Wall: Brick.

   The wall will change appearance.

- Click Properties.

- Click Edit/New.

- Click Duplicate.

- Enter **Brick over Block**.

- Click OK.

8.12 Click the Preview button at the lower-left corner of the Edit Assembly dialogue.

The preview pane toggles between plan and section view. You can zoom and pan in this view to examine the structure of your wall, roof or floor.

- Click on the drop-down list in the View control at the lower right of the preview pane.

- Pick Section: Modify Type Attributes.

   The Modify Vertical Structure controls activate when Section view is current in the preview (see Figure 3.106).

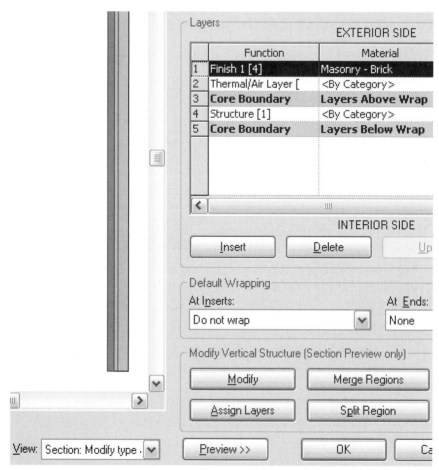

Figure 3.106 *Prepare to modify the vertical structure by opening a section preview*

Layer 1 is highlighted black in the assembly pane and red in the preview pane. This is the layer you will augment.

8.13 Click Split Region in the Modify Vertical Structure section of the dialogue.

- In the Preview, place the cursor over the red section on the left of the wall.

  The Tooltip and Status Bar will identify the layer as Layer 1: Masonry – Brick. A temporary dimension will appear. As you move the cursor up and down, the dimensions will snap to the base or top of the wall. You can zoom using the scroll wheel and pan by using the scroll bars on the Preview pane. As you zoom in closer, the dimension snap value decreases in distance.

- When the temporary dimension reads approximately 9' 0" [3000] from the bottom, click to place a split, as shown in Figure 3.107.

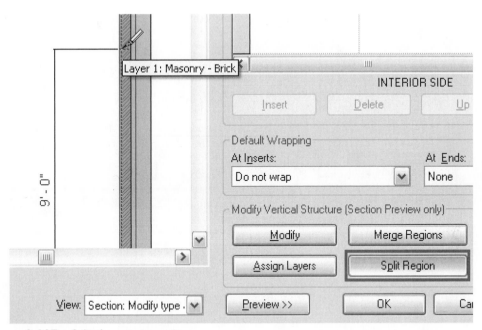

Figure 3.107   *Split the exterior region*

8.14 Click Modify to terminate the Split tool.

- Zoom in so you can see the split line clearly.

- Place your cursor over it; the Tooltip and Status Bar will identify it as Border Between Layer 1 and Layer 2.

- Click on the border.

    Temporary dimensions will appear. You may have to zoom out to read the dimension field.

    Click in the field.

    Enter **13' 4" [4064]**.

    Now that you have split a region, you need to create a new layer to hold the material for the new region.

8.15 Click on Layer 2 in the assembly pane.

- Click Insert.

    The new layer will have default values for Function, Material and Thickness.

- Set the Structure For Layer 2 to Finish [1].

- Click in the Material field for Layer 2.

- Set the Material to Masonry – Concrete Masonry Units [Masonry – Concrete Blocks].

- Click OK.

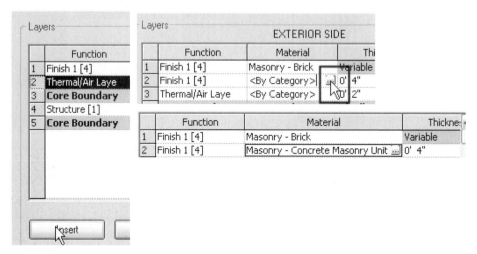

Figure 3.108 *Insert a layer and set its properties*

8.16 Click Layer 2 if it is not highlighted in the assembly pane.

- Click Assign Layers.

- Place the cursor over the left side of the wall in the preview.

  The Tooltip and Status Bar will identify the material. This will be the same in both the upper and lower regions.

- Click in the lower region.

  This will change the material to that of Layer 2, Concrete Masonry.

- Click Layer 1 in the assembly pane to highlight that region red in the preview (see Figure 3.109).

  The lower region will be white. There are now two materials divided at the split line. They will have the same thickness. Basic wall types do not allow layers with different thickness. If you need to represent thicker block than brick, for instance, use a Stacked Wall type.

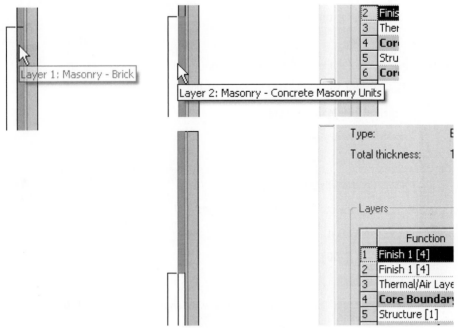

Figure 3.109 *Assign a layer to a split region to apply different materials*

Sweeps are profiles that pass along a path. If the path is the face of a wall or roof, you can create a molding, cornice, fascia or gutter. You can also create free-standing sweeps when making Mass Objects or In-Place Families. You will work with these in later exercises. In the next few steps, you will place sweeps using two different profiles on the face of this wall type so that it resembles the CAD elevation.

ADD SWEEPS

8.17 Click Sweeps in the Modify Vertical Structure (Section Preview only) area of the dialogue.

The profiles you need to create the sweeps you measured previously are not in this project yet.

- In the Walls Sweeps dialogue, click Load Profile.

- Navigate to the Profiles folder in the Imperial or Metric library.

- Click *Wall Sweep–Brick Course.rfa* [*M_Wall Sweep–Brick Course.rfa*].

- Click Open.

- Click Load Profile.

- In the Profile folder, click on *Wall Sweep-CMU Course.rfa* [*M_Wall Sweep-CMU Course.rfa*].

You are using these profiles for their size only, not because you will set the material according to the profile names.

- Click Open.

8.18 Click in the Profile field for the sweep.

- Select Wall Sweep-Brick Course: 1 Brick.

- Click in the Material field.

- Select Masonry – Stone.

- Click OK.

- Click in the Distance field.

- Set the value to **3' 0" [900]** from Base.

- Click in Offset.

- Enter **-2" [-50]**.

- Click Apply.

The sweep will appear in the preview, as shown in Figure 3.110.

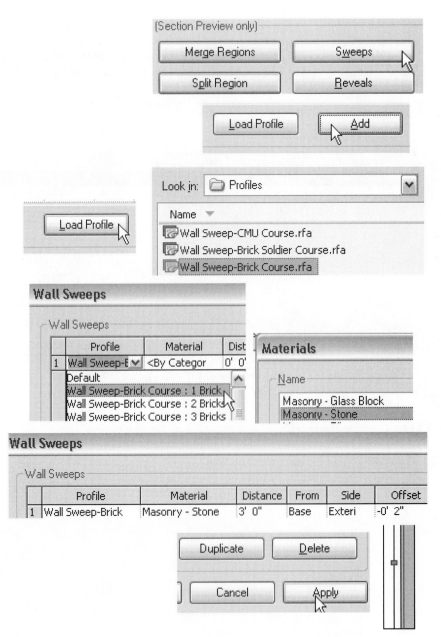

Figure 3.110 *Create a sweep, load profiles and define the sweep properties.*

8.19 Click the number 1 to the left of the sweep so the entire row highlights. Click Duplicate three times to create new sweeps. In the Distance fields, enter **5' 4"**, **7' 8"** and **13' 4"** [**1524, 2235, 4064**]. Click Apply. The layers reorder themselves by height. Set the Offset for Layer 1, the CMU Course, to **-4" [-100]** (see Figure 3.111).

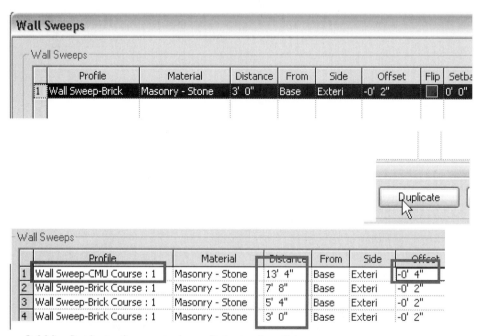

Figure 3.111 *Duplicate the sweep, then edit properties of the new sweeps*

8.20 Click OK to exit all dialogues. The wall face will update.

APPLY THE NEW TYPE TO OTHER WALLS

8.21 Zoom to Fit.

- On the Toolbar, click Match Type.

  The cursor changes to an eyedropper. This will allow you to change components from one type to another quickly, without using the Type Selector list.

- Click on the face of the new wall type you just created.

  The eyedropper fills.

- Click on wall segments, as shown in Figure 3.112.

- Match the CAD elevation with your picks.

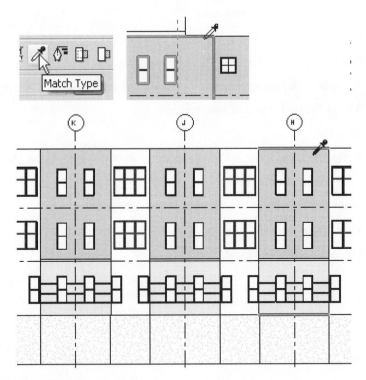

Figure 3.112　*Change walls to the new type*

CREATE A METAL PANEL WALL TYPE WITH CUSTOM SURFACE PATTERN

8.22 Zoom in to the right side of the model.

- Change the wall segment at Grid B to Brick over Block.

- Click Properties.

- Click Edit/New.

- Click Duplicate.

- Enter **Metal Panel over Block**.

- Click OK.

- Click Edit in the Structure field.

- Click in the Material field for Layer 1.

- Select Finishes - Exterior – Metal Panel in the list.

- In the Surface Pattern section, click the 3-dot icon to open a further dialogue.

8.23 In the Fill Patterns dialogue, select Model.

- Click New.

- In the Add Surface Pattern dialogue, enter **Metal Panel** in the Name field.

- Enter **90** for the Angle.

- Enter **3' 0" [900]** in Line Spacing 1 (see Figure 3.113).

- Click OK to exit all dialogues.

  The wall segment will update.

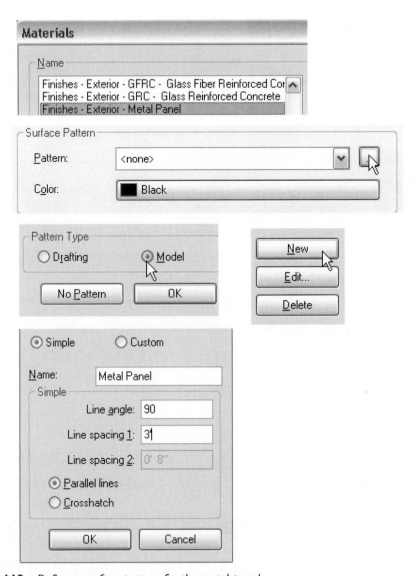

Figure 3.113   *Define a surface pattern for the metal panel*

8.24 Click Match Properties on the Toolbar.

- Change all the wall segments with triple windows to the new type.

8.25 Click on the CAD import.

- Type **VH** at the keyboard to hide it in the view. Zoom to Fit.

8.26 Open 3D view SW – Existing.

8.27 Save the file.

EXERCISE 9. EXERCISE 9: EDIT A WALL PROFILE

You can create walls with sloped or curved top and bases. You can also place holes of any shape in walls by editing their sketches in elevation, section or 3D views.

SET UP VIEWS

9.1 Open or continue working in the file from the previous exercise.

9.2 Open Elevation view West.

- Right-click and pick View Properties.

- Set the Phase to New Construction.

- Set the Phase Filter to Show Complete.

- Click OK.

9.3 Click the Show Crop tool on the View Toolbar.

- Click on the double-arrow drag control on the bottom edge.

- Pull it up to conceal the foundation walls and simplify the view.

- Pull the left and right crop close to the model (see Figure 3.114).

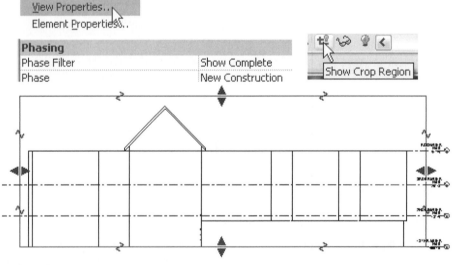

Figure 3.114 *Set up the elevation view*

9.4 Click the Show Crop tool to toggle off the crop display.

- Zoom to Fit.

 You will return to this view to change the shape of a new wall.

9.5 Open Floor Plan 1ST FLR.

- CTRL + select a green Reference Plane and a grid.

- Enter **VH** at the keyboard to turn off their display.

- Right-click and pick View Properties.

- Set the Phase to New Construction.

- Set the Phase Filter to Show Complete.

- Set the Underlay to 2ND FLR.

 You will refer to the walls on the floor above your work.

- Click Edit in View Range.

- Set the View Depth Level to Associated level (1ST FLOOR) (see Figure 3.115).

- Click OK.

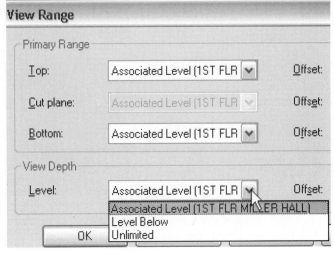

Figure 3.115 *Prepare the view so it will be easy to place a wall*

9.6 Zoom in to the left side of the North-South wing.

The walls above will be light gray. You can select and move them, so pick carefully.

9.7 Click Wall on the Basics Tab of the Design Bar.

- In the Type Selector, select Basic Wall: Generic 12" [Basic Wall: Generic 300mm].

- Set the Height to 2ND FLR.

- Set the Location Line to Finish Face Exterior.

- Draw the wall from the lower-left corner of the walls one floor up straight up, as shown in Figure 3.115.

  A warning will appear. This new wall overlaps with walls above it.

- Click Modify.

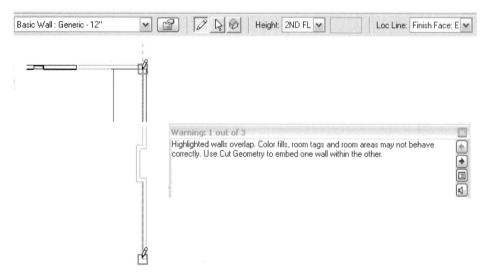

Figure 3.116 *Create the wall you will edit.*

9.8 Open Elevation view West.

9.9 Place the cursor over Level 2ND FLR.

- Move the cursor along the level line until the wall highlights.

- Click to select it.

  Controls will activate.

9.10 Pull the top down to meet the bottom of the wall above, which you previously set to a negative base offset from the level.

- Click the padlock.

You are now working in New Construction phase. You have placed a wall to indicate one side of the new ground floor condition. This upper stories of this wing will become elevated hallways and the first floor will now allow open air cross traffic. This wall will support the walls above and hold openings for paths. You will sketch these openings in the wall profile.

9.11 Click Edit Profile from the Options Bar.

The outline sketch becomes available and the Design Bar switches to Sketch mode. You will center an archway under the central bay of the wall above.

9.12 Click Ref Plane on the Sketch Tab of the Design Bar.

- Place two Reference Planes, as shown in Figure 3.117, extending down from wall edges.

9.13 Click Lines on the Design Bar.

- Click the Arc passing through three points option.

- Start the arc at the intersection of level 1ST FLR and one Reference Plane.

- Click the intersection of level 1ST FLR and the other Reference Plane.

- Pull the cursor up to define a 180° arc (see Figure 3.117).

- Click Modify.

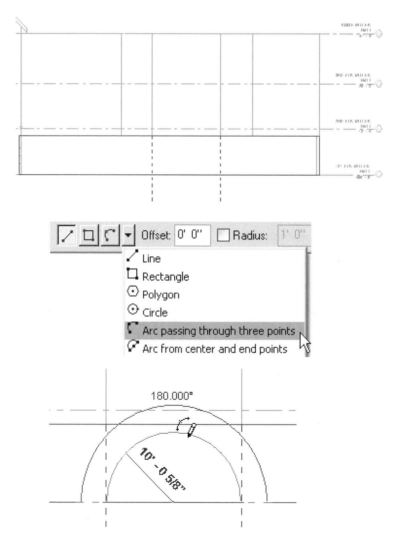

Figure 3.117 *Use Reference planes to place an arc inside the wall profile sketch*

9.14 Select the new arc.

- Click Copy on the Toolbar.

- Click a point on one of the vertical walls as the start point.

- Pull to the left horizontally and click the corresponding point two walls to the left.

9.15 Click Split from the Toolbar.

- On the Options Bar, check Delete Inner Segment.

- Click at the left base point of the left arch.

- Click at the right base point of the same arch to remove the line between the pick points.

- Repeat on the right arch (see Figure 3.118).

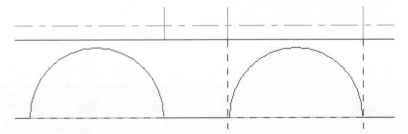

Figure 3.118   *Remove lines inside the arches by using Split*

9.16 Click Finish Sketch.

- Select the Wall. In the Type Selector, click Brick over Block.

  A warning will appear, because this wall instance does not have enough height to create the top sweep.

9.17 CTRL + click the left, middle and right upper wall segments.

- Set their type to Brick.

9.18 CTRL + click the narrow upper wall segments.

- Set their type to Metal Panel.

9.19 Zoom to Fit.

- Set the Model Graphics Style to Shading with Edges.

- Click the Shadows control to the right of the Model Graphics control on the View Control Bar.

- Select Shadows On (see Figure 3.119).

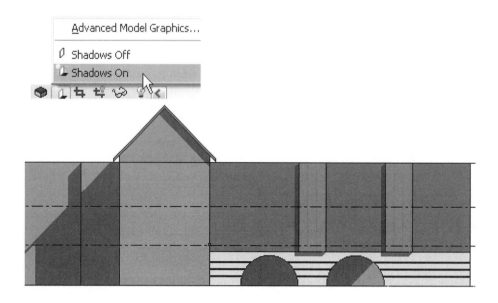

Figure 3.119 *Examine the new wall by turning on shading and shadows*

9.20 Save the file.

SUMMARY

Congratulations. You have created a new Revit Architecture file, imported files into it, and referred to the CAD content of these files while you created a site plan and the exterior shell of a building model using Revit Architecture elements and components. You have begun the process of detailing the shell of this sizable structure, and created a new wall with unique openings. In the process, you have worked with many of the basic design and editing tools within Revit Architecture, and some techniques that are a little more advanced than the basic level. In the next chapter, you will use the model created in this chapter, plus an imported site plan sketch, to set up a multibuilding, multifile campus development project.

REVIEW QUESTIONS – CHAPTER 3

MULTIPLE CHOICE

    1. To control visibility of layers in an imported CAD file

        a. use link instead of import

        b. invert the layer colors

        c. use the DWG/DWF/DGN tab in the Visibility Graphics dialogue

        d. You can't control layers in an imported file.

2. Phases in Revit Architecture projects control

    a. the display of objects in views

    b. how many objects of a certain type the file can contain

    c. the voltage the computer uses while working in Revit Architecture

    d. the date display in titleblocks

3. Reference Planes

    a. can be named

    b. can be turned on and off in views

    c. are excellent alignment tools

    d. all of the above

4. Sketch mode on the Design Bar is active

    a. when creating or editing a roof

    b. when editing a wall elevation profile

    c. whenever no other command is active

    d. a and b, but not c

5. Revit Architecture's default Wall and Window families

    a. can't be edited

    b. are made to be edited

    c. can be edited if you pay a special fee

    d. fight all the time, like most families

TRUE/FALSE

6. True _ False _   The View Properties dialogue controls the appearance of nearly everything in Revit Architecture.

7. True _ False _   Wall and Window types cannot be edited during placement.

8. True _ False _   The Align tool will snap to walls, but not to windows.

9. True _ False _   Plan views can be duplicated in the Project Browser, but Elevations have to be copied in Plan views to create new Elevation views.

10. True _ False _   An object's Phase Created and Phase Demolished properties work with View Phases and Phase Filters to control display of the object.

 Answers will be found on the CD.

CHAPTER 4

Design Modeling with Mass Objects

INTRODUCTION

This section continues a series of exercises using basic Revit Architecture techniques to create related, interconnected project files. The exercises in Chapter 3 showed how to use 2D imported CAD file information as the basis for a Revit Architecture model. You created the model using building elements (walls, roofs) and components (doors, windows). For your hypothetical multi-building project, the starting point is an existing building. You modeled this building in Chapter 3. In this chapter's exercises, you will start modeling three structures for Phases 1, 2 and 3 of this campus development. You will create mass objects to define the shape, size and location of these structures.

To get started, you will import a file holding a site topography/building model similar to the one you created in Chapter 3 and a sketch of a proposed site plan into a project file, and then correlate these two sources. You will trace building footprints on top of the raster image and copy/paste those footprints into three additional empty project files. Each of those project files will become a building project file, using 3D massing components to start a building. You will link those building model files back into the first file you created, which will hold the overall view of this extensive development.

OBJECTIVES

- Create mass models using a variety of techniques in three separate files
- Link Revit Architecture files together; share locations and coordinates
- Create perspective views of mass models for client presentation

REVIT ARCHITECTURE COMMANDS AND SKILLS

File save settings

File link

Shared levels

Coordinates—acquire; relocate project

File import attributes

Model lines—line, arc, fillet arc

Edit commands—move, copy, trim, rotate, copy to Clipboard, paste

Add/Cut Mass—Extrusion, Sweep, Blend

View Camera

EXERCISE I. IMPORT PROJECT DATA—ACQUIRE COORDINATES

1.1 Launch Revit Architecture.

It will open to an empty project file.

- From the File menu, pick File>Close.

- Select File>Open from the File menu.

- Navigate to the folder that holds the files for Chapter 4.

- Select *Chapter 4 start imperial.rvt* [*Chapter 4 start metric.rvt*]. Click Open.

The file will open to a plan view.

- From the File menu, pick File>Save As.

- In the File Save Option dialogue, make the maximum number of backup(s) value 1. Choose OK.

 Note: This value is set in each Revit Architecture file. It is **not** global. Do this step for each file you create in this and other exercises.

If you are working in a class situation, set the value as directed by your instructor. If you are working on your own on a computer that may have limited storage space, you will want to make this a habit when first saving project files.

1.2 Save the file as **Chapter 4 Overview.rvt**, in a location specified by your instructor if you're in a class, or in a folder on a drive with adequate space and read-write permission.

1.3 Pick File>Import/Link>RVT from the File menu, as shown in Figure 4.1.

- Navigate to the folder containing *Chapter 4 Miller Hall imperial .rvt* [*Chapter 4 Miller Hall metric.rvt*]; select the appropriate file name so it appears in the File name field.

- Accept the default positioning option: Automatically place Center-to-center.

- Click Open.

The building outline will appear. The linked file is a single object that you can move, rotate or copy.

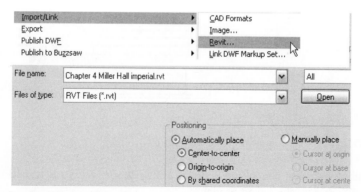

Figure 4.1　*Link a Revit Architecture file*

1.4 Open View Elevations: East.

- Type **VG** to open the Visibility/Graphics Overrides dialogue for this view.

- In the Model Categories Tab, clear the Topography check box, as shown in Figure 4.2.

- Click OK.

  The Toposurface under the building will disappear.

- Zoom to Fit.

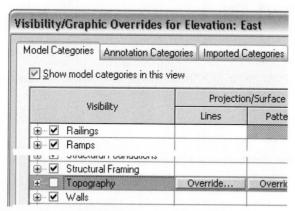

Figure 4.2　*Turn off the topography for clarity*

1.5 Zoom in to the level lines at the lower left of the view.

- Click on the line for Level 1 (not the bubble).

- Click the check boxes that appear to hide the bubble on the right end and show the bubble on the left end (see Figure 4.3).

- Repeat for Level 2.

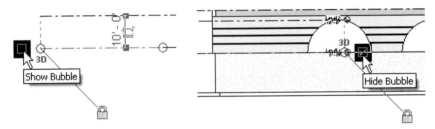

Figure 4.3 *Display the Level bubble at the left end only*

1.6 With either level line selected, use the drag control at the right end to extend the level lines past the linked model.

- Drag the left ends of the level lines close to the model.

1.7 Click on Level 1 to select it. In the Type Selector, pick Level: Shared.

You are changing the level type so it reads shared coordinates. You created a Shared Level type in a previous exercise. You will update the elevation value in the next steps.

Revit Architecture does not use an "exposed" coordinate system, meaning there is no way to specify an absolute position for entities in a file independent of other entities. Revit Architecture does track and use the center point of each file, which changes as the building model footprint changes. Revit Architecture also provides a coordinate location point (unseen by the user) with each project file, so that linked files can synchronize positions. Generally the project file that contains site information becomes the basis for shared coordinates, and building models will be located according to the topography.

1.8 From the Tools menu, pick Tools>Shared Coordinates>Acquire Coordinates, as shown in Figure 4.4.

The Status Bar will read Select a linked project from which to acquire shared coordinate system.

- Select the elevation of the linked file.

Level 1 will read the value of 1ST FLR MILLER HALL level in the linked file.

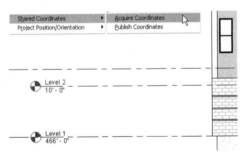

Figure 4.4 *Acquire coordinates from the link*

1.9 Click Align from the Toolbar.

- Click on 2ND FLR MILLER HALL in the linked file.

- Click on Level 2 to align it at the elevation of the linked level.

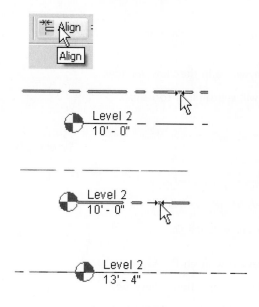

Figure 4.5 *Adjust the Elevation of Level 2*

1.10 Zoom to Fit.

- Open View Floor Plans: Level 1. Save the file.

EXERCISE 2. SKETCH BUILDING FOOTPRINTS FROM AN IMPORTED IMAGE

2.1 Open or continue working with the *Chapter 4 Project Overview.rvt* file from the previous exercise.

- Type **VV** to open the Visibility/Graphic Overrides dialogue.

  This is an alternate keyboard shortcut. You have previously used **VG**.

- On the Annotation Categories Tab, clear Elevations.

- Click OK.

  Only the building outline will now be visible in the View window.

2.2 From the File menu, select File>Import/Link>Image, as shown in Figure 4.9.

- Navigate to the folder that contains the file *proposed site plan.jpg*.
- Select the file name.
- Click Open.

  Four blue control dots indicating the boundary of the image will appear around the cursor.

- Left-click anywhere in the View window.

  The image will appear. It is too small.

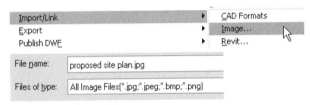

Figure 4.6 *Add a raster image to the file*

2.3 Pick the Properties icon on the Options Bar.

- Clear the Lock Proportions check box.
- Change the Width value to **1140' [349000]**.
- Change the Height value to **925' [278500]** (see Figure 4.10).
- Click OK.

Parameter	
Dimensions	
Width	1140' 0"
Height	925' 0"
Horizontal Scale	349.276596
Vertical Scale	349.912434
Lock Proportions	☐

Figure 4.7 *Adjust the size of the image to fit the model*

 Note: These values have been determined by trial and error. You will have to adjust the size of other images you may import or link into project files experimentally.

The image will change size. It will still be selected—the cursor is the double-headed double arrow, indicating that you can drag the selected item to a new position.

- Pick Move from the Toolbar.

2.4 Zoom in Region in the image file where the hand lettering reads **Existing Building 3 Story**.

- Left-click at one of the corner points of the drawn building outline, then left-click at the corresponding corner of the linked building model to move the image over the model (see Figures 4–8 and 4–9).

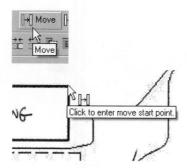

Figure 4.8 *Start moving the image*

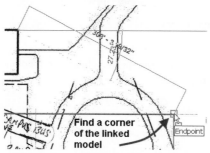

Figure 4.9 *The second pick for moving the image*

2.5 Zoom in Region to the right side of the building model.

The walls of the model will show as white over the black raster image pixels. The image will still be selected.

- Move the image as necessary to get a careful alignment with the walls (see Figure 4.10).

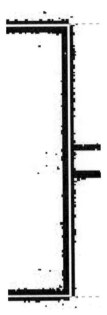

Figure 4.10 *Careful alignment is possible*

 Tip: Use the arrow keys to move selected objects small distances. This will work better than dragging with the cursor.

- When you are satisfied, Zoom to Fit and study the site plan sketch.

2.6 Click on the raster image so it highlights.

- Click Pin from the Toolbar, as shown in Figure 4.11.

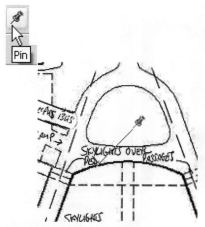

Figure 4.11 *Lock the location of the image before drawing over it*

A blue pushpin symbol will display. The image is now locked in place.

2.7 Zoom in Region around the Phase 1 building outline sketch.

- Right-click and pick View Properties from the context menu.

- Set the Phase Filter to Show All.

- Click OK.

  Additional walls will appear in the linked file. These walls will be demolished, and until now they have been hidden in this view. You will use a wall intersection to start your sketch in the next steps.

In CAD applications, everything you create is line-based. Revit contains many model object placement tools, in which you designate a location for an object that has already been defined. In the next exercises, you will practice drawing lines and manipulating them. These lines will become the basis for 3D model objects. In later exercises, you will draw and edit lines to provide outlines for floors, roofs and stairs.

SKETCH BUILDING OUTLINES USING THE IMPORTED IMAGE

You will now start to sketch over this very rough image.

2.8 Pick Lines from the Basics Tab of the Design Bar.

These are Model lines, and will be visible in other views.

- Verify that Lines is selected in the Type Selector.

2.9 Sketch a line starting from the intersection of the walls of the linked model as shown.

- Draw straight up a distance of 18'-0" [5400 mm], as shown in Figure 4.12.

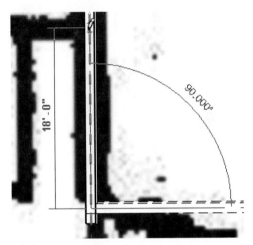

Figure 4.12 *The first sketch line*

2.10 From the File menu, pick File>Manage Links.

- On the Revit Tab, select the name of the *Chapter 4 Miller Hall.rvt* file so that it highlights (see Figure 4.15).

- Select Unload.

- Click Yes in the question box.

- Click OK.

  The building model will disappear.

Figure 4.13 *Manage Links to unload the building model*

2.11 Continue sketching from the upper endpoint of the line you just drew. Draw a line segment 44'-0" [13200 mm] to the left. Draw a line down 18'-0" [5400 mm] to match the first line.

2.12 Draw five lines around the footprint of the Phase 1 building, as shown in Figure 4.16.

The line lengths will be 272'-0" [81600 mm], 76'-0" [22800 mm], 184'-0" [45200 mm] and 40'-0" [12000 mm].

- Do not add the dimensions; they are for guidance only.

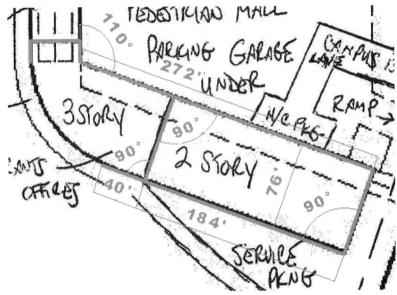

Figure 4.14 *Draw the simple building outline*

2.13 To draw the arc at the lower left of the sketch outline, pick the Fillet Arc icon on the Options Bar, as shown in Figure 4.15. Click the Radius check box. Enter **100'** **[30000]**.

Revit Architecture will prompt you to pick two lines near the endpoints where you want the fillet to be applied.

- Pick the two lines shown in Figure 4.15 at the approximate locations shown.

  Revit Architecture will place an arc between the lines you picked.

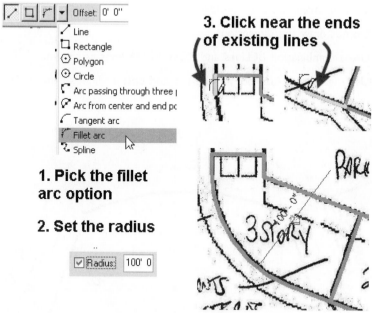

Figure 4.15 *The Fillet arc tool with a predefined radius*

2.14 With the Lines command still active, change the Type Selector to <Overhead>, as shown in Figure 4.20. On the Options Bar, click the straight line and Chain Options

- Draw the two lines shown in Figure 4.20 over the dashed lines representing a building overhang.

- Do not draw the offset dimension of 20'-0" [6000 mm].

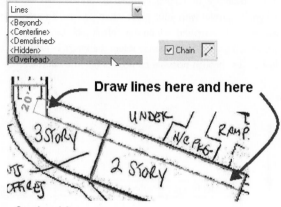

Figure 4.16 *The new Overhead lines*

2.15 Continue adding sketch lines, as shown in Figure 4.17.

The raster image has been removed in this figure for clarity.

- Do not draw the dimensions.

- Use a combination of Lines and <Overhead> lines as shown.

- Use the 3-Point Arc tool for the top and bottom lines of the outline.

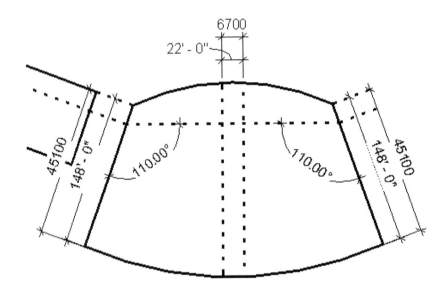

Figure 4.17 *The central building outline—the raster image has been turned off for clarity*

 Tip: Try to take advantage of editing commands rather than drawing all parts of a symmetrical design. Be careful with your picks when you are zoomed in close; the raster image that provides the background will be the default pick. Learn to read the cursor (see Figure 4.18) so that you do not inadvertently select the image when trying to do something else. The tab key cycles through possible choices under the cursor.

2.16 Zoom to Fit.

Figure 4.18 *The cursor Tooltip and appearance let you know when you have made a selection*

- Pick Modify from the Basics Tab of the Design Bar to terminate the Lines command.

- Click the raster image to select it.

- Right-click and select Hide in View>Elements to turn the raster image off in the view.

2.17 Zoom in Region around the sketch lines.

- Select the lines as shown in Figure 4.19.

- Select Mirror from the Toolbar.

- Choose the Draw option.

- Select the midpoint of the horizontal line in the central building outline for the first point of the mirror axis, then pull the cursor straight down (or up) to define a vertical mirror axis at the middle of the outline.

- Click to mirror the selection.

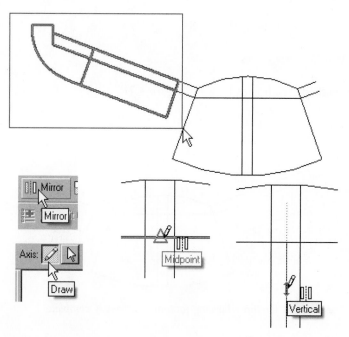

Figure 4.19 *Select the lines to mirror*

2.18 Edit the top and curved side of the right-hand building outline sketch to eliminate the gap, as shown in Figure 4.20.

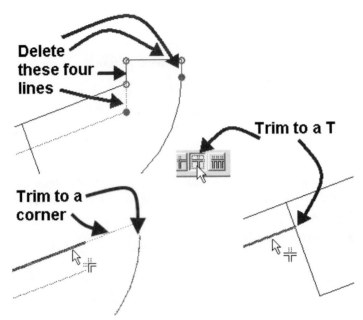

Figure 4.20 *Steps for editing lines*

2.19 The finished sketch lines look like Figure 4.21.

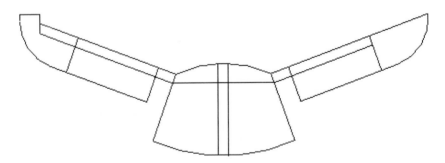

Figure 4.21 *The linework for three building footprint outlines is complete*

2.20 Click Modify on the Design Bar.

- Zoom to Fit.

- Click the Reveal Hidden Elements (lightbulb) control on the View Control Bar.

  Hidden elements will be displayed. The view will show a red border, and the lightbulb icon will have a red background.

- Click on the raster image.

- Right-click and select Unhide in View>Elements.
- Click the Hidden Elements icon to toggle it off (see Figure 4.26).

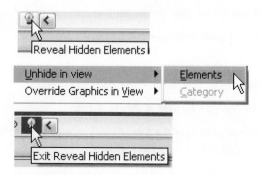

Figure 4.22 *Make the image visible to check your sketch against it*

2.21 Check your sketch for accuracy and make any necessary adjustments.

- Select the raster image. Click the Delete icon on the Toolbar (see Figure 4.23).

Figure 4.23 *Revit's Delete tool*

A warning will appear because you have deleted a pinned object. You can ignore the warning. It will disappear.

- Zoom to Fit.
- Save the file.

Now that you have used the existing model file and a sketch of the proposed site plan to capture design intent information accurately, you are ready to set up files for the individual buildings that will constitute this campus development. First you will move the sketched lines into separate files for each building model, and then link those model files back into the Overview file. The building models will be developed and revised extensively during the course of this (or any) project. The Overview file provides a central place for designers or project managers to check progress and design correlation.

EXERCISE 3. TRANSFER SKETCH LINES INTO NEW PROJECT FILES.

3.1 Select File>Open from the File menu.

- Navigate to the folder that holds the files for Chapter 4.

- Select *Chapter 4 start imperial.rvt* [*Chapter 4 start metric.rvt*].

- Click Open.

- From the File menu, select File>Save As.

- Save the new file as **Chapter 4 Phase 1.rvt**.

- Use the Options button in the Save As dialogue to set the backup count for this new file to 1, or as specified by your instructor.

3.2 From the Window menu, pick *Chapter 4 Project Overview.rvt – Floor Plan: Level 1* from the list of available views in the bottom section of the menu to return to that view (see Figure 4.27).

Your Project list may look different from the figure.

Figure 4.24 *Return the view to the Overview file*

3.3 With no command active, pick two points around the building outline on the left side of the View window.

- Pick Edit>Copy to Clipboard from the Edit menu or Toolbar, as shown in Figure 4.25.

  You can also use CTRL + C, the Windows keyboard shortcut.

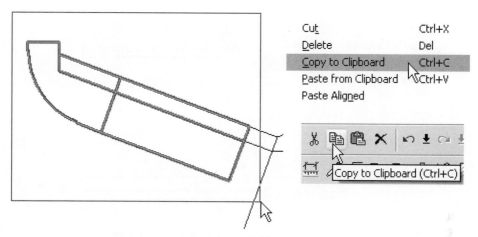

Figure 4.25 *Select the Phase I building outline linework*

3.4 CTRL + TAB to cycle the view to the other project file.

This is a Windows keyboard shortcut.

- Pick Edit>Paste, from the Edit menu or Toolbar, as shown in Figure 4.30.

  You can also use CTRL + V as a keyboard shortcut.

- Place the copied lines inside the elevations.

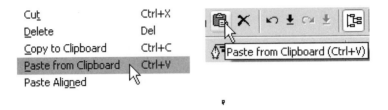

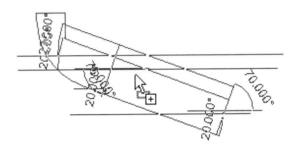

Figure 4.26 *Edit>Paste from the Clipboard*

3.5 Choose Rotate from the Toolbar.

- Pick a point to the right of the rotation icon that appears in the center of the pasted content.

- Pull the cursor up until the temporary angle dimension reads 20°.

- Click to rotate the highlighted elements counterclockwise together (see Figure 4.27).

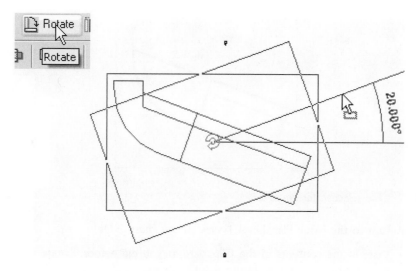

Figure 4.27   *Rotate the new lines to horizontal orientation*

3.6 From the File menu, select File>Save.

- From the File menu, select File>Close.

3.7 From the File menu, select *Chapter 4 start imperial.rvt* [*Chapter 4 start metric.rvt*] from the list of recently opened files at the bottom of the menu (see Figure 4.28).

Figure 4.28   *Reopen a recent file from the menu list*

- From the File menu, select File>Save As.

- Save the new file as **Chapter 4 Phase 2.rvt**.

  The original topography for the referenced building model shows that there are different elevations where the concept plan sketch locates the new buildings. These steps set the new files at different absolute elevations. In the next exercise, you will correlate them at correct vertical and horizontal placement in the overview file.

3.8 Return to the Floor Plan: Level I view of the Project Overview file, as you did before.

- Select the right-hand building outline, as shown in Figure 4.29.

- Copy the selection to the Clipboard.

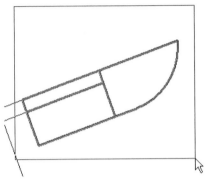

Figure 4.29   *The selection for Phase 2*

3.9 Return to the Floor Plan: Level 1 view of the *Phase 2* file.

- Paste in the contents of the Clipboard as you did before. Rotate the contents 20° clockwise this time (see Figure 4.30).

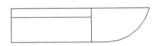

Figure 4.30   *The building outline placed horizontally*

3.10 Save the new file.

- Close the file.

3.11 From the File menu, select *Chapter 4 start imperial.rvt* [*Chapter 4 start metric.rvt*] from the list of recently opened files at the bottom of the menu, as you did in Step 3.7.

- From the File menu, select File>Save As.

- Save the new file as **Chapter 4 Phase 3.rvt**.

3.12 Return to the Floor Plan: Level 1 view of the Project Overview file, as you did before.

- Select the middle building outline, as shown in Figure 4.31.

- Copy the selection to the Clipboard.

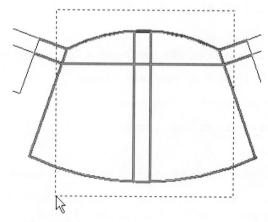

Figure 4.31 *Copy the middle outline to the Clipboard—note the right-to-left selection window*

3.13 Return to the Floor Plan: Level I view of the new file.

- Paste in the contents of the Clipboard as before.

Revit Architecture allows the user to copy and paste from file to file, as you have just done, and also allows copy/paste operations within a project, with alignment as an option. A particular arrangement of walls or furniture components can easily be copied from floor to floor of a multi-level project in exact alignment. Revit Architecture's capacity to align objects with each other or with Reference Planes is sophisticated and extremely useful for the busy designer.

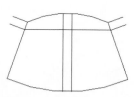

Figure 4.32 *The new file with Elevations in place*

3.14 Save the new file.

- Close the file.

EXERCISE 4. LINK THE NEW PROJECT PHASE FILES TO THE OVERVIEW FILE

4.1 Open or continue working with the *Chapter 4 Project Overview* file.

- Zoom to Fit in the Floor Plan: Level 1 view.

- Right-click and select View Properties.

- Set the Phase Filter to Show Complete.

- Click OK.

4.2 From the File menu, select File>Import/Link>Revit, as shown in Figure 4.33.

- In the Look in: section of the Add Link dialogue, navigate to the folder that contains the file *Chapter 4 Phase 1.rvt* that you created earlier.

- Select that file name so that it appears in the File Name field in the Add Link dialogue box.

- In the Positioning section of the dialogue, select Manually Place>Cursor at origin.

- Click Open.

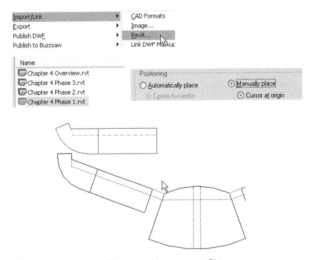

Figure 4.33 *Link a Revit Architecture file into the current file*

Linework from the linked file will appear in the View window. The cursor Tooltip and Status Bar will read Left-Click to place Import Instance.

4.3 Place the linked instance to the left of the View window near the linework you pasted into the link file.

4.4 Repeat the Add Link process for files *Chapter 4 Phase 2.rvt* and *Chapter 4 Phase 3.rvt*.

- Place the Phase 2 file to the right of the View window.

- Place the Phase 3 file at the bottom of the View window.

- Zoom to Fit.

4.5 Select the Phase 3 link instance.

It will highlight red and a boundary will appear around it. The cursor will become the double-headed edit arrow when you move it over the edges of the link.

- Move the linked file using appropriate pick points so that it sits directly on top of the linework in the current file (see Figure 4.34).

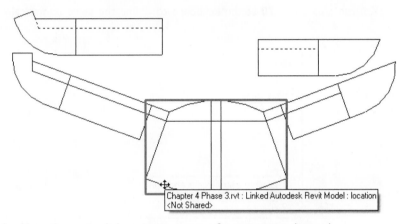

Figure 4.34 *Move the center link so it sits on top of its originating linework*

4.6 Select the Phase 1 link.

- Move it so that the upper-left corner of the link sits directly on the corresponding corner of the linework in the current file (see Figure 4.35).

- Choose Rotate from the Toolbar.

The Rotation icon will appear in the center of the link Instance.

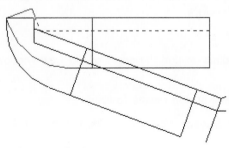

Figure 4.35 *Move the link so corners align*

4.7 Place the cursor over the icon so that it turns black.

- Click and hold the mouse button down.

- Drag the Rotation icon to the corner you used for your alignment point.

- Pull the cursor vertically and pick to start a rotation angle vector.

- Pull the cursor down to the right (clockwise), so that the angular listening dimension reads 20° (see Figure 4.36).

- Click to enter the angle.

    You can also enter **20** at the keyboard to define the angle numerically.

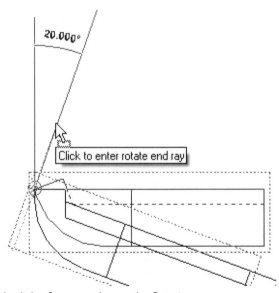

Figure 4.36   *Rotate the link after you relocate the Rotation center*

4.8 Select the Phase 2 import Instance.

- Repeat Move and Rotate (20° counterclockwise) so that the import sits directly on top of the linework.

- Zoom to Fit.

4.9 Choose two points around the edges of the View window to select everything visible.

- Click the Filter button on the Toolbar.

- In the Filter dialogue, clear RVT Links, as shown in Figure 4.37.

- Click OK.

    You have selected only the lines.

- Delete the lines from the project.

  Lines will no longer be colored, and as you move your cursor over the visible entities, it will only read the import instances.

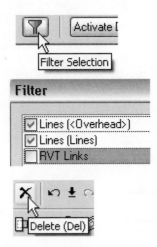

Figure 4.37 *Filter out links from the selection before deleting lines*

4.10 From the File menu, select File>Manage Links.

- On the RVT Tab in the Manage Links dialogue, choose *Chapter 4 Miller Hall.rvt* so that it highlights.

- Select Reload.

- Click OK.

- Zoom to Fit.

  The building outline will appear at the upper left (see Figure 4.38).

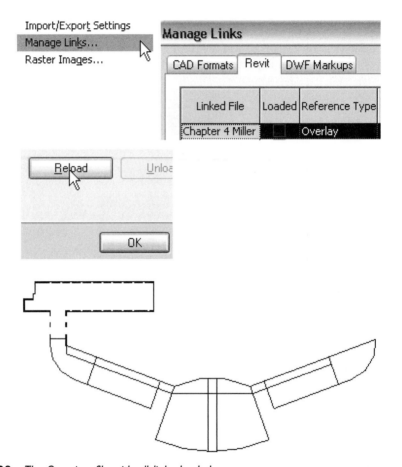

Figure 4.38 *The Overview file with all links loaded*

4.11 Enter **VG** at the keyboard.

- In the Visibility/Graphics Overrides dialogue, on the Annotations Tab, check Elevations.

- Click OK.

4.12 Select and move Elevations around the links (see Figure 4.39).

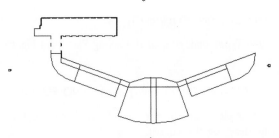

Figure 4.39 *Move elevations so you will be able to see linked models as they develop*

4.13 Zoom to Fit.

- Save the file.

- Close the file.

This file is now set up. As the designs of the other project building models develop and evolve, a project manager or senior designer can monitor overall progress. In the next exercise, you will start mass models in the three associated project files, and then return to this file to create a perspective view of the combined facades for a hypothetical client presentation.

EXERCISE 5. CREATE A BUILDING MODEL USING MASSING—EXTRUSIONS

5.1 Launch Revit Architecture if it is not running.

- Close any open files.

- Open file *Chapter 4 Phase 1.rvt*.

- Open View Elevations: East.

- Zoom in so you can see the Level heads clearly.

5.2 Select Level 1.

- Rename Level 1 *1ST FLOOR PHASE 1*.

- Click Yes to rename the corresponding Views.

- Right-click and pick Maximize 3D Extents.

- Select Level 2.

- Rename Level 2 *2ND FLOOR PHASE 1*.

- Click Yes to rename the corresponding Views.

- Right-click and pick Maximize 3D Extents.

- Zoom to Fit.

5.3 Select Level from the Basics Tab of the Design Bar.

- On the Options Bar, click Plan View Types.

- In the Plan View Types dialogue, clear Ceiling Plan (see Figure 4.40).

- Click OK.

- Add a Level at **13' – 4" [4000 mm]** above 2ND FLOOR.

  You can type the elevation for the new level at the keyboard to enter it directly rather than by dragging the cursor.

- Click in the name field for the level.

- Rename it *3RD FLOOR PHASE 1*.

- Click Yes in the rename dialogue.

5.4 Add a Level at **14' – 4" [4300 mm]** above the new Level.

- Click the Level name and enter **ROOF PHASE 1**.

- Click Yes.

- Click Modify.

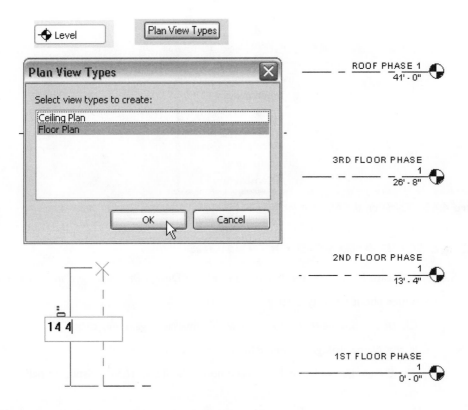

Figure 4.40 *New and renamed Levels without Ceiling Plans associated*

5.5 Open the Site View.

- If the Design Bar does not display a Massing Tab head in the stack at its bottom, put the cursor over the Design Bar and right-click to bring up the list of available tabs.

- Check Massing, as shown in Figure 4.41.

  That tab will open up in the Design Bar.

Figure 4.41 *Toggle on the Massing Tab*

CREATE SOLID AND VOID EXTRUSIONS

5.6 Pick Create Mass from the Massing Tab of the Design Bar, as shown in Figure 4.42.

A notice about visibility settings will display.

- Check the box next to Don't show this message again and click OK.

- In the Name dialogue, enter **Phase 1 Shell** and click OK.

The Design Bar will switch to Mass mode. The linework will display as halftone gray.

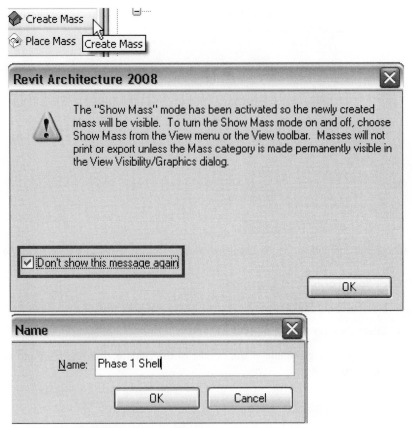

Figure 4.42　*Create and name a mass*

5.7　Choose Solid Form>Solid Extrusion from the Mass Bar.

It will switch to Sketch mode, with the Lines tool active. The cursor will change to a pencil outline. The Options Bar will show linework controls and a Depth control default of 20' – 0" [5000].

5.8　Set the Depth value for this extrusion to **28' [8400]**.

- Select the Pick icon (arrowhead) on the Options Bar.

Line Options will disappear from the Options Bar once you make that selection. The cursor shows the selection arrow icon.

- Select the lines that make up the right half of the footprint of the building.

Each will highlight magenta as it is selected, so that you can start to modify the profile sketch you are creating (see Figure 4.43). The top line will extend the length of the outline.

- Choose Trim and trim lines as necessary.

- When you have a complete outline selected and highlighted, click Finish Sketch on the Sketch Tab of the Design Bar.

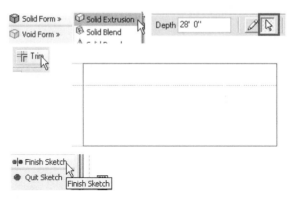

Figure 4.43 *Select the outline for the Extrusion sketch—trim to make a simple rectangle*

5.9 Click Solid Form>Solid Extrusion.

- Set the Depth value for this extrusion to **41' [12300]**.

- Select Pick on the Options Bar.

- Start by picking the line in the middle of the outline, then the lines and arc to its left (see Figure 4.44).

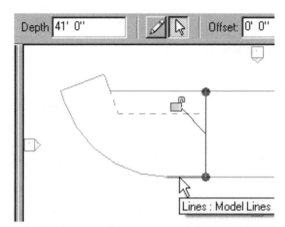

Figure 4.44 *The first picks for the second extrusion*

5.10 Use the Trim command to trim back the upper and lower outline lines as necessary to make a complete outline with no overlapping lines (see Figure 4.45).

- Select Finish Sketch from the Sketch Tab on the Design Bar.

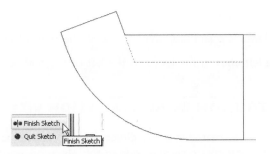

Figure 4.45   *The complete, trimmed outline for the second extrusion*

The lines will disappear under the masses.

5.11 Enter **WF** at the keyboard to shift to Wireframe display mode.

5.12 Pick Void Form>Void Extrusion from the Mass Tab of the Design Bar.

- Set the Depth value of the Cut to **12' [3600]**.

- Select the Pick icon on the Options Bar.

5.13 Pick the two Overhead lines inside the building outline, as shown in Figure 4.46.

- Pick the top and right edge lines of the outline.

- Trim as necessary to make a complete outline.

- Select Finish Sketch from the Sketch Tab on the Design Bar.

- Select Finish Mass from the Mass Tab.

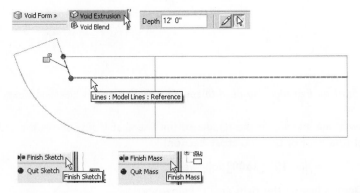

Figure 4.46   *Create a Void Extrusion. It will cut the previous masses to create a building overhang.*

5.14 Type **HL** to return the current View to Hidden Line mode.

The View will not change.

5.15 Open Elevation view East.

- Select a Level line and drag the bubbles to the right of the building outline.

- Drag the left ends of the levels to the left of the outline as well. Zoom to Fit.

CREATE AN EXTRUSION IN AN ELEVATION VIEW

Extrusions can be created in other directions than vertical and other in views than plans. You will now create an extrusion in this elevation, using one face of the mass object you have created as your work plane.

5.16 Select Create Mass from the Design Bar.

- Name the new mass **Atrium Roof**.

- Click OK.

5.17 Select Solid Form>Solid Extrusion.

- Select Pick a Plane in the Work Plane dialogue.

- Click OK.

- Place the cursor over the left side of the building outline and press the TAB key until the outline highlights, as shown in Figure 4.47.

- Click to make the face of the mass the new work plane.

Figure 4.47 *Start an Extrusion. You need to specify a Work Plane in Elevation views.*

5.18 Draw a line starting in the middle of the line that is just above the 3RD FLOOR level (the roof of the two-story section).

- Make the line **12' [3600]** long vertically.

- Draw a line to the right at 30° below vertical.

  You do not need to extend it to the roof line.

- Click Modify.

- Select the new line.

- Click Mirror from the Toolbar.

- Mirror the line to the left using the first line as the mirror axis.

5.19 On the Design Bar, click Lines.

- Draw a line along the roof line.

- Use Trim to make a complete triangle outline.

- Delete the first sketch line.

- Select Finish Sketch. See Figure 4.48.

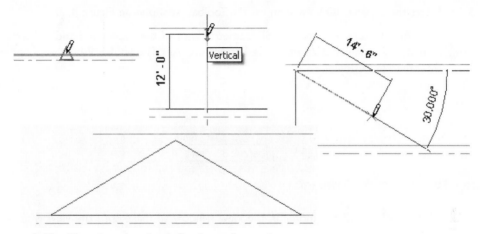

Figure 4.48 *The triangular sketch for the roof extrusion*

5.20 Open the Site View.

The Extrusion has been created, but it extends away from the building rather than over the lower half.

- Select the Extrusion and its control arrows activate.

- Select the right arrow and drag it to the left until it snaps to the left side face (see Figure 4.49).

The depth indicator will read a negative value, meaning that the extrusion extends away from the plane of the view used to create it.

- Select Finish Mass.

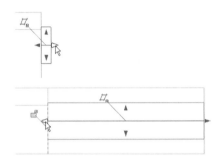

Figure 4.49 *Adjust the Extrusion proportions using the Shape Handle*

5.21 Select the Default 3D View icon on the Toolbar, as shown in Figure 4.50.

Revit Architecture will open an SE isometric view of the model and add a View named {3D} under 3D Views in the Project Browser. The void will be visible under the other extrusions.

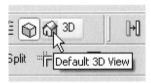

Figure 4.50 *The Default 3D View icon*

5.22 Click anywhere off the model in the View window to make sure the view is active.

  • Select View>Orient>Northeast from the View menu, as shown in Figure 4.51.

Figure 4.51 *Orient the 3D View to the Northeast*

5.23 Select View>Orient>Save View from the View menu, as shown in Figure 4.52. Enter **Northeast** in the Name box and choose OK.

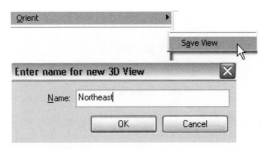

Figure 4.52   *Save the view orientation under a name*

5.24 Pick Shading with Edges from the View Control Bar.

- Select the View Mass toggle.

  The mass will disappear.

- Select the lines and delete them (see Figure 4.53).

- Toggle the View Mass control back on.

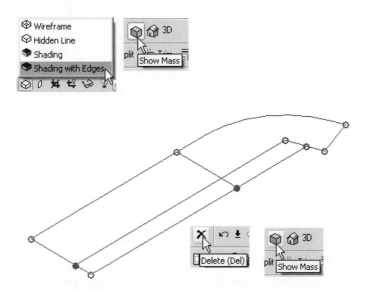

Figure 4.53   *Toggle the mass display off, erase the lines, toggle the mass back on*

TRANSFORM THE MASS INTO BUILDING ELEMENTS

The Massing Tab contains tools to create building elements—walls, floors, roofs and curtain systems—directly from mass faces (see Figure 4.54). The Wall by Face tool creates a wall each time you click. All the others are two-step operations. Floor by Face requires floor faces inside masses.

Figure 4.54 *Tools for creating building elements from masses*

5.25 Select the mass object, as shown in Figure 4.55.

- Do not select the atrium roof.

- Select Floor Area Faces from the Options Bar.

- In the dialogue that opens, select 1ST FLOOR and 2ND FLOOR.

- Pick OK.

  The mass now contains faces at the selected levels.

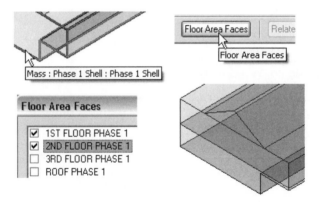

Figure 4.55 *Creating Floor Area Faces*

5.26 Select Floor by Face on the Massing Tab.

- Select the two floor faces.

- Select Create Floors from the Options Bar (see Figure 4.56).

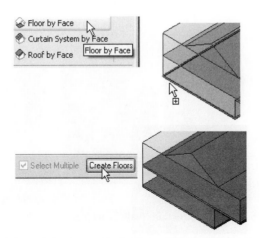

Figure 4.56   *Create two floors using the Floor by Face tool*

5.27 Select Roof by Face.

- Select the two flat roof faces.

- Choose Create Roof from the Options Bar.

5.28 Select Curtain System by Face.

- Select the slanted faces and end face of the triangular extrusion.

- Select Create System on the Options Bar (see Figure 4.57).

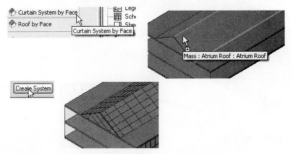

Figure 4.57   *Create a curtain system on the atrium roof extrusion*

5.29 Select Wall by Face.

- Set the Type Selector to Exterior: Brick on Mtl Stud.

- Pick vertical faces, as shown in Figure 4.58, to create three brick walls.

  Each click will place a wall. If you click twice on the same face, a warning will appear. In this case, click Undo one time to remove the last wall, and the Wall by Face tool will stay active.

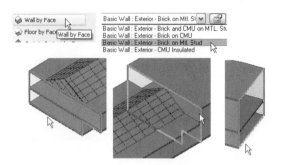

Figure 4.58 *Create Brick Walls by Face*

5.30 Set the Type Selector to Curtain Wall: Storefront.

- Pick Faces, as shown in Figure 4.59, to create four glass curtain walls with predefined panel sizes.

  You will define your own curtain wall type in a later exercise.

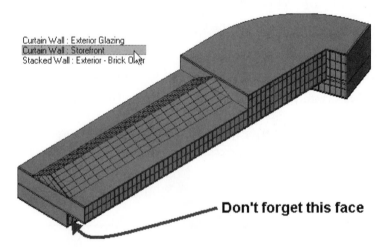

Figure 4.59 *Curtain Walls by Face*

5.31 Open the default 3D view.

- Create three Walls by Face using Curtain Wall: Storefront along the south and west sides of the building (see Figure 4.60).

- Click Modify to terminate wall placement.

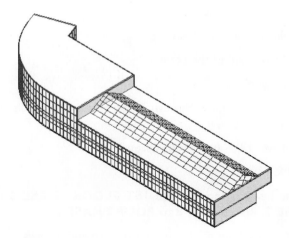

Figure 4.60   *Curtain walls on the curved side*

5.32 Return to View Northeast.

- Toggle the View Mass control off.

- Zoom to Fit.

- Save the file.

- Close the file.

EXERCISE 6. CREATE AN EXTRUDED MODEL WITH SWEEPS AND REVOLVES

CREATE SOLID AND VOID EXTRUSIONS

6.1 Open the file *Chapter 4 Phase 2.rvt.*

- Open the West Elevation View.

- CTRL + click Levels 1 and 2.

- Right-click and pick Maximize 3D Extents.

- Zoom to Fit.

- Click Level 2.

- Set the Elevation to **13'-0" [3900]**.

When you create levels by using Copy or Array, the new levels do not have floor plan or ceiling plan views associated with them by default. The new level heads are black on the screen, indicating they are non-story (no associated view).

6.2 Click Copy from the Toolbar.

- Click Multiple on the Options Bar.

- Click the left end of the Level.

- Pull the cursor up.

- Click to add 5 Levels (3 through 7) at a standard offset distance of **13' – 0" [3900]** from the Level below.

- Add a Level **14' – 0" [4200]** above Level 7.

- Click Modify.

- Rename Levels 1 through 7 to **1ST FLOOR PHASE 2** through **7TH FLOOR PHASE 2**. Rename level 8 to **ROOF PHASE 2**.

- Click Yes each time to rename the corresponding Views (see Figure 4.61).

    The story levels will not trigger the rename dialogue.

Figure 4.61   *Levels in the Phase 2 file*

6.3 Open the Site view.

- Select the Massing Tab head on the Design Bar to activate that tab.

- Select Create Mass from the Massing Tab of the Design Bar.

- Name the Mass **Phase 2 Block**.

- Click OK.

6.4 Choose Solid Form>Solid Extrusion.

- Set the Depth Value for the first extrusion to **28' [8400]**.

- Select the Draw icon on the Options Bar.

- Select the Rectangle option.

- Pick two corner points to create the rectangular shape shown in Figure 4.62.

- Select Finish Sketch.

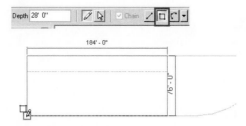

Figure 4.62   *The outline for the first extrusion*

6.5 Type **WF** to set the View mode to Wireframe.

- Select Void Form>Void Extrusion on the Massing Tab of the Design Bar.

- Set the Depth to **12' [3600]**.

- Select Rectangle on the Options Bar.

- Pick two corner points to create the rectangular shape shown in Figure 4.63.

- Select Finish Sketch.

- Select Finish Mass.

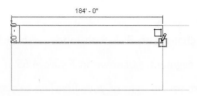

Figure 4.63   *Define a void sketch on the upper part of the mass*

6.6 Select Create Mass.

- Name the Mass **Phase 2 Tower**.

- Select Solid Form> Solid Extrusion.

- Set the Depth to **92'-0" [27600]** (the height of the Roof level).

- Select Pick on the Options Bar.
- Choose the vertical line and then the lines and fillet arc to the right of that line to complete the outline.
- Trim lines as necessary (see Figure 4.64).
- Select Finish Sketch.

Figure 4.64 *The outline for the second extrusion*

6.7 Open the default 3D view.

CREATE A VOID SWEEP

The tower mass has been started. You will now create setbacks at each floor level to create a slimmer profile for the tower.

- Select Void Form>Void Sweep on the Massing Tab of the Design Bar.

The Design Bar changes to Sketch mode.

A Sweep consists of a Profile moved along a Path. You first create a path, then the profile. You can edit the path or profile any time later. The profile can be an outline you sketch, as you will do in the next steps, or you can load a profile family and use that. The same technique applies whether you are creating something as large as a building setback, as you are doing in the current exercise, or something as small as a window frame or decorative stop.

6.8 Select Pick Path.

The Design Bar changes to Pick Path mode, with Pick selected.

- Click the line segment, as shown in Figure 4.65; a profile locator will appear.
- Choose the arc to the right of your first line selection to define the rest of the path.
- Click Finish Path.

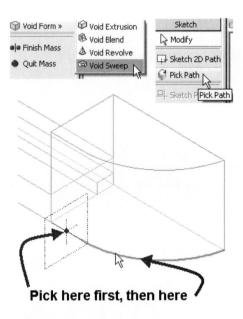

Pick here first, then here

Figure 4.65 *Void sweep—pick the path segments carefully*

6.9 Select Sketch Profile on the Sketch Tab, as shown in Figure 4.66.

- Open the West Elevation View.

- Choose Lines from the Sketch Tab.

- Click the Chain option on the Options Bar.

- Start a profile sketch at the right side of the building outline at the top of the Block mass, just above 3RD FLOOR.

- Make the first line **12' [3000]** long to the left as shown.

- Pull straight up to 4TH FLOOR and click.

- Pull to the left 12' [3000] and click.

- Repeat the previous two steps until the line reaches the top level.

- Draw lines to the right and down to create a profile outline, as shown in Figure 4.66.

- Do not draw the dimensions; they are for your guidance only.

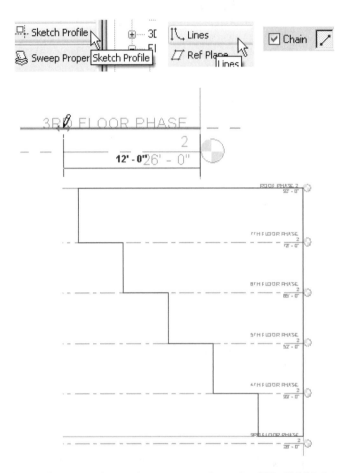

Figure 4.66 *Start a Sketch Profile on the mass just above the 3RD FLOOR level*

6.10 When the Sweep Profile is a complete loop, select Finish Profile.

- Choose Finish Sweep.
- Open the default 3D view to check the results, as shown in Figure 4.81.
- Choose Finish Mass.

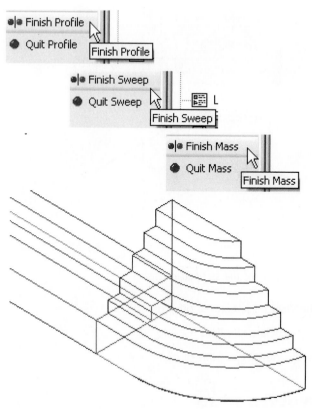

Figure 4.67   *Finish the profile, finish the sweep, finish the mass*

6.11   Select View>Orient>Northwest from the Menu Bar.

- Select View>Orient>Save View.

- Name the view Northwest and choose OK (see Figure 4.68).

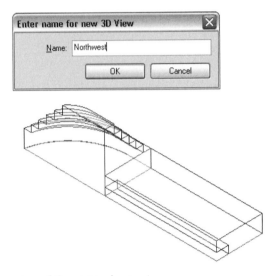

Figure 4.68 *Northwest view of the stepped extrusion*

    6.12 Toggle the View Mass control off.

- Select and delete the lines.

- Toggle the View Mass control on.

CREATE A REVOLVED SOLID

A Revolve consists of a profile spun around an axis. Since you are working in a 3D view, you will first define the work plane, then locate the axis, and then draw a profile as before.

    6.13 Select Create Mass from the Massing Tab on the Design Bar.

- Name the mass **External Elevator**.

- Click OK.

- Select Solid Form>Solid Revolve.

    The Design Bar shifts to Sketch mode.

    6.14 Select Set Work Plane.

- Select Pick a Plane in the dialogue and choose OK.

- Put the cursor over the edge of the tower profile and TAB until the outline of the tower highlights, as shown in Figure 4.69. Click to select the face of the tower as the work plane.

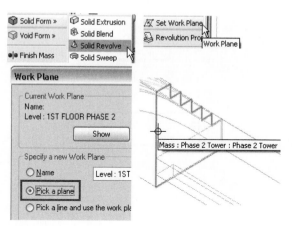

Figure 4.69 *Set the work plane*

6.15 Choose Axis from the Sketch Tab.

- Draw a vertical line from the roofline of the Block mass to the Roof of the Tower mass, as shown in Figure 4.70.

    You will adjust it precisely in the next step.

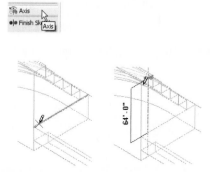

Figure 4.70 *Draw a vertical axis on the work plane*

6.16 Open view West Elevation.

- Select Dimension from the Sketch Bar.

- Place the dimension between the axis line and the left side of the tower.

- Select the axis.

- Change the dimension value to **5' [1500]**, as shown in Figure 4.86.

- Check to make sure that the axis line terminates at the upper edges of the two masses.

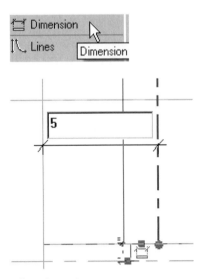

Figure 4.71   *Locate the axis with a dimension*

You are going to create a bowed shape, so you will sketch an arc of specific dimensions. First you will create a temporary construction line.

6.17 Choose Lines from the Sketch Tab to start drawing the profile.

- Draw a line starting at the midpoint of the axis.

- Make the line **3'6" [1050]** long, horizontal to the left (see Figure 4.72).

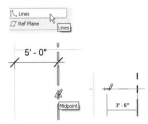

Figure 4.72   *Place a short construction line at the midpoint of the axis*

6.18 Hit ESC once.

This will enable you to start the next line freely.

- Select the 3-point Arc tool from the Lines options on the Options Bar.

- Start the arc at the left side of the tower by the lower end of the axis.

- The second point will be the upper-left end of the tower, and the middle of the arc will be the endpoint of the horizontal sketch line, as shown in Figure 4.73.

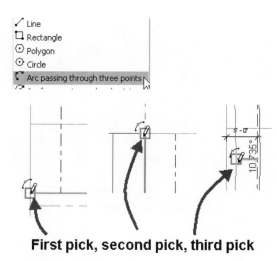

First pick, second pick, third pick

Figure 4.73 *Picks for a 3-point arc*

6.19 Select the Straight Line tool.

- Finish the profile, as shown in Figure 4.74.

  The right side of the profile sketch is on top of the axis.

- Click Modify.

- Delete the horizontal line.

- Select Revolution Properties on the Sketch Tab.

- Change the End Angle value to **-180**.

  This will make the revolution travel in a half circle around the axis toward you. This value was determined by trial and error.

- Select Finish Sketch.

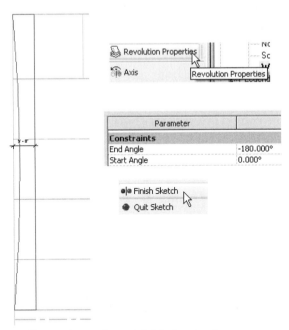

Figure 4.74 *Finish the profile, set the angle, and finish the sketch*

6.20 Open 3D view Northwest to check your revolved mass.

- Click Finish Mass.

CREATE BUILDING COMPONENTS FROM THE MASSES

You will now create floors, walls, roofs and curtain systems as you did in the previous exercise.

6.21 Select the tower.

- Click Floor Area Faces.

- Check levels 1ST FLOOR through 7TH FLOOR.

- Click OK.

- Select the block.

- Click Floor Area Faces.

- Check 1ST FLOOR and 2ND FLOOR.

- Click OK.

6.22 Click Floor by Face.

- On the Options Bar, set the Offset to **1' [300]**. This will place the edges of the floors inside the mass faces by the offset distance, eliminating interference with walls.

- Select all the floor faces.

Tip: You can window-select rather than pick individual items.

- Click Create Floors.

- Set the view display to Shaded with Edges, as shown in Figure 4.75.

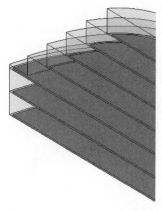

Figure 4.75   *New floors with offset edges*

6.23 Click Roof by Face.

- Select the roof of the block.

- Click Create Roof.

- Carefully select roof faces on the tower, as shown in Figure 4.76.

- If you have difficulty selecting horizontal faces in this view, open the default 3D (Southeast) view to continue the command.

- Select all the exterior horizontal segments of the tower.

- Click Create Roof.

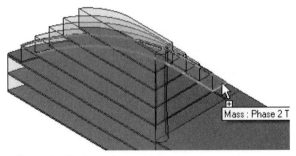

Figure 4.76 *Create roofs on the block and tower*

6.24 Click Wall by Face.

- Place instances of wall type Exterior: Brick on Mtl Stud and Curtain Wall: Storefront, as shown in Figure 4.77.

- Use the default 3D view to place curtain walls on the setbacks.

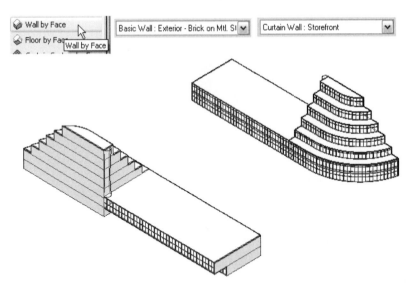

Figure 4.77 *Place walls by face. Shading has been removed for clarity.*

6.25 Open the Northwest view.

- Click Curtain System by Face.

    The default curtain grid spacing will not fit on the elevator tower profile.

- Select the Properties icon.

- Click Edit/New.

- Click Duplicate.

- Name the new Curtain System Type **2' x 2' [600 x 600]**.

- Click OK.

- Set its grid values to **2' [600]** each way.

- Choose OK twice to finish the settings.

- Select the face and top of the elevator mass.

- Select Create System (see Figure 4.78).

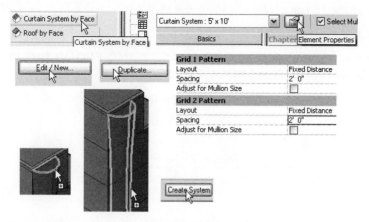

Figure 4.78   *Define a curtain system and apply it to the elevator mass*

6.26 Zoom to Fit in the view.

- Toggle the View Mass control off.

6.27 Save the file.

- Close the file.

EXERCISE 7. CREATE A MASS MODEL USING BLENDS

So far you have used extrusions, sweeps and revolves to create building shapes. These all use a single profile that is processed to make a 3D shape. A blended Mass in Revit Architecture uses two profiles, one each for the top and bottom, and creates a volume between the two. Blends can create surfaces that curve in two directions.

You will start by sketching the bottom profile, then the top.

7.1 Open the file *Chapter 4 Phase 3.rvt*.

- Open View Elevations: East.

- CTRL + select Levels 1 and 2.

- Right-click and pick Maximize 3D Extents.

- Zoom to Fit.

- Select Level 2.

- Set the Elevation to **12'-0" [3600]**.

7.2 Click Copy from the Toolbar.

- Check Multiple on the Options Bar.

- Click anywhere to set a start point.

- Pull the cursor up 12'-0" [3600] and click to create a story level.

- Pull the cursor up 16' 0" [4800] and click.

- Click Modify.

7.3 Rename the Levels **1ST FLOOR PHASE 3**, **2ND FLOOR PHASE 3**, **3RD FLOOR PHASE 3**, and **ROOF PHASE 3**.

- Click Yes to accept View name changes (see Figure 4.79).

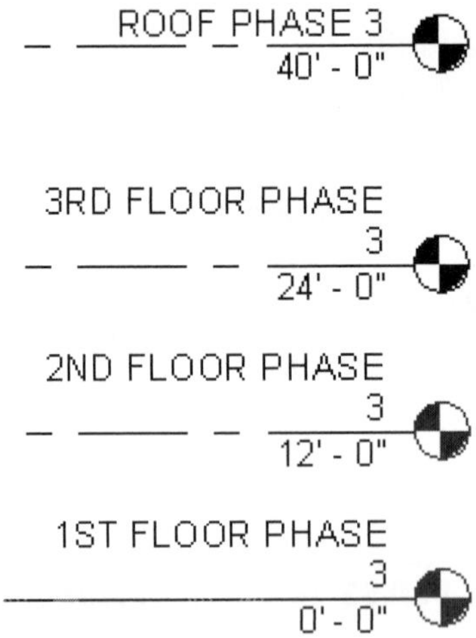

Figure 4.79 *Levels in the Phase 3 file*

7.4 Open the Site Floor Plan View.

- Select Create Mass from the Massing Tab.

- Name the new mass **Phase 3 Shell**.

- Click OK.

- Select Solid Form>Solid Blend.

  The Design Bar will change to Sketch mode.

7.5 Set the Depth value to **40' [12000]**.

- Select the Pick icon and pick the sides of the building outline.

- Select the Draw icon and draw horizontal lines between the endpoints of the magenta lines you created by your picks (see Figure 4.95).

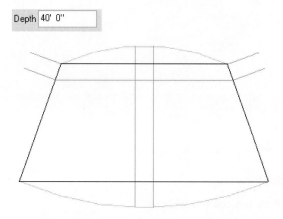

Figure 4.80 *The outline for the blend bottom*

7.6 Pick Edit Top in the Sketch Tab.

- Select the Pick icon and pick the outline of the building footprint (see Figure 4.81).

- Select Finish Sketch.

Figure 4.81 *Edit the top and pick the profile*

7.7 Click the Default 3D View icon on the View Toolbar.

- Set the display controls to Shaded with Edges (see Figure 4.82).

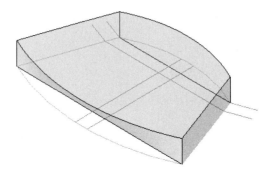

Figure 4.82 *View the Blend*

7.8 Open the Site View.

- Type **WF** to see the model lines.
- Select Void Mass>Void Extrusion on the Massing Tab.

7.9 Set the Cut Depth to **23' [6900]**.

- Select the Pick icon on the Options Bar.
- Pick the Overhead line near the top of the figure.
- Pick the arc at the top of the building outline.
- Select the Draw icon and trace lines connecting the endpoints of the magenta lines created by your previous picks (see Figure 4.83).

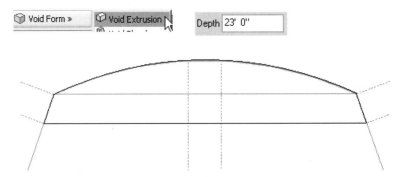

Figure 4.83 *Outline for the Extrusion cut*

7.10 Click Finish Sketch.

- Examine the results in the 3D View.
- Return to the Site View.

7.11 Select Void Mass>Void Blend from the Massing Tab.

7.12 Set the Depth to **40'**.

- Select the Pick icon from the Options Bar.
- Pick the two Overhead lines running vertically down the middle of the outline.
- Pick the upper and lower horizontal lines as with the previous blend.
- Trim the lines to complete the outline of the base (see Figure 4.84).
- Select Edit Top.

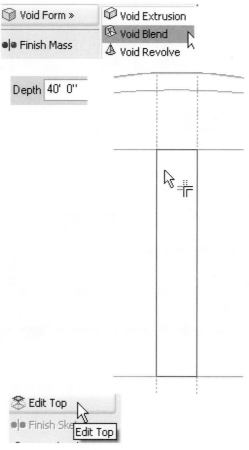

Figure 4.84 *Outline for the Blend cut bottom*

7.13 Select Pick.

- Make the Offset value **4' [1200]**.

- Pick the two Overhead lines so that the offset lines appear to the outside of each (wider).

- Pick the upper and lower arcs so that the offset lines appear to the inside.

- Trim the arcs back to the vertical lines (see Figure 4.85).

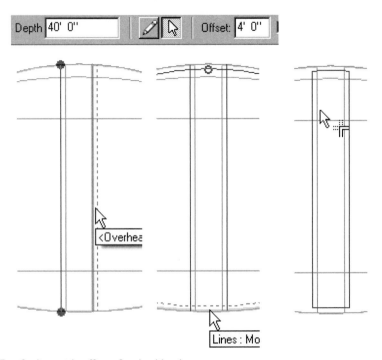

Figure 4.85 *Outline with offsets for the blend top*

7.14 Select Finish Sketch.

- Select Finish Mass.

- Examine the results in the 3D view, as shown in Figure 4.86.

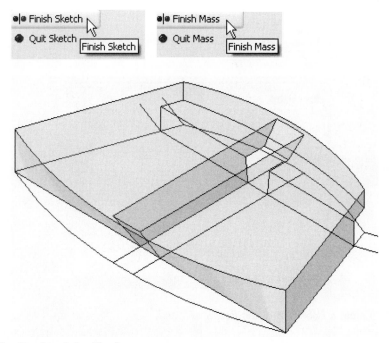

Figure 4.86 *Results of the Blend cut*

CREATE BUILDING COMPONENTS

7.15 Activate the View Tab of the Design Bar.

- Click Floor Plan.

- In the New Plan dialogue, select ROOF PHASE 3, as shown in Figure 4.87.

- Click OK.

  That plan view will open.

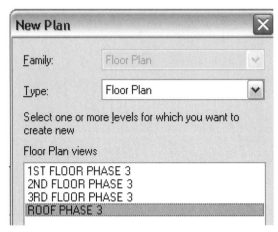

Figure 4.87 *Create a Floor Plan at the Roof level*

7.16 Make the Basics Tab active.

- Select Roof>>Roof by Footprint.

7.17 Choose Lines on the Sketch Tab.

- Select Pick from the Options Bar.

- Check Defines Slope.

- Carefully select the outer sides of the Blend cut.

- Clear Defines Slope.

- Select the upper and lower ends of the Blend cut (see Figure 4.88).

- Select Finish Roof.

  The roof will appear cut because of the view settings.

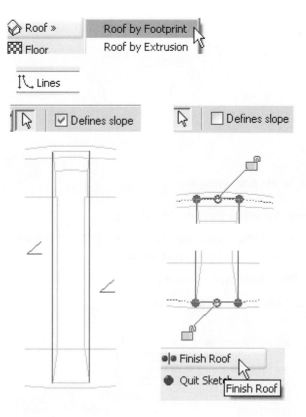

Figure 4.88 *Profile for the atrium roof*

7.18 Open the 3D View to see the roof you just created.

7.19 Open the 2ND FLOOR PHASE 3 Floor Plan View.

- Zoom to the upper left.

- Select Roof>>Roof by Footprint.

- Choose Lines from the Sketch Tab.

- Select the Pick icon from the Options Bar.

- Put a check next to Defines Slope.

- Pick the two Overhead lines at the upper-left of the building outline.

7.20 Clear Defines Slope.

- Select Draw.

- Draw two lines between the endpoints of the lines you just created (see Figure 4.89). Select Finish Roof.

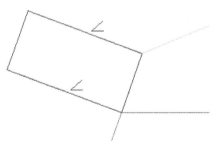

Figure 4.89 *Profile for the passageway roof*

7.21 Select Roof Properties from the Sketch Tab.

- Set the Base Offset from Level value to **11' [3300];** this will align the roof with the height of the void cut you made earlier.

- Select Finish Roof (see Figure 4.90).

    The roof will not be visible in the view.

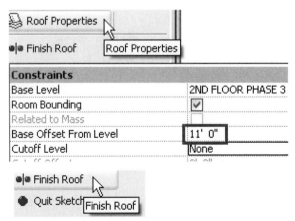

Figure 4.90 *Set the height offset and finish the roof*

7.22 Open the Site view.

- Select the new roof you just created.

- Choose Mirror from the Toolbar.

- Select the Draw icon.

- Pick the midpoint of the building front, as shown in Figure 4.91.

- Pull the cursor straight up or down to create a vertical mirror axis.

- Click to create the copy on the right side of the building.

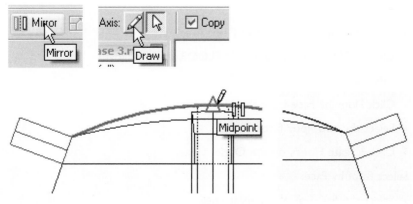

Figure 4.91 *Mirror the left-hand roof to the right side*

7.23 Open the 3D view.

- Toggle the View Mass control off.

- Select and delete the lines.

- Toggle the View Mass control on.

 Tip: Select everything in the view and use the Filter Selection tool from the Options Bar. Clear Roofs from the list in the Select dialogue, and what remains in your selection set will be the two types of lines. The Filter Selection tool speeds up selections.

7.24 CTRL + Select the three roofs.

- Use the Type Selector to change them to Sloped Glazing.

- They will change appearance, as shown in Figure 4.92.

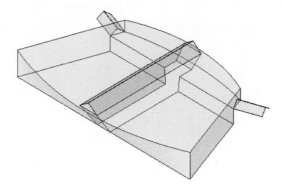

Figure 4.92 *The new roofs are defined as glass, but without any grids yet*

7.25 Select the shell mass.

- From the Options Bar select Floor Area Faces.

- Choose 2ND FLOOR and 3RD FLOOR in the dialogue.

- Select OK.

- Click Floor by Face from the Massing Tab.

- Select the two floor faces.

- Click Create Floors on the Options Bar.

7.26 Select Roof by Face.

- Select the flat roof of the mass shell.

- Select Create Roof.

7.27 Select Curtain System by Face.

- Select the arc blend (south) face of the building.

- Select Create System (see Figure 4.93).

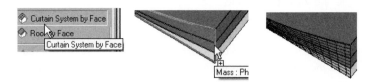

Figure 4.93 *Create a curtain system on the south wall*

7.28 Select Wall by Face.

- Set the Type Selector to Brick on Mtl Stud.

- Select the right face to create a wall (see Figure 4.94).

  If a warning appears, ignore it.

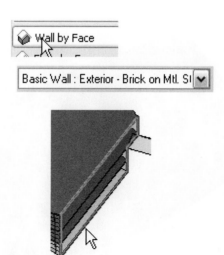

Figure 4.94   *Create a wall on the right side*

7.29 From the View menu, select View>Orient>Northwest.

- Wall by Face is still active. Select the right side face.
- Pick Curtain System by Face.
- Select the Blend faces (there are two, so pick carefully).
- Click Create System.
- Click Modify.
- Toggle View Mass off (see Figure 4.95).

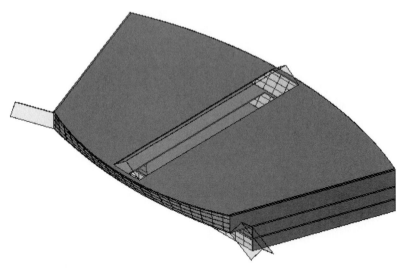

Figure 4.95 *The completed model*

7.30 Save the file.

- Close the file.

EXERCISE 8. VIEW THE NEW MODELS IN THE OVERVIEW FILE

8.1 Open the file *Chapter 4 Project Overview.rvt*.

The models have been linked in plan, but their elevations have not been adjusted to take into account site levels.

8.2 Open the South Elevation View.

- Enter **VG** at the keyboard.

- Clear Topography on the Model Categories Tab.

- Click OK.

- Zoom to Fit.

8.3 Select the Phase 3 link. Click Move on the Toolbar. Click anywhere for a start point. Pull the cursor straight up and type **4' [1200]**.

8.4 Select the Phase 2 link. Click Move. Click anywhere. Pull the cursor straight up and type **8' [2400]** (see Figure 4.96).

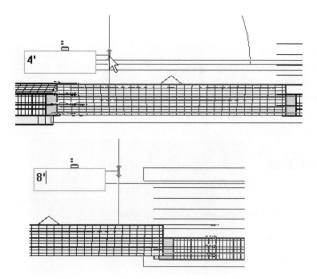

Figure 4.96 *Move two links so they will fit the topography*

8.5 Open the Default 3D View.

- Orient the view to the Northeast and examine the model.

- If your mouse is equipped with a scroll wheel, hold down the wheel and the SHIFT key together and you can spin the view to other angles (see Figure 4.97).

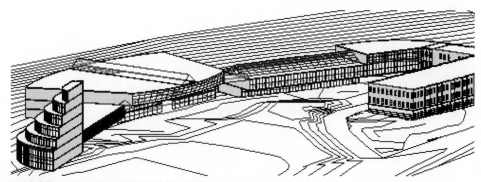

Figure 4.97 *A low-angle view of the linked models*

8.6 From the menu bar select View>Orient>Save View.

- Name the View **Northeast**.

- Set the view display mode to Shading with Edges.

8.7 Save the file.

- Close the file.

SUMMARY

You have seen in this series of exercises how Revit Architecture accepts information from other sources, including CAD files, images and Revit Architecture files. You used simple modeling techniques based on Massing to create building shells for a multi-building project, and linked them into a central file so as to view them all at once.

REVIEW QUESTIONS

MULTIPLE CHOICE

1. Typed shortcuts that deal with Views include

    a. VV, VG, VP

    b. WF, HL, SD

    c. F8

    d. all of the above

2. To create a Blended Massing Form, you define

    a. a Top Profile

    b. a Bottom Profile

    c. both a and b

    d. both a and b, plus a Profile Angle

3. Sharing Coordinates by Acquire or Publish between files

    a. allows synchronized elevations

    b. requires Administrator privileges on the network

    c. takes too much system memory to be useful

    d. all of the above

4. A Shared Level

    a. is easily created from a default Level by changing the Project Base for the Level

    b. takes its elevation information from a file with Shared coordinates

    c. always has to be a different color from regular Levels

    d. a and b, but not c

5. Edit/Copy and Edit/Paste

    a. are only for use between files, not within a project file

    b. can only be used on door and window objects, not model lines

    c. can take advantage of Paste Aligned

    d. can only be used once per file

TRUE/FALSE

6. **True** _ **False** _   Erasing a linked Revit Architecture file does not remove the link from the host.

7. **True** _ **False** _   Linked images can't be deleted.

8. **True** _ **False** _   The Solid Form and Void Form types are Extrude, Revolve, Sweep and Blend.

9. **True** _ **False** _   Sketches for Roofs and Massing always have to be drawn—there is no option to Pick lines or edges.

10. **True** _ **False** _   Model Line types include Medium, Thin and Wide lines, plus Overhead, Demolished and Hidden.

 Answers will be found on the CD.

Subdivide the Design: Interiors, Worksets, Groups and Phases

INTRODUCTION

This chapter continues with model building techniques. You will learn how to create floors of various types, including ones that slope. You will work with curtain walls and their component grids, mullion and panels. You will create layouts of internal walls and furniture components.

We have mentioned and concentrated on the teamwork aspects of design projects. The ability to share information directly from a building design model with collaborators, consultants, clients and officials will only grow in importance for designers. In Chapter 3, you examined how Revit Architecture imports from and exports to CAD files. In Chapter 4, you explored how to link Revit Architecture model files together to coordinate multi-building site developments.

In this chapter, you shall look at Revit Architecture's Worksets, Phasing and Group mechanisms. Worksets are designed to allow real-time collaborative team design. A Workset is simply a named, segregated collection of elements in a model file. When a project model is divided according to Worksets, more than one user can work simultaneously on different parts of the model. All Worksets are part of the same building model file, so when design developments are published from a local copy to the Central File and reloaded by other users, changes to the model are propagated and coordinated tightly.

Worksets are a professional toolset, and their proper and efficient use takes planning and adherence to common-sense best practices. There isn't enough room in this book to discuss advanced uses of Worksets (editing at risk, rolling back changes or restricting references). Worksets provide the ability to open only part of a building file's contents. Many Revit Architecture users divide nearly all their projects—even those that only one person will work on—into Worksets according to a standard setup, to take advantage of the reduced load on their computer system resources.

Phases are Revit Architecture's way of putting depiction of time span into a building model. All building projects have at least two implicit phases (existing conditions and new construction); all remodeling projects have at least three phases (existing, demolition and new); many projects have four or more phases (existing, demolition, temporary construction, removal of temporary work, new construction phase 1 and so forth). Revit Architecture allows the user to define any number of phases for purposes of viewing a project at any stage along its projected timeline, which can be of significant value in a collaborative setting.

Groups are named collections of objects. They can be copied throughout a design. Editing one group instance updates all instances, which makes them efficient for populating repetitive designs. Groups can be saved as independent files, which can then be linked into other projects. Links can be bound into host files as groups. This ability provides another mechanism for collaboration, since bound groups can be edited and re-exported.

Phases and Worksets are not designed explicitly for exploring alternative options or variations on a design. The Design Options feature in Revit Architecture will be explored in Chapter 10.

OBJECTIVES

- Activate, create and use Worksets in a model file
- Work with Floor and Wall objects
- Work with Furniture objects
- Examine Worksets in linked files
- Set-up and use Phases in a model file
- Work with Groups

REVIT ARCHITECTURE COMMANDS AND SKILLS

Worksets—rename, new

Save to Central

Save Local

Create walls from mass elements

Floor surface offset

Floor material variable thickness

Slope Arrow

Split walls

Attach walls

Opening

Convert to Curtain wall

Add and Edit Curtain Grid

Manage Links

Phasing—create Phases

View Properties for Phases

Add Components in Phases

Create Groups

Edit Groups

Save Group to Library

Load File as Group

EXERCISE I. ACTIVATE AND CREATE WORKSETS IN A PROJECT FILE

1.1   Launch Revit Architecture.

- Open the file *Chapter 5 start imperial.rvt [Chapter 5 start metric.rvt]*.

It will open to a floor plan view that shows exterior walls, interior walls and a collection of toposurfaces. In your hypothetical project, the design for the building and site has thus far been prepared by a single person working in the file. You will now prepare the file so that a team will be able to work in the file simultaneously.

1.2   From the File menu, pick File>Worksets, as shown in Figure 5.1.

The Worksets dialogue will appear. This names the default Worksets that will be created automatically.

- Click OK to accept the default Workset names.

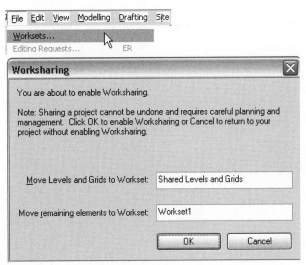

Figure 5.1   *Activating Worksets from the file menu*

The Worksets dialogue appears. It shows the active Workset (Workset1 by default) and the properties of the User-Created Worksets now in the file. They are all Opened and Editable. Your Revit Architecture username will appear next to the Workset name in the Owner column (see Figure 5.2).

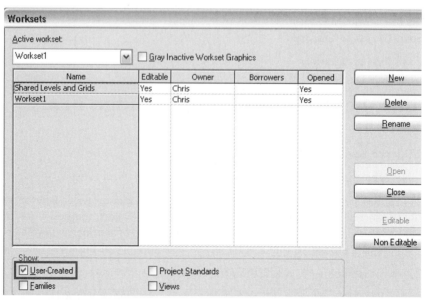

Figure 5.2 *The Worksets dialogue with default Worksets*

1.3 Take a moment to study the Worksets that Revit Architecture creates, other than the User-Created ones.

- Clear User-Created in the Show section of the Worksets dialogue.

- Check Families.

- Note the Revit Architecture object categories that have been moved into Worksets, as shown in Figure 5.3.

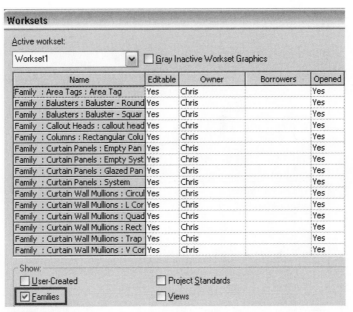

Figure 5.3   *Family Worksets*

1.4 Clear Families.

- Check Project Standards (see Figure 5.4).

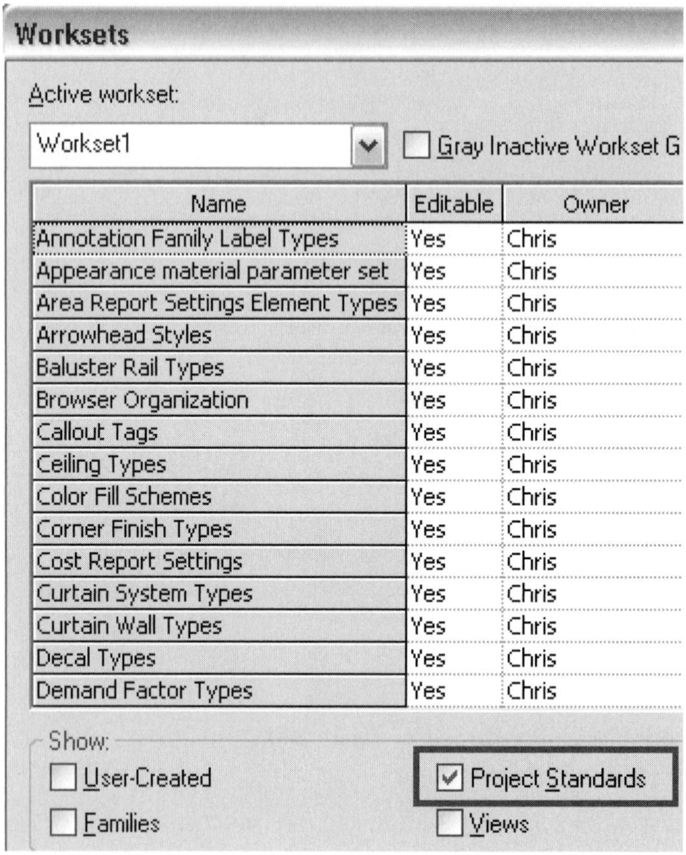

Figure 5.4 *Project Standards Worksets*

- Clear Project Standards.
- Check Views (see Figure 5.5).

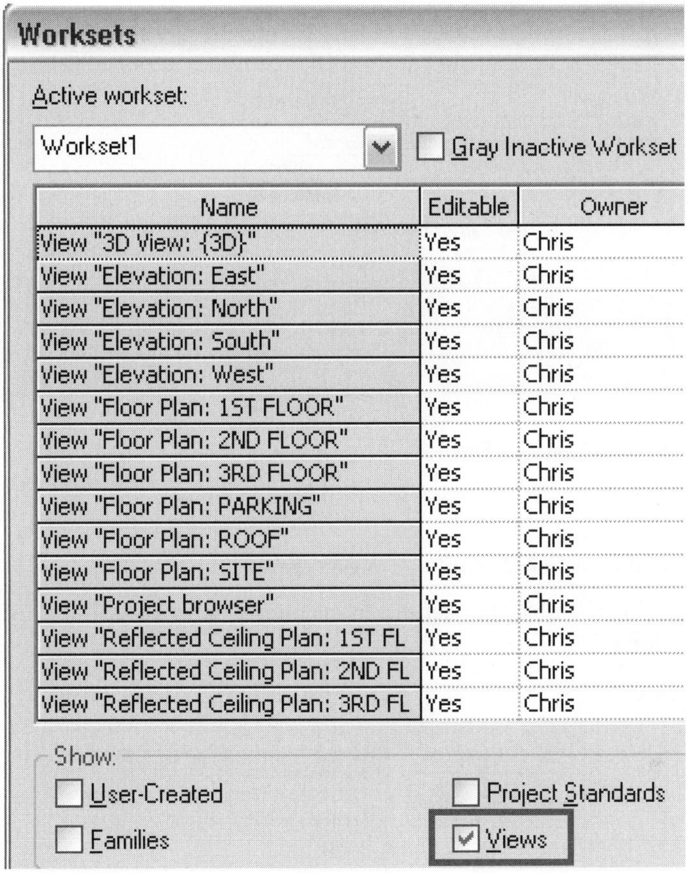

Figure 5.5   *Views Worksets*

1.5 Clear Views and check User-Created.

- Select Workset1 in the Name field of the dialogue, and pick Rename.

- In the Rename dialogue, enter **Exterior** (see Figure 5.6).

- Click OK.

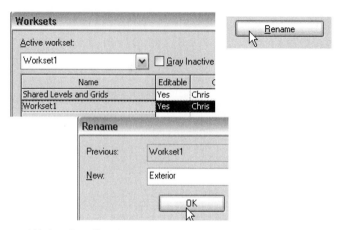

Figure 5.6 *Rename Workset1 to Exterior*

1.6 Select New.

- Enter the name **Interior** (see Figure 5.7).

  Note that Visible by default in all views is checked—this is appropriate for interior walls.

- Click OK.

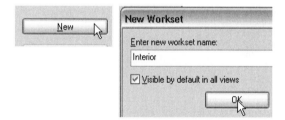

Figure 5.7 *A new Workset named Interior*

1.7 Select New.

- Enter the name **Site**.

- Clear Visible by default in all views, as shown in Figure 5.8.

- Click OK.

- Select New.

- Enter the name **Furniture**.

- Clear Visible by default in all views.

- Click OK.

- Click OK in the Worksets dialogue to close it.

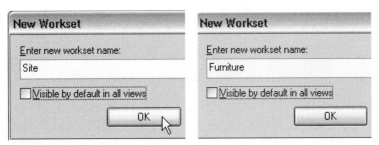

Figure 5.8 *The Site and Furniture Worksets will not be visible in all views*

1.8 Place the cursor anywhere over the Toolbar.

- Right-click to bring up the list of available toolbars.

- Check Worksets.

  The Worksets toolbar will appear, with Exterior in the drop-down field, indicating that Exterior is the current Workset. Any elements created now will appear in the Exterior Workset (see Figure 5.9).

  The Options Bar now shows an Editable Only check option. With this option checked, you will not be able to select elements that are in a non-editable Workset.

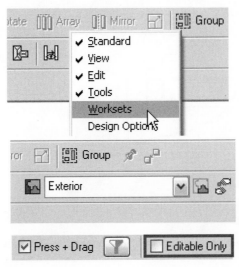

Figure 5.9 *Turn on the Worksets toolbar*

PLACE THE MODEL CONTENTS INTO WORKSETS

1.9 Zoom to Fit.

- Pick two points to select the entire visible model.

- Select the Filter icon.

- In the Filter dialogue, pick Check None, then check Topography, as shown in Figure 5.10.

- Click OK.

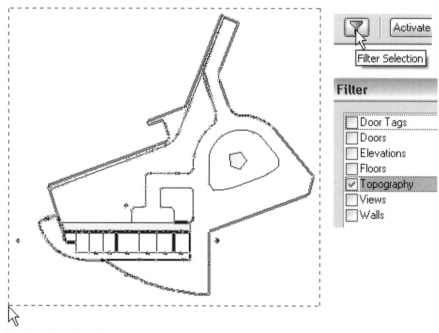

Figure 5.10 *Filter the selection*

1.10 Pick the Properties icon.

- In the Element Properties dialogue, change the Workset Value to Site, as shown in Figure 5.11.

- Click OK.

  The highlighted topography will disappear, as the Site Workset is not visible in all views by default.

Figure 5.11   *Change the selected topography elements to the Site Workset*

1.11 Zoom to Fit.

- Select all interior doors and walls, as shown in Figure 5.12.

- Pick the filter icon and clear Door Tags.

- Select the Properties icon.

- Change the Workset Value to Interior.

  The doors and walls will not disappear, as the Interior Workset is visible in all views by default. Door Tags have automatically been placed in a View workset.

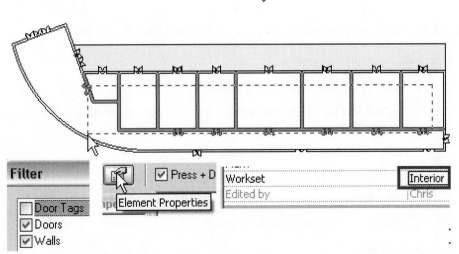

Figure 5.12   *Select interior doors and walls in the 1st FLOOR view and place the selection in the Interior Workset*

1.12 Place the cursor over an exterior wall.

- TAB until the edge of the floor highlights and the Tooltip and Status Bar read Exterior: Floors: Floor.

- Click to select the floor.

- Right-click and pick Select All Instances from the context menu.

  This is a quick way to collect all items of a certain type so you can modify them all at once.

- Click Properties.

- Place all floors of this type into the Interior Workset (see Figure 5.13).
- Click OK.

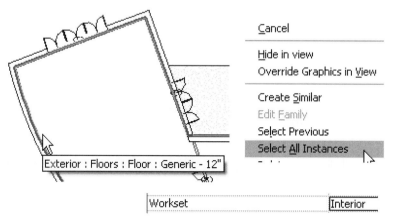

Figure 5.13 *Select the floor, select all floors of the type and place them in the Interior Workset*

1.13 Open the 2ND FLOOR plan view.

- Select all interior doors and windows.

  You will find that a combination of right-left crossing window selection and CTRL + individual selection will work best.

- Filter out the Door Tags, as in Step 1.13.
- Right-click.
- Select Element Properties.
- Change the Workset value to Interior.
- Click OK.

  You have already placed the floor in a workset, and Door Tags are automatically assigned a workset.

1.14 Open the PARKING plan view.

- Pick two points around the model to select everything visible.

  Only the Elevations and the green trapezoidal shape (sketch lines for the Parking Garage) will highlight red.

- Use the Filter to clear Elevations and Views, leaving Lines (Lines) checked.
- Click OK to exit the Filter.
- Hold down the CTRL key and select the two lines to the right of the trapezoid.

  These are also sketch lines to use in creating a ramp (see Figure 5.14).

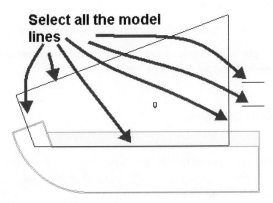

Select all the model lines

Figure 5.14   *Pick the two types of model lines*

1.15 Select the Properties icon.

- Place the lines in the Site Workset.

- Click OK.

    The lines will disappear, as the Site Workset is not visible in all views by default.

- Zoom to Fit.

1.16 Open the SITE plan view.

- Type **VV** to enter the Visibility/Graphic Overrides dialogue, which now contains a Worksets Tab.

- Make the Worksets Tab active.

- Clear Furniture and Interior.

- Check Site (see Figure 5.15).

- Click OK.

    Topography is now visible around the building outline.

Visibility/Graphic Overrides for Floor Plan: SITE

| Model Categories | Annotation Categories | Imported Categories | Filters | Worksets |

☑ Exterior
☐ Furniture
☐ Interior
☑ Shared Levels and Grids
☑ Site

Figure 5.15   *Change the Workset visibility in the SITE plan view*

1.17 Open the PARKING view.

- Type **VG** to enter the Visibility/Graphic Overrides dialogue.

- Repeat the previous checks from Step 1.16 on the Worksets Tab to make the Site Workset visible, and Interior not visible, in this view.

- Click OK.

1.18 Open the 1ST FLOOR view.

- Zoom to Fit.

The Central File is created automatically the first time a file is saved after Worksets are enabled. Location of this file is important—all team members who will use the file should have access to it, using the same network path.

EXERCISE 2. CREATE A CENTRAL FILE

2.1 From the File menu, pick File>Save As.

2.2 In the Save As dialogue, select Options.

- In the File Save Options dialogue, notice that the option to Make this a Central lfile after save is checked and grayed out—this will automatically happen the first time, and then become optional for later saves.

- Change the number of backups to **1**, or as specified by your instructor (see Figure 5.16).

- Click OK.

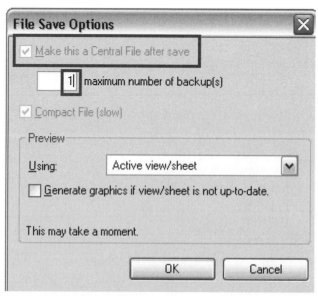

Figure 5.16 *The backup and Make this Central options*

2.3 In the Save As dialogue, give the file you are about to save the name ***Chapter 5 central.rvt***, and save it in a location specified by your instructor.

- If you are working this exercise on your own, or outside a network environment, save the Central File on your hard drive.

- Click Save.

INSTRUCTOR'S NOTE If at all possible, have the students save their central files to a shared network location with a drive path that will be the same for all. If your network security policy mandates private folders for student work, create a temporary public folder for the duration of this exercise. Have the students save the Chapter 5 central file with their first initial and last name as part of the file name, as in **Chapter 5 central J Lopez.rvt**. This will facilitate sharing of central files in the second part of the exercise. The central file storage folder will need to be on a drive with adequate space (see the next exercise step).

2.4 Open the File Manager or Windows Explorer (My Computer) on your computer and navigate to the folder where you just saved the Central File.

Note that Revit Architecture has created a folder named *Chapter 5 central_backup*. Open that folder. There will be files with extension *.dat* and *.rws*. Revit Architecture will not create *Chapter 5 central.000x.rvt* files for the Central File as it does for non–Central Files. Each local copy that is made from the Central file will have its own backup folder.

Caution: If you ever move a Central File after creating it, the backup folder and its contents should be moved with it. Make no changes to the contents of the backup folder at any time.

⊟ 📁 Exercise files		
📁 Chapter 5 central C Fox_backup		
📄 wperms.dat	1 KB	DAT File
📄 users.dat	1 KB	DAT File
📄 requests.dat	1 KB	DAT File
📄 incrementtable.0044.dat	1 KB	DAT File
📄 eperms.dat	399 KB	DAT File
📄 contents.0001.dat	4 KB	DAT File
📄 Chapter 5 central C Fox.slog	2 KB	SLOG File
📄 basicfileinfo.0010.dat	1 KB	DAT File
📄 formats.0039.dat	69 KB	DAT File
📄 0_15706.rws	1,431 KB	RWS File
📄 preview.0042.dat	2 KB	DAT File
📄 partitiontable.0042.dat	7 KB	DAT File
📄 history.0042.dat	9 KB	DAT File
📄 global.0042.dat	25 KB	DAT File
📄 elemtable.0042.dat	71 KB	DAT File

Figure 5.17 *A typical central_backup folder and contents*

2.5 Close the File Manager without making any changes to it or in the *Chapter 5 central_ backup* folder.

2.6 Before closing the new Central File, release the Worksets so they will be accessible to other users who will use this file to create local files to develop the design.

- Pick the Worksets icon on the Toolbar you made visible earlier.

- In the Worksets dialogue, select the top Workset name field, hold down the left mouse button, and drag down to select all the Worksets, as shown in Figure 5.18.

- Pick Non-Editable.

- Your name will be removed from the Owner fields and the Editable values become No.

  In this dialogue you can also manage visibility of Worksets by the Open/Close controls.

- Click OK.

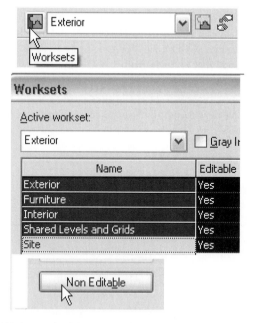

Figure 5.18 *Relinquish Workset editability in the Central File*

2.7 Pick File>Close from the File menu.

CREATING AND USING LOCAL FILES

At this point, you are about to create a copy of the Central File for your local use. You will need to check out at least one Workset to make edits. To publish your design changes so others can see them, you should save work frequently (every 30 minutes is recommended) and save centrally—a different step—every one to two hours. After your last central save, remember to release Worksets as you just did, then save locally and close the file.

In the next part of the exercise, you will create two local files and work in each as a separate user, accessing the same project Central File. If you are working individually without a network you can open a separate session of Revit Architecture for each file (depending on system resources you may want to limit yourself to two open sessions) to see the changes propagate.

2.8 Pick File>Open from the File menu.

- In the File Open dialogue, use the Open Worksets drop-down list and select Specify.

- Select the *Chapter 5 central.rvt* file you created (see Figure 5.19).

- Click Open.

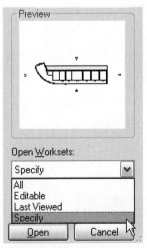

Figure 5.19   *The choices for opening Worksets*

In the Opening Worksets dialogue, you will choose which Worksets to make open (visible) in the copy of the Central File you are about to create. Since you did not close Worksets when closing the file, all are open by default.

In this file you will work on the site.

2.9 Select the Furniture and Interior Worksets, as shown in Figure 5.21.

- Pick Close.

- Click OK.

  You can also close Worksets at any time after the file is opened.

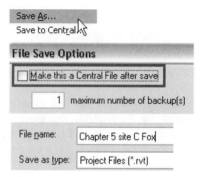

Figure 5.20 *Select the Worksets you do not need to see*

The *Chapter 5 central.rvt* file opens. No interior walls or doors are visible.

2.10 Select File>Save As from the File menu.

- Click Options.

  The option to Make this a Central file after save is now available.

- Make sure it is not checked (see Figure 5.21).

- Click OK to close the options dialogue.

- Save the file as **Chapter 5 site.rvt** on your local drive, or as specified by your class instructor.

- Click Save.

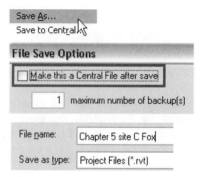

Figure 5.21 *Do not make this the Central File*

Note: Once you have saved the file as a central copy, Revit Architecture activates the icon for the Save to Central function on the Toolbar (see Figure 5.23). Set the save reminder in the Settings>Options dialogue as appropriate so you can keep your work safe and your design changes published to the Central File regularly. Remember—Revit Architecture **does not** have an autosave.

This file is now your local copy to work in. It will maintain a connection to the Central File so that certain functions and conditions (Workset editability) are constantly monitored, while other changes update when local files are saved and Worksets reloaded.

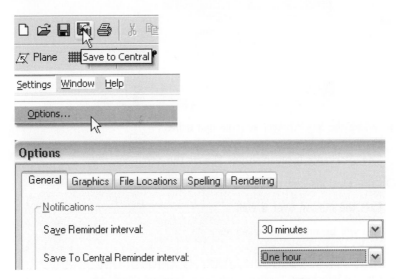

Figure 5.22 *The Save to Central icon appears in Workset-activated files*

2.11 Pick the Worksets icon on the Toolbar.

Certain Worksets are open (visible), but none have been made editable (checked out of the Central File).

- Select the Site Workset.

- Pick Editable.

- Your name appears as the Owner, as shown in Figure 5.24.

- Use the Active Workset drop-down list to make Site the active Workset.

- Click OK.

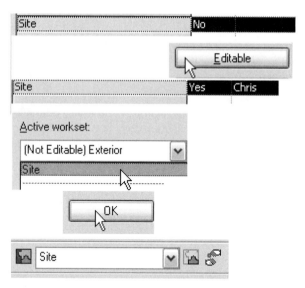

Figure 5.23 *Make the Site Workset editable and active*

The Worksets Toolbar shows Site as the active Workset.

2.12 On the Options Bar, check Editable Only.

- Pass the cursor over a wall.

  You will be unable to select it.

2.13 Open the SITE view.

- On the Worksets Toolbar, click Gray Inactive Worksets.

  Note the change in the view. Walls will go gray, while toposurfaces remain black.

- On the Options Bar, check Editable Only.

- Pass the cursor over a wall (see Figure 5.24).

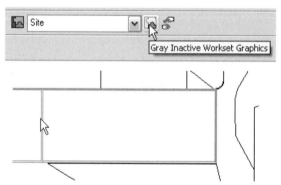

Figure 5.24 *You can adjust visibility and selectability of worksets*

2.14 Open the 1ST FLOOR view.

- On the Worksets toolbar, click Gray Inactive Worksets to toggle wall visibility back to normal.

  The Editable Only option does not persist—the check box is now empty. The active workset is Site. Exterior is not editable, but no other user has checked it out.

2.15 Select a wall.

The Tooltip will identify its type. The workset icon with Tooltip message Make element editable appears.

- Right-click.

  The context menu includes options to make the element and workset editable (see Figure 5.24).

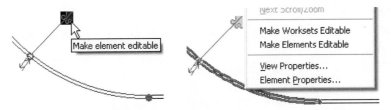

Figure 5.25 *Select a wall and note the workset options*

- Use the Type Selector to change the wall type to Basic Wall: Exterior: Generic - 12" [300 mm].

  A warning may appear, depending on which wall you have selected.

- Click the Worksets icon on the toolbar to open the Worksets dialogue (see Figure 5.26).

Warning

Two elements were not automatically joined because one or both is not editable.

Worksets

Active workset:

Site ☐ Gray Inactive Workset Graphics

Name	Editable	Owner	Borrowers
Exterior	No		Chris

Figure 5.26 *Changing a non-editable but not owned element makes you the borrower*

The Exterior workset now shows you as the Borrower, because you changed an element. There is no owner, since you are the only one working in a local file for the central file you created and you did not previously make the workset editable.

2.16 Select Exterior.

- Click Non-Editable.

  A warning appears (see Figure 5.27). You cannot make this change without a Save To Central or other file update action.

  Click OK.

- Click Editable.

  Your name now appears as owner. This is the same as picking the Make Worksets Editable option from the right-click menu with a wall selected.

- Click OK.

- Click No in the question box about making Exterior the active workset.

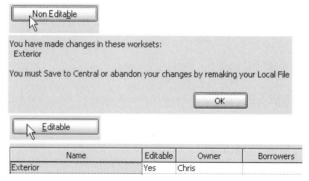

Figure 5.27   *Revit shows a warning message about this workset*

2.17 Click Undo from the Toolbar to return the wall to its original type.

2.18 Open the Worksets dialogue.

- Select Exterior.

- Click Non-Editable.

- Click OK.

  You are no longer the owner (see Figure 5.28).

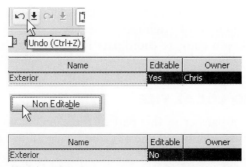

Figure 5.28 *Undoing the change allows you to change the workset status*

Revit allows you to make changes in worksets without owners as a borrower, or to become the owner of those worksets.

EXERCISE 3. WORKING SIMULTANEOUSLY TO DEVELOP THE MODEL

If you are working this exercise on your own, at this point you will open a second session of Revit Architecture and run the second session under a different username. If you are working in a class situation or with a partner, skip to Step 3.3.

3.1 Minimize but do not end the current session of Revit Architecture.

- Start another session using the desktop icon or Start menu.

- Close the file that opens.

- From the Settings menu, pick Settings>Options, as shown in Figure 5.29.

Figure 5.29 *Reach Revit Architecture's Options with no project open; change your Worksets Username*

3.2 On the General tab of the Options dialogue, enter **User2** in the Worksets Username value field.

- Click OK.

The Workset Username value field will hold the last entered value. After completing this exercise and closing all open Revit Architecture sessions, change the Worksets Username back to your Windows Login name to coordinate Revit Architecture with your usual login for future work.

CREATE A SECOND LOCAL FILE

If you are working with a partner in this part of the exercise, you will both have created a Central File on the network. Choose one of those Central Files to use for the rest of this exercise. The open local file from the selected Central File will stay open—that user will now be called User1, and the user who creates a second local file in the next steps will be called User2.

USER2:

3.3 Close any open Revit Architecture files.

- Open the Central File *Chapter 5 central.rvt* (open the file created by your partner for this exercise if you're working with a partner).

- Accept the Last Viewed option for Visible Worksets.

- Select File>Save As from the File menu.

3.4 In the Save As dialogue, click Options.

- Make sure that Make this the Central Location after save is **not** checked.

- Save the file on your local hard drive as **Chapter 5 exterior.rvt**.

    You are now working in a local file. Revit Architecture places a Save to Central icon on the Toolbar so design changes can be published to the Central File and shared with your partner.

3.5 Place the Worksets icon on the Toolbar as before.

- Open the Worksets dialogue.

- Pick Exterior in the Name field to highlight it.

- Click in the Editable column; change the value from No to Yes.

    Your username (which will be different from User2 if you are working the exercise with a partner) will appear in the Owner column (see Figure 5.30).

- Click OK.

Name	Editable	Owner
Exterior	Yes ⌄	User2
Furniture	No	
Interior	No	
Shared Levels and Grids	No	
Site	No	Chris

Figure 5.30   *Making the Exterior Workset Editable*

USER1:

You are still working in the file *Chapter 5 site.rvt.*

3.6 Open the Worksets dialogue.

The Exterior Workset will show that User2 has checked it out. The Editable status of that Workset in your file is No.

- Try to change the status of the Exterior Workset to Yes.

    Revit Architecture will display a warning that the Workset is being edited by another user, as shown in Figure 5.31.

- Click OK to close the warning.

- Click Cancel to exit the Worksets dialogue.

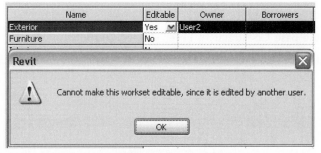

Figure 5.31   *Revit Architecture knows this Workset is checked out*

 INSTRUCTOR'S NOTE The next steps in this exercise for both User1 and User2 introduce new skills other than Worksets. We recommend that you have users complete both sections of this exercise. Once both users have completed their relative exercises 4 or 5, have them return to Step 3.3, open the copy of Chapter 5 central.rvt NOT used the first time through, and work the exercise steps (site or exterior) that they did not complete the first time.

USER2:

3.7 Move to Step 5.1 to work by yourself—you will continue to work in the file *Chapter 5 exterior.rvt.*

3.8 When your partner notifies you of an editing request, go to Step 4.22.

USER1:

Start Exercise 4:

EXERCISE 4. CREATE WALLS AND FLOORS

CREATE FOUNDATION WALLS AND A FLOOR

4.1 Open the PARKING plan view.

This view shows the level above as the Underlay, so you see the walls and topography at ground level.

- Pick Wall from the Basics tab of the Design Bar.

4.2 Set the Type value of the selected walls to **Foundation – 12" Concrete [Foundation – 300mm Concrete]**.

Foundations walls are created from the top down. By default, Revit will place them at the level of the current view extending down.

- Select Properties.
- Set the Base Constraint to Level **1ST FLOOR**.
- Set the Top Constraint to **1ST FLOOR** (see Figure 5.32). Click OK.

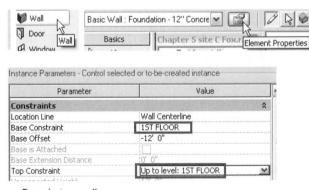

Figure 5.32 *Create Foundation walls*

4.3 Click the Pick option.

- Place the cursor over a green model line.

- TAB until all four lines highlight.
- Click to place walls.
- Click Modify.

4.4 Pick Floor on the Basics Tab of the Design Bar.

- The Design Bar will change to Sketch mode. Pick Walls will be selected by default on the Design Bar.
- Clear Extend into wall (into core) on the Options Bar.
- Place the cursor over a wall.
- TAB + select the walls you just created (see Figure 5.30).
- Pick Finish Sketch on the Design Bar.

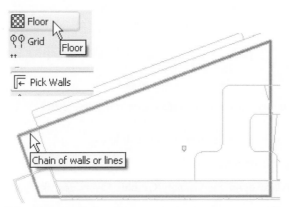

Figure 5.33   *Select walls to create the new floor*

PLACE AN OFFSET POINT IN THE FLOOR SURFACE

You will now locate a point for a floor drain.

4.5 Place the cursor over a wall.

- TAB until the floor edge highlights.
- Click to select it.
- On the Options bar, pick Draw Points.
- Set the Elevation to **-6" [-150]**.
- Place the new point in the center of the floor to the left of the elevation marker, as shown in Figure 5.34.

- Click to place a single point.
- Click Modify.

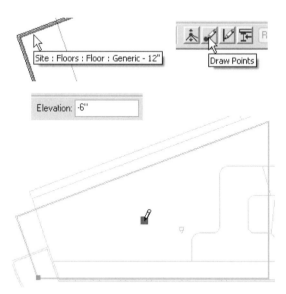

Figure 5.34 *Place a point in the floor surface with a negative offset*

The floor surface updates with break lines.

4.6 Open Section View Section I.

- Type **VG** to open the Visibility/Graphic Overrides dialogue.
- On the Worksets tab, check Site.
- Click OK.

4.7 Zoom in until you can see the floor clearly.

- Both the top and bottom edges have been depressed. This is not how concrete floors with drain slopes are created.
- Select the floor.
- Click Properties.
- Click Edit/New.
- Click Edit in the Structure field.
- Check Variable for Layer 2 in the Floor Assembly (see Figure 5.35).

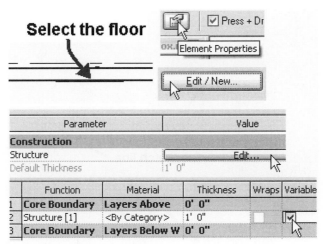

Figure 5.35 *Giving the floor variable thickness*

- Click OK to exit all dialogues.

  The bottom edge of the floor is now straight, and the floor will display correctly in section views.

  You will now create a second floor that will slope from the PARKING level up to 1ST FLOOR, to provide a ramp down from the surface to the underground garage.

CREATE A SLOPED FLOOR USING THE SLOPE ARROW

4.8 Open PARKING floor plan view.

4.9 Zoom in to the right side of the new walls.

- Pick Floor on the Design Bar.

- The Design Bar will change to Sketch mode with Pick Walls the default option.

- Select the left side of the wall, as shown in Figure 5.35.

- Select Lines on the Sketch Tab of the Design Bar.

- Select the Pick icon (arrow) on the Options Bar.

- Two Model lines have previously been placed in the Central File to indicate edges of a ramp from the driveway on the 1ST FLOOR level down to the PARKING level.

- Select those two lines to start the outline of the ramp.

- Select Draw on the Options Bar. Draw a vertical line to create the right edge for the floor outline.

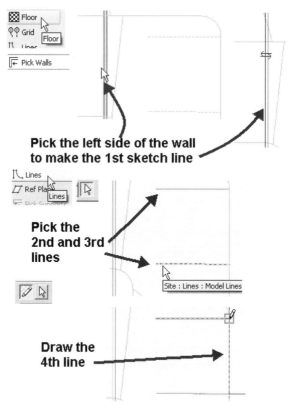

Figure 5.36 *Pick a wall, pick two lines, draw one line*

4.10 Trim the four new lines to complete a rectangular outline, as shown in Figure 5.37.

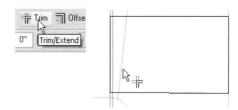

Figure 5.37 *Trim to finish the sketch*

4.11 Pick Slope Arrow from the Sketch Tab of the Design Bar.

- Pick the midpoint of the left side of the outline as the start point.
- Pull the cursor horizontally to the right to define the Slope Arrow, as shown in Figure 5.32.
- Pick the midpoint of the right side of the floor outline for the end point of the arrow.

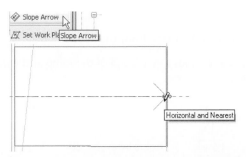

Figure 5.38 *Draw the slope arrow from the left-side line to the right-side line*

4.12 Select the new Slope Arrow.

- Pick the Properties icon from the Options Bar.

- Set the Level at Tail value to **PARKING**, with a Height Offset of **0' 0" [0]**.

- Make the Level at Head value **1ST FLOOR** (see Figure 5.39).

- Click OK.

- Pick Floor Properties.

- Make the Type value **Concrete walkway 4" [In situ concrete 225 mm]** (see Figure 5-39).

- Click OK.

- Click Finish Sketch. Pick No in the question about joining to the walls.

    The ramp will display according to the cut line for the view.

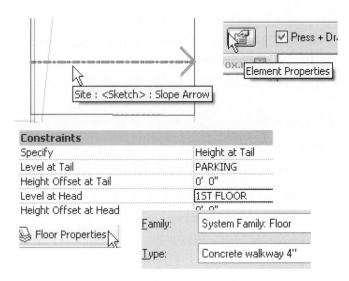

Figure 5.39 *Properties for the head and tail of the ramp Slope Arrow*

4.13 Select the Split tool from the Toolbar.

- Check Delete Inner Segment from the Options Bar.

- Split the wall that crosses the left end of the ramp at the edges of the ramp, as shown in Figure 5.40.

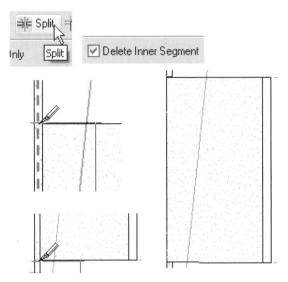

Figure 5.40 *Splitting the wall to remove a section*

CREATE SLOPED WALLS FOR THE SLOPED FLOOR

4.14 Select Wall from the Basics Tab of the Design Bar.

- From the Type Selector drop-down list, select **Basic Wall: Retaining – 12" Concrete [300 mm Concrete]**.

- Click Properties. Set the Base Constraint and Top Constraint to 1ST FLOOR, as before.

- On the Options Bar, set the Location Line to Finish Face Exterior.

- Clear the Chain Option.

- Draw a horizontal wall **47' [14100 mm]** long along the upper edge of the ramp.

  Be sure the wall is on the ramp, not beside it.

- Set the Location Line to Finish Face Interior.

- Place a corresponding wall on the bottom edge of the sloped slab (see Figure 5.41).

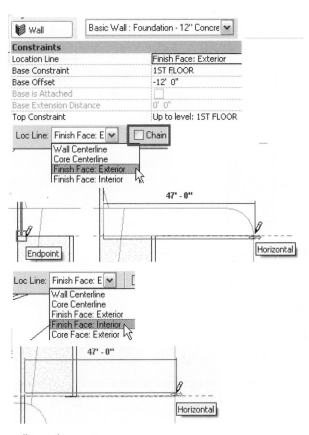

Figure 5.41   *Draw walls on the ramp*

4.15 Select Modify.

- Pick the two new walls, then select Attach from the Options Bar, as shown in Figure 5.42.

- Choose Base.

- Pick the Ramp.

    The walls will change appearance and end at the cut line, consistent with the ramp.

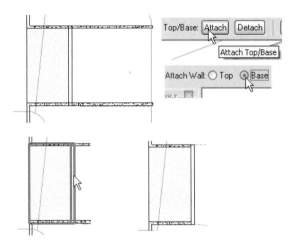

Figure 5.42 *Attach the bottom of the new walls*

4.16 Zoom to Fit in the PARKING plan view.

- Open the Elevations: South view.

- Type **VG**.

- On the Model Categories Tab, make sure Topography is cleared.

- On the Worksets Tab, make sure that Exterior, Shared levels and Grids and Site are checked.

- Click OK.

4.17 Zoom in Region to the right side of the view.

- Examine the sloped floor and side walls you have just created (see Figure 5.43).

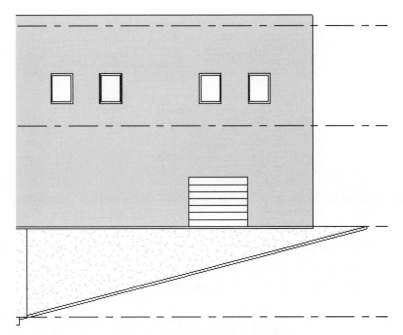

Figure 5.43 *A side view of the walls and floor they are attached to*

SHARING WORKSETS: EDITING REQUESTS

4.18 Open the Floor Plan View 1ST FLOOR.

- Type **VP**.

- In the View Properties dialogue, make the Underlay value **PARKING**.

- Pick Edit in the Visibility field.

- On the Worksets Tab, check **Site** (see Figure 5-44).

- Click OK twice.

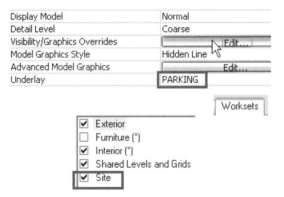

Figure 5.44 *Change the Underlay in this view and make the Site workset (jot visible by default) visible*

4.19 Zoom in Region to the lower-right corner of the now-visible parking garage walls.

- On the Worksets Toolbar, make Exterior the current Workset, even though it is not editable in this local file (see Figure 5.45).

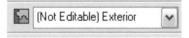

Figure 5.45 *Make the Exterior Workset the active one*

4.20 Make the Modeling Tab active on the Design Bar.

You are going to create an opening in the walkway for an elevator stair down to the parking garage.

- Pick Opening>Opening by Face.

  The cursor will change shape.

- Select the exterior walkway, as shown in Figure 5-46.

  An error message appears.

- Click Place Request.

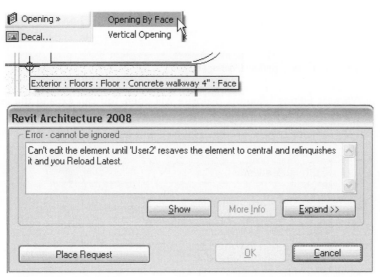

Figure 5.46   *Use the Opening tool to cut a hole in the walkway, which is a floor*

Revit Architecture will display a confirmation dialogue with options to continue work or to check for a grant to your editing request (see Figure 5-47).

Figure 5.47   *The dialogue to check your request*

4.21 If you are working this exercise alone, move to the second session of Revit Architecture that has the file *chapter 5 exterior.rvt* open. If you are working with a partner, notify her or him that you have reached this point in your exercise.

USER2:

4.22 You are working in the file *chapter 5 exterior.rvt*. When your partner notifies you (or if working alone, when you have made the editing request in Step 4.20), pick the Editing Requests icon from the Worksets Toolbar (see Figure 5.48).

Figure 5.48 *Open the Editing Request dialogue*

4.23 The Workset Editing Requests dialogue will open. You can expand the list to see who is requesting what change in the Worksets you control from your local file. Select the one from your partner (or yourself) and click Grant (see Figure 5.49).

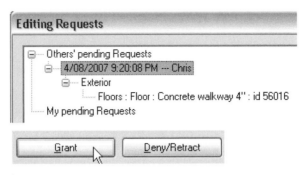

Figure 5.49 *Grant the request*

4.24 Click Close and continue with your work in Exercise 5.

USER1:

4.25 Once you (or your partner) have completed Steps 4.22 to 4.24, pick Check Now in the Check Editability Grants dialogue. Click OK in the confirmation dialogue (see Figure 5.50). Revit Architecture will create the opening in the walkway.

Figure 5.50 *Check the request and finish the sketch*

You could have picked Continue in the Grant Check dialogue and continued with other work in your project file, rather than requesting an immediate grant. In a real-world situation, the demands of work flow, and a possibly far-flung design team, may very well mean that requests accumulate while other work goes on.

4.26 Open the Worksets dialogue.

- Note that your name now appears as a Borrower(s) of the Exterior Workset (see Figure 5.51).

- Click Cancel to exit the dialogue.

Name	Editable	Owner	Borrowers	Opened
Exterior	No	User2	Chris	Yes
Furniture	No			No
Interior	No			No
Shared Levels and Grids	No			Yes
Site	Yes	Chris		Yes

Active workset: Site — Gray Inactive Workset Graphics

Figure 5.51 *You are now a borrower of the Exterior Workset in this local file*

4.27 Click Opening>Opening By Face.

- Select the exterior walkway.

- Select the rectangle option from the Options Bar.

- Sketch a 6' x 10' [1800 x 3000] rectangle, as shown in Figure 5.52, starting at the intersection of the walkway edge and the wall below.

- Click Finish Sketch.

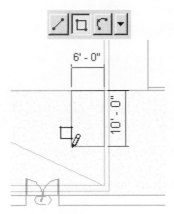

Figure 5.52 *Sketch the Opening you requested*

4.28 Open the default 3D view. Type **VG**.

- On the Worksets Tab, check **Site**.

- Click OK.

- From the View menu, pick View>Orient>Northeast (see Figure 5.53).

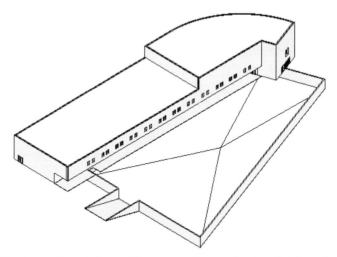

Figure 5.53 *A Northeast view of the parking garage floor, walls, sloped slab and opening cut*

4.29 Pick the Save to Central icon from the Toolbar. From the File menu, pick File> Close. In the dialogue that opens, click Relinquish & Save.

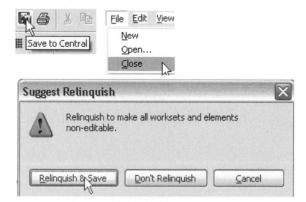

Figure 5.54 *Save to Central and close, relinquishing worksets*

4.30 Move to Step 5.1 unless you have already worked Exercise 5.

USER2:

If User1 has Saved to Central before you get to this step, Revit Architecture will ask you to reload Worksets before continuing work. Certain views may therefore look different from the illustrations in this chapter.

EXERCISE 5. CREATE AND EDIT A CURVED CURTAIN WALL

You are working in file *Chapter 5 exterior.rvt.*

    5.1 Open the default 3D view.

- Pick the Worksets icon on the Toolbar.

- Close Site.

- Make Interior Editable.

- Click OK.

- Click No in the question, as shown in Figure 5.55.

    This will keep Exterior the active workset.

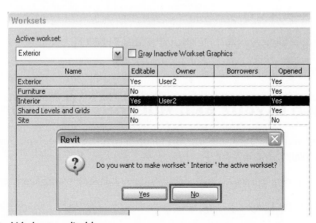

Figure 5.55 *Make Worksets editable*

    5.2 From the View menu, pick View>Orient>Southwest.

- Zoom in Region at the curved wall section.

    5.3 Select the Split tool.

    As you move the cursor over the building exterior, the curved wall segment will highlight.

- Move the cursor up the wall, as shown in Figure 5.56.

- Pick a point near but below the roof to split the wall into two segments vertically.

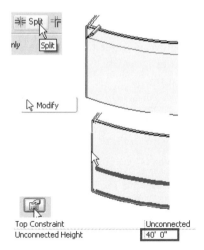

Figure 5.56 *Using the Split on a vertical wall*

5.4 Select Modify from the Basics Tab of the Design Bar.

- Pick the large curved wall at the floor line.

- Choose the Properties icon.

- Make the Unconnected Height value **40' [12000]**.

- Click OK.

CREATE A GRID LAYOUT AND MULLIONS FOR A CURTAIN WALL

5.5 With the wall still selected, use the drop-down list in the Type Selector to select Curtain Wall: Curtain Wall 1.

- The wall displays as a single straight (not curved) transparent panel.

- Click Properties.

- Click Edit/New.

- In the Vertical Grid area, set the Layout value to Maximum Spacing.

- Set the Spacing to **5'-0" [1500]**.

- In the Horizontal Grid area, set the Layout Value to Fixed Distance.

- Set the Spacing to **8'-0" [2400]** (see Figure 5.57).

- Click OK to exit the dialogues.

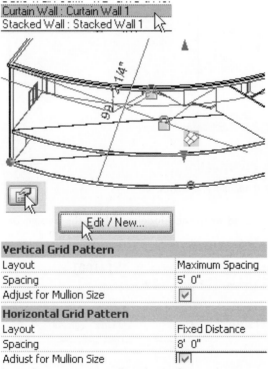

Figure 5.57 *Change the wall to a curtain wall, and specify a grid pattern*

5.6 Open the South Elevation view. Zoom in Region to the right side of the curtain wall.

5.7 Place the cursor over the first (rightmost) vertical curtain grid.

- TAB so that it highlights with a dashed green line and select it.

  The Options Bar will change its options.

- Select Add or Remove Segments.

- Pick the lowest segment of the curtain grid.

  It will highlight with a dashed green line, as shown in Figure 5.58.

- Click anywhere off the walls.

  The grid segment will disappear.

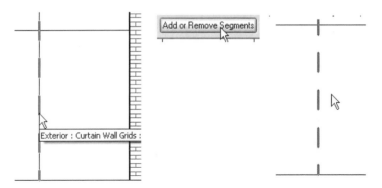

Figure 5.58 *Removing a curtain grid segment*

5.8 Select Curtain Grid from the Modeling Tab of the Design Bar.

- Choose One Segment from the Options Bar.

- Place a horizontal curtain grid at **7' [2100]** elevation in the open grid you just created, as shown in Figure 5.59.

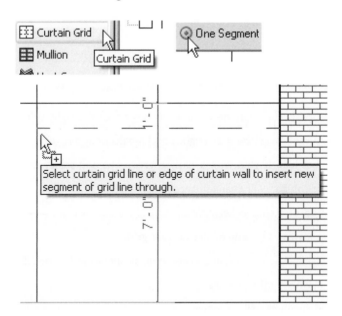

Figure 5.59 *Adding a curtain grid segment*

5.9 Zoom out until you can see the entire curtain wall. Pick Mullion from the Modeling Tab of the Design Bar. From the Options Bar, choose All Empty Segments. Accept the default Mullion type in the Type Selector. Put the cursor over the curtain wall and the whole wall will highlight. Click to create mullions throughout the curtain wall (see Figure 5-60).

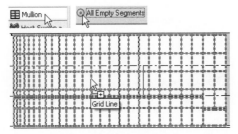

Figure 5.60   *Create mullions on the entire curtain wall*

ADD A DOOR TO THE CURTAIN WALL

5.10 Expand the Families section of the Project Browser.

- Expand the Curtain Panels Section under Families.

  Note that it contains four types of panels: Empty Panel, Empty System Panel, Glazed Panel, and System Panel. You need a door to go in a curtain wall panel.

5.11 From the Modeling Tab, select Door.

- On the Options Bar, choose Load from Library.

- Navigate to the Revit *Content/Imperial Library/Doors [Content/Metric Library/Doors]* folder.

- Select *Curtain Wall-Store Front-Dbl.rfa [M_Curtain Wall-Store Front-Dbl.rfa]* (see Figure 5.61).

- Click Open.

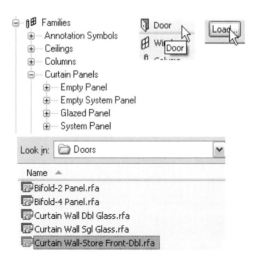

Figure 5.61 *The door type in the Imperial Library*

5.12 Revit Architecture will attempt to place a door, but not the type you just loaded, and the new door type will not be in the Type Selector list.

- Click ESC.

    The Families/Curtain Panels section of the Browser will now show **Curtain Wall-Store Front-Dbl** (see Figure 5.62). The Door family has been loaded into the file, but does not work with the Door tool. There is another way to specify the correct door type in the curtain wall panel.

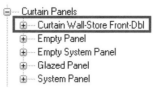

Figure 5.62 *The new curtain panel type in the Project Browser*

5.13 Zoom in Region around the open panel you created earlier.

- Place the cursor over a mullion at the edge of the panel.

- Hit TAB until the four sides of the panel pre-highlight and the Tooltip identifies the System Panel (see Figure 5.63).

- Click to select it.

- From the Type Selector drop-down list, choose Curtain Wall-Store Front-Double Door.

- The panel will update with a door to fit.

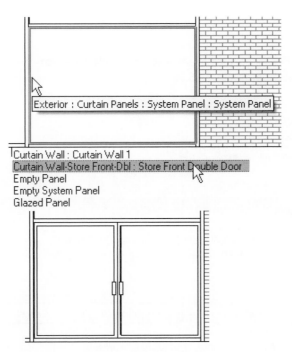

Figure 5.63 *Select the System Panel and change it to a door*

5.14 Open the 3D View. From the View control Bar, change the display to Shading with Edges. A warning will display (see Figure 5.64).

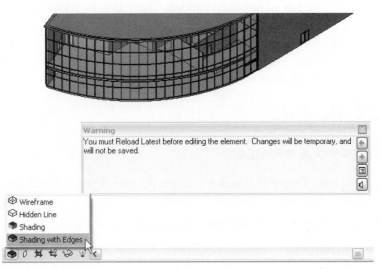

Figure 5.64 *A shaded view of the curtain wall*

5.15 Pick Save to Central from the Toolbar. The view will revert to the original orientation and graphics style.

5.16 From the File menu, pick File>Close. In the dialogue that opens, click Relinquish & Save, as shown in Figure 5.65.

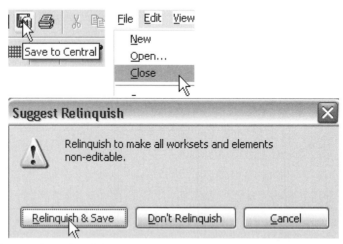

Figure 5.65 *Save to Central and close, relinquishing worksets*

This ends the tandem-user section of exercises in this chapter. For the rest of the exercise steps, there will be no usernames specified. Close all open files. If you have worked the tandem-user exercises by yourself, close one session of Revit Architecture and use the menu to choose Settings>Options to set the Worksets Username to its previous value.

The next exercise will concentrate on showing changes to a building over time. You have already worked with the Phase and Phase filter of views to show that existing walls can be demolished and new ones created. Now you will create phases to show how one area of a building will change more than once.

EXERCISE 6. PHASING

ADD PHASES TO A PROJECT

6.1 Open the file *Chapter 5 phases imperial.rvt [Chapter 5 phases metric.rvt]*. The file will open to a plan view of the 2ND FLOOR.

6.2 From the Menu bar, pick Settings>Phases, as shown in Figure 5.66.

On the Project Phases Tab of the Phasing dialogue, two default Phases will display: (1) Existing and (2) New Construction.

• Select the Name field for Phase 2 and change it to **Building Shell**.

- Click the After button, in the Insert section of the dialogue box, to add a new line.

- Revit Architecture will create Phase 1 on line 3.

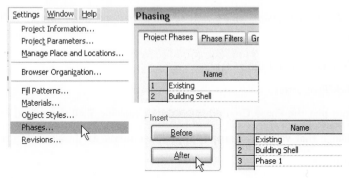

Figure 5.66 *Rename a default phase, add another*

6.3 Make the Name for Phase 1 **Temporary classrooms**.

- In the Description field, type **2nd floor used for classes during Miller Hall renovations**.

- Click Insert After again.

- In the Name field for Phase 1 (on line 4), type **Offices**.

- Enter the following in the Description field: **2nd floor offices and labs** (see Figure 5.67).

	Name	Description
1	Existing	
2	Building Shell	
3	Temporary classrooms	2nd floor used for classes during Miller Hall renovations
4	Offices	2nd floor offices and labs

Figure 5.67 *Two new phases with descriptions*

6.4 Select the Phase Filters Tab.

- Study the default Phase display settings for a moment.

- Select the Graphic Overrides Tab and study it so that you gain an understanding of the display characteristics for the different Phase conditions.

- Click OK to exit the dialogue.

APPLY PHASE TO OBJECTS

6.5 CTRL + select all the interior walls.

You can use a crossing window to select doors and door tags as well.

- Click Filter.

- Clear Doors and Door Tags.

- Click Properties.

  Note that the Phase Created value shows Building Shell, the edited name for Phase 2.

- Change that value to **Temporary classrooms**, as shown in Figure 5.68.

- Click OK.

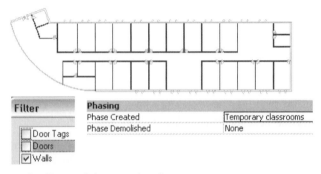

Figure 5.68   *Change the Phase of doors and walls*

The walls and doors will disappear.

6.6 Type **VP** to enter the View Properties dialogue.

The Phase for this View is Building Shell, which predates the phase you just assigned to the walls.

- Change the Phase Value for the View to Temporary classrooms, as shown in Figure 5.69.

- Click OK.

- The walls and doors appear.

Phasing	
Phase Filter	Show All
Phase	Temporary classrooms

Figure 5.69   *Change the Phase of the view*

6.7 Select walls and doors, as shown in Figure 5.70.

- Use the Filter to select only walls. Doors will assume the phase of their hosts.

- Select the Properties icon.

- Change the Phase Demolished value to Offices.

- Click OK.

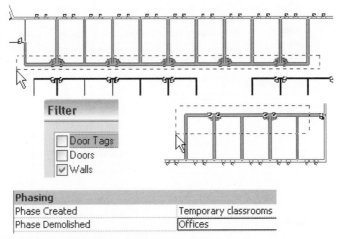

Figure 5.70   *Select walls and doors to designate as temporary*

APPLY PHASE AND PHASE FILTER TO VIEWS

6.8 Right-click on 2ND FLOOR in the Project Browser.

- Select Rename.

- Name the view **2ND FLOOR CLASSROOMS**.

- Click No in the rename level query.

6.9 Right-click on 2ND FLOOR CLASSROOMS in the Project Browser.

- Choose Duplicate View>Duplicate.

- Rename the new view **2ND FLOOR OFFICES**.

- Click OK.

- Place the cursor in the view, right-click and pick View Properties.

- Change the Phase of the View to Offices.

- Change the Phase Filter to Show Previous + New, as shown in Figure 5.71.

- Click OK.

The walls that you designated to be removed in the Office construction Phase disappear. The Walls constructed in the previous phase (Temporary Classrooms) that remain in the phase for this view appear halftone.

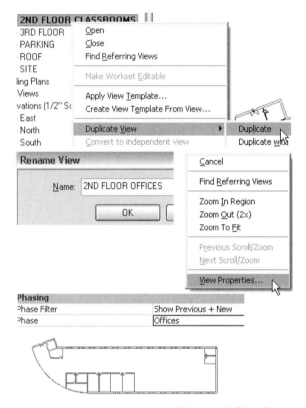

Figure 5.71 *Rename a view, duplicate it and set the Phase and Filter for the new view*

Your task is to create 20 offices along the upper wall, and fill the open space that remains with workstation cubicles.

6.10 Zoom to the area, as shown in Figure 5.72.

- Select the vertical wall. Click Array on the Toolbar.

- On the Options Bar, enter **21** for the Number.

- Verify that Move to: 2nd is selected.

- Check Constrain.

- Click anywhere to start a distance vector.

- Pull the cursor to the right until the dimension reads 11' [3300] and click to establish the distance.

You can also type the distance directly. The array count field will display.

- Click anywhere off the array to accept the number.
- Click OK in the warning.

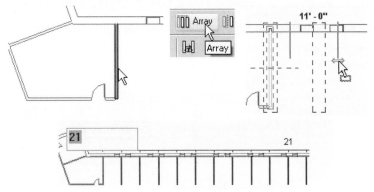

Figure 5.72 *New walls by array*

6.11 Select the horizontal wall, as shown in Figure 5.73.

- Click Create Similar on the Toolbar.
- Draw a wall to the right of the end of the one you just selected.

  The wall segment can be any length.

- Click Trim to make a corner with the rightmost vertical wall of the array.

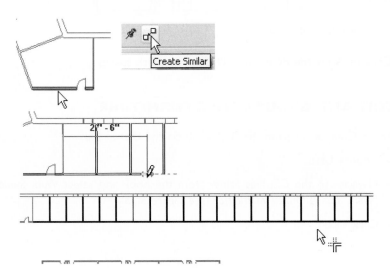

Figure 5.73 *Create a new wall to match an existing one*

6.12 Select the door in the horizontal wall you matched previously.

- Click Create Similar.

- Place two doors, as shown in Figure 5-74, 2' 6" [750 mm] from the walls.

- Use the Spacebar and Flip controls to manage hinge and swing.

- Click Modify.

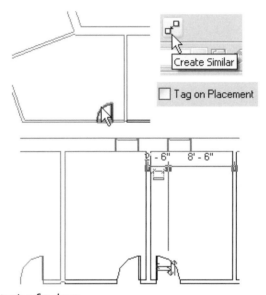

Figure 5.74 *Typical spacing for doors*

6.13 Copy or Array the doors into all the new office spaces.

PLACE, EDIT AND GROUP OFFICE FURNITURE

6.14 Select Component from the Basics Tab on the Design Bar.

- Choose Load.

- Navigate to the CD files folder from the book and select *Work Station Cubicle.rfa* *[Work Station Cubicle metric.rfa]* (see Figure 5.75).

- Click Open.

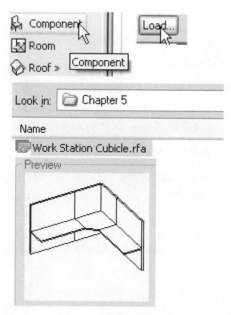

Figure 5.75 *Find a Work Station Cubicle component to insert*

6.15 Locate an instance of the workstation, as shown in Figure 5.76.

- Do not draw the dimensions.

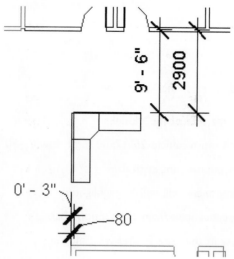

Figure 5.76 *Locate the first cubicle*

6.16 Choose Modify from the Design Bar to terminate component placement.

- Select the new workstation.

- Pick Array from the Toolbar.

- Clear Group and Associate on the Options Bar.

- Enter **24** for the Number.

- Check Constrain.

- Click at a convenient point to start the distance, and pull the cursor to the right until the temporary dimension reads 8'-2" [2490].

- Click to accept the distance.

  Revit will snap to points on the cubicle. You can also enter **8'2" [2490]** at the keyboard (see Figure 5-77).

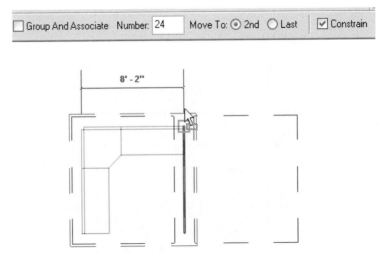

Figure 5.77 *Getting ready to array the workstation*

6.17 Zoom to the right side of the model.

- Copy the rightmost cubicle 25' [7500 mm] down (–90°).

  The exact location is not critical.

  The new component will still be highlighted.

6.18 Select the Properties icon from the Options Bar.

- Pick Edit/New in the Type Properties dialogue.

- Click Duplicate.

- In the Name box, enter **108" x 108" [2700 x 2700]**.

- Click OK.

- In the Type Properties dialogue, change the Panel1 Length and Panel2 Length values to **6'-0" [1800]**, as shown in Figure 5.78.

- Click OK twice to exit the edit dialogues.

  The cubicle will grow.

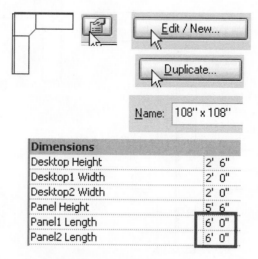

Figure 5.78   *Adjust the panel length in the new larger cubicle type*

6.19 Mirror the workstation along one edge. Mirror the two instances to make a four-sided cubicle arrangement, as shown in Figure 5.79.

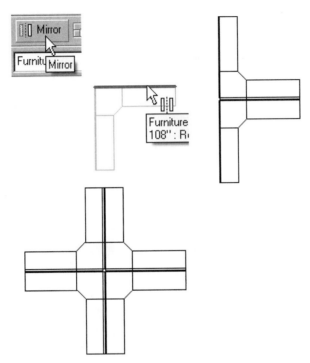

Figure 5.79 *The large cubicles arranged back to back to back*

6.20 Select all four cubicles. Choose Group from the Toolbar. Enter **Quad Workstation Long**. Click OK.

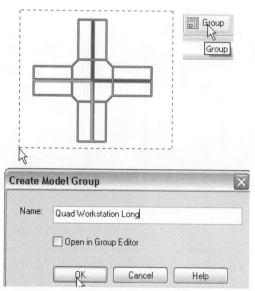

Figure 5.80 *Rename the Group*

CAD Manager Note: Groups are a very useful tool in Revit Architecture. When combined with Worksets they introduce a level of complexity that needs to be carefully managed. Groups can be created with elements from different Worksets; this can create headaches later, as designs develop and teams interact.

The new group displays with control points.

6.21 Select Rotate from the Toolbar.

- Click to the right of the group, then pull to create a 45° rotation (see Figure 5.81).

352

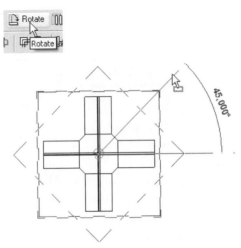

Figure 5.81 *Rotate the workstation*

6.22 Pick Copy.

- Clear Constrain on the Options Bar.

- Check Multiple.

- Copy the Group to the left and right at a distance interval of **25' [7500 mm]** four times, as shown in Figure 5.82.

- When the copies have been created, hit ESC twice to terminate the Copy tool.

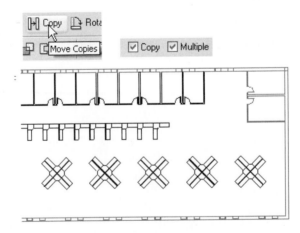

Figure 5.82 *Workstations in place in the Furniture Plan*

EXPORT AND IMPORT GROUPS

6.23 Zoom to Fit.

- From the File menu, select Save to Library>Save Group.

- Navigate to a convenient location to hold the new file.

- Note that Revit will save the file using the Group name unless you enter another name (see Figure 5-83).

- Enter **Quad Workstation Export** as the file name. This will save confusion when we import the same file in the next step.

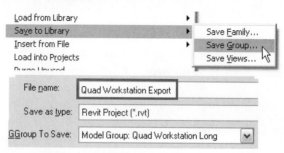

Figure 5.83   *Save the Group as a Library file*

- Click Save.

6.24 From the File menu, Select Load From Library>Load File as Group.

- Navigate to the folder where you saved Quad Workstation Export.

- Select the name of the new file (see Figure 5.84).

- Click Open.

- Click Yes in the question box.

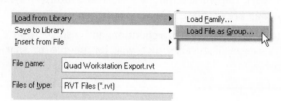

Figure 5.84   *Load the new group*

6.25 In the Project Browser, expand the list next to Groups.

- The new group Quad Workstation Export (actually the same as the one in your project, but renamed) is now in the list.

- Click the name of the new group. Hold down the cursor button and drag it into the view window.

- Click in the open area at the left, as shown in Figure 5.85, to place an instance.

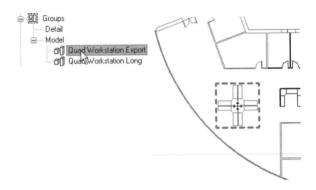

Figure 5.85 *Place a copy of the imported group*

- Click Modify.

6.26 Save the File. Close the file.

SUMMARY

This chapter's exercises have introduced you to the uses of Worksets Phasing and Groups, important mechanisms in Revit Architecture for organizing a building model project file. Worksets divide a project's contents according to whatever system you decide works best for you. The Save to Central/Save Local function allows numerous users to work on a project simultaneously and coordinate their efforts. Many firms find that using Worksets in projects without multiple users is worthwhile for the control it gives designers over component visibility and organization.

Phases provide a way to show changes in a project over time. Phases, together with their associated Filters and Graphic Overrides, give complete control over view visibility characteristics—new work, old work, temporary work and future work can all be displayed or hidden according to your preferences.

Groups allow you to take advantage of repetition, and also provide a way to export and import portions of a design. This is another way for people to share design information between or within projects.

REVIEW QUESTIONS - CHAPTER 5

MULTIPLE CHOICE

1. A Workset can be

   a. Open but not Editable

   b. Editable but not Open

   c. Active but not Editable

   d. all of the above

2. To select elements in a Workset-enabled file,

   a. at least one Workset must be editable

   b. Editable Only must be cleared on the Options Bar

   c. either a or b

   d. both a and b

3. Properties of Slope Arrows for floors include

   a. Width and Rotation

   b. Height at Tail, Height at Head, Offsets

   c. Thickness

   d. all of the above

4. When placing gridlines in a curtain wall,

   a. you can set vertical and horizontal lines in any order

   b. you must work right to left for vertical lines

   c. you must work top–down for horizontal lines

   d. Revit Architecture disables snaps

5. To place a door in a curtain wall,

   a. Create a hole in the curtain wall and place a standard door

   b. Create a section of standard wall inside the curtain wall

   c. Load the curtain wall door family you want to use, and edit the curtain wall panel to the correct door type

   d. You can't place doors in curtain walls

TRUE/FALSE

6. True _ False _ In order to make a change in a model object that is part of a Workset already being edited by another user, you must make an Editing Request of that user.

7. True _ False _ The user needs to create Worksets for annotation types, as Revit Architecture does not create those Worksets automatically.

8. True _ False _ When a Workset-enabled file is linked into a host file, Revit Architecture displays only the Worksets from the link that were originally set to display in all views.

9. True _ False _ When you Attach the bottom of a wall to a roof or floor below the wall, Revit Architecture will split the wall at the floor line and you have to delete the extra section by hand.

10. True _ False _ Revit Architecture will let you create as many Phases in a project as you want.

 Answers will be found on the CD.

CHAPTER 6

Managing Views and Annotation

INTRODUCTION

So far you have been working on the design tools integrated into Revit Architecture. In previous chapters, you created masses to represent buildings and added walls, doors and windows. You have linked models and sites and shared models with others. The main purpose of building a model is to communicate to the construction team what the design team wants to build. Traditionally, design communication has been in the form of construction documents: printed collections of plans, sections, elevations and specifications. This may not be the only way to deliver or communicate design intent in the future, but documents of a certain widely accepted appearance are today's standard and will not soon be completely replaced. If a wonderful model of a wonderful building cannot be understood by those who have to approve, pay for or build it, the building won't be built. Proper annotation is essential to effective design practice.

Revit Architecture is comprehensive building information modeling software for the full spectrum of architectural practice, so you shall now explore its tools for documentation of your designs. This chapter will contain several sections to illustrate different types of standard documentation and the management of each technique. In the following exercises, we will demonstrate techniques for adding, editing and managing annotation. Please note that many of these exercises are organized to show you techniques for setting up and managing annotations, and therefore a certain similarity will become evident. In a working environment, once annotations have been set up in a project, the settings can be saved in company standard templates and therefore will not have to be repeated.

OBJECTIVES

- Create interior elevations and plan callout views
- Manage Room objects
- Create, insert, manage and modify tags—Room, door, window, sections, elevation and grids
- Insert, manage and modify annotation—Notes and dimensions
- Transition entire set from Schematic annotation style to Design Development annotation style

REVIT ARCHITECTURE COMMANDS AND SKILLS

Creation and customization of:

Room Tags, Section Indicators, Elevation Symbols, Titleblocks, Dimensions, Units, Font Styles, Notes and Grids

Place Elevation

Place Section

Place Callout

Working with Phases

View Templates

Project Browser as a Tool

Views and Sheets

View Titles

Tag All Not Tagged

One of the great benefits of using Revit Architecture is its coordination of all objects in the model, including tagging. Because the tag is merely a view or "looking glass" into one specific piece of information about the object it annotates, the coordination of that information is handled by the object and the computer. Very little user input is necessary, which eliminates opportunities for user error. There are tags for just about any Revit Architecture object you can think of, and customizing them to match company standards is a simple task.

Revit's Rooms are model objects, not simply annotation devices. You associate rooms with bound spaces in a plan view. An area enclosed by three walls or room boundary lines will hold a room. Rooms are three dimensional and can extend higher than one story if you wish. Rooms read their perimeter, wall area and volume and can therefore be used to calculate and schedule floor finish area, baseboard length and paint. Rooms hold occupant and phase information, and so can be used for extensive facilities management well after construction is complete. This capacity of a building model is only now starting to be utilized by design and architectural service firms as a source of revenue.

Sustainable building design criteria will become increasingly important for designers over the next few years. Energy performance analysis is a major component of successful sustainable design. In Revit MEP rooms hold energy information such as heat loss, airflow and lighting load, and as such form the basis for energy performance analysis. Much will be made of these room properties in the very near future.

Annotation of a building project transforms a collection of views of a model into a set of construction instructions and therefore a legal document. There are standardized formulas for arrangements of text notes, symbols and dimensions on printed pages to make instructions clear to the many people who need to view and use a document set.

Governments, professional organizations and companies all have long-established drafting standards. As a student or member of a firm implementing Revit, you will spend time setting up Revit's views, annotations, dimensions, text, tags and symbols to match pre-existing standards. Once setup is complete, managing annotation takes very little time and effort.

In many cases you will want your annotations to convey a different image at different times. You may want to show or not show certain information at different stages of design. The best way to do that is through the customization of the tag that reports that information. In the following exercise, you will create some custom tags for a Schematic Design look.

Note: To demonstrate the management of annotation and its appearance in different phases of design, we will assume that you are in our project's Schematic Design phase during the first portion of the exercises. Annotation during this phase will be loose in appearance. You will be using the Comic Sans MS font style (standard on Windows PCs) to illustrate this. Midway through the chapter you will switch to a Design Development phase appearance with more formal "hard-line" tags and annotations. You will make this change in a few short steps.

EXERCISE 1. ROOMS AND ROOM TAGS

PLACE ROOMS

1.1 Launch Revit Architecture.

- From the File menu, select File>Close.

- From the File menu, select File>Open.

1.2 Navigate to the folder that holds the CD Library files for Chapter 6.

- Open the file *Chapter 6 start imperial.rvt [Chapter 6 start metric.rvt]*.

    It will open to the 1ST FLOOR – FLOOR PLAN view that shows exterior and interior walls.

- Save the file as ***Chapter 6 annotation.rvt*** in a convenient location or as determined by your instructor.

    At the right side of the building, rooms have tags visible.

1.3 In the view window, pass the cursor near the room tags until the diagonal indicators of a room display, as shown in Figure 6.1.

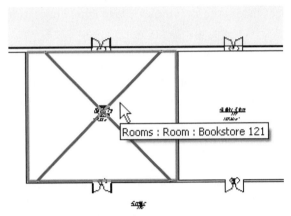

Figure 6.1 *A room object highlights as the cursor passes over it*

1.4 When a room outline displays, click to select it.

- Right-click to open the context menu.

- Select Element Properties.

- Note the room properties (see Figure 6.2).

- Click Cancel.

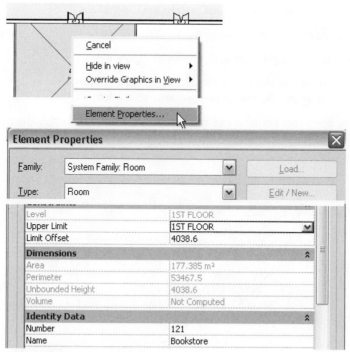

Figure 6.2 *Properties of a room*

1.5 From the Basics Tab of the Design Bar, click Room.

- Existing rooms on the right side of the building will highlight cyan.

- On the Options Bar, clear Tab on Placement.

- Place the cursor in a white area, as shown in Figure 6.3.

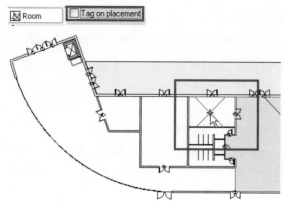

Figure 6.3 *Place a room without a tag*

A room outline and diagonal marker lines will show.

- Click to place a room without a tag. You will place tags in a later step.

1.6 Place a total of eight rooms, as shown in Figure 6.4.

- Place rooms in the two toilet blocks and the janitorial space between them, in the elevator shaft, and the large open area at the left.

The open area continues to the right, but a room separation line has been placed to divide it for placing rooms.

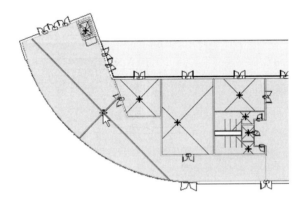

Figure 6.4 *New rooms in place*

1.7 Click Modify.

The new rooms will disappear, but will highlight under the cursor as in Step 1.3.

CUSTOMIZE A ROOM TAG FAMILY

- From the File pull-down menu, select File>Open.

- In the dialogue that opens, navigate to the CD files *Library/Chapter 6* folder.

- Select *Room Tag With Area Imperial.rfa [Room Tag with Area – Metric.rfa]* from the *Library* folder (see Figure 6.5).

- Click Open.

- Select File>Save As from the File menu.

- Save the file as **Room Tag with Number - Area - SD.rfa** (the SD in this file name stands for Schematic Design) in a convenient location or as designated by your instructor.

Room name **Room name**

| 101 |

101

150 m2   150 SF

Figure 6.5 *The Room tag file before changes: metric left, imperial right*

1.8   Select and delete the lines that form the box around the room number.

Note: Pre-highlighting one of the lines and selecting TAB will select all the lines in the box.

The text objects you see in this file are all labels. Labels are not simply text; they read and report (display in text form) the content of parameters. These labels report the properties of rooms that you looked at in Step 1.4: name, number and area. If the name, number or area of a room changes during design development, the label for that room will update automatically.

1.9 Select the Room name label and pick the Properties button in the Options Toolbar.

- Select Edit/New and change the Text Font to Comic Sans MS.

- Check the Bold, Italic and Underline checkboxes, as shown in Figure 6.6.

- Click OK twice to see the results.

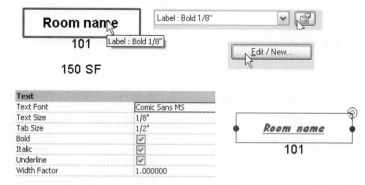

Figure 6.6 *Change the font for the Room name label*

1.10 Repeat the preceding steps with the Room Number label.

- Change the Text Font in the Room Number label to Comic Sans MS.
- Select the Italic checkbox.
- Click OK twice to see the change.

   Both the Room Number and Room Area label text fonts change. They are the same Type of label.

1.11 Type **VG** to open the View Graphics dialogue.

- Select the Annotation Categories Tab.
- Check the Reference Planes checkbox.
- Click OK.

   The intersection of the Reference Planes represents the insertion point of the tag.

1.12 Move the Room Number and Room Area labels closer together so they look more like a single text item than individual lines (see Figure 6.7).

 Tip: Use the nudge tool to move the text closer together by tapping the arrow keys once you have selected the text. You will note that the closer you are zoomed in, the less the text actually moves. The move distance is controlled by the snap distance, which is zoom-dependent.

Room name
101
150 SF

Figure 6.7 *Room tag with room number and area set up for Schematic Design annotation*

1.13 Save the file.

- From the Family Tab on the Design Bar, pick Load Into Projects.
- Click Yes in the question box to overwrite the existing tag type with your newer customized version.

Figure 6.8 *Load the customized tag into the project file and overwrite the previous version*

1.14 From the Window menu, select Room Tag with Area – Number – SD to return to the family file.

- Close the family file.

Note: You can open a family file directly from a Revit Architecture project by selecting an instance of the family and picking Edit Family on the Options Bar.

PLACE ROOM TAGS IN A PROJECT

1.15 From the Drafting Tab on the Design Bar, select the Room Tag tool.

The rooms will display and highlight cyan.

- Use the Type Selector to make Room Tag with Number - Area – SD the active tag.

- Place a room tag in all of the rooms that do not already have one, plus the open curved area at the left of the plan (see Figure 6.9).

- Select Modify to terminate placement.

The room highlights will disappear.

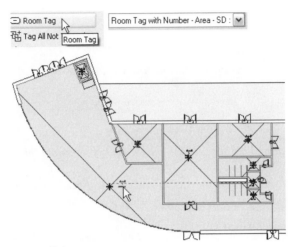

Figure 6.9 *Place instances of the new room tag*

1.16 Zoom in Region to the far left of the plan.

- Select the Room Tag in the large curved area, click on the Room label text, and it will become editable.

- Change the name to **Lobby**.

- Do the same for the Number label and change it to **101**.

- Continue from left to right and name the rooms as follows:

  152 – Elevator

  102 – Office

  103 – Office

  104 – Storage

  105 – Male

  106 – Female

  151 – Janitorial (see Figure 6.10).

  Actual room names and numbers are not critical.

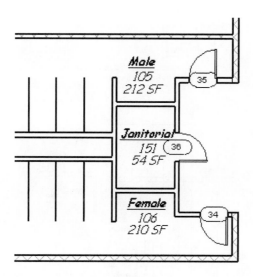

Figure 6.10 *Room tags with room numbers and area*

1.17 Highlight and select any room object in a space where you placed a tag.

- Open its Properties dialogue.

- Note that the Name and Number are the values you just entered.

- Click Cancel.

- Save the file.

EXERCISE 2. CUSTOMIZE VIEW TAGS

Documentation views contain indicators of other standard views, i.e., elevations, sections and callouts. These indicators denote where other views are placed in the model, which portion of the model they show, and the document address of the view (sheet number and detail number). This provides a trail through the document set. You can control and customize the display of view tags in many ways.

SECTION INDICATORS

2.1 Open or continue working in the file from the previous exercise.

2.2 From the Basics Tab of the Design Bar, choose the Section tool.

- In the Type Selector, pick Section: Building Section from the list.

- Place three sections, as shown in Figure 6.11.

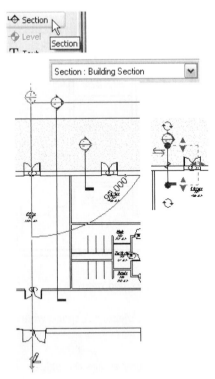

Figure 6.11 *Section indicators—controls allow you to flip the section head, split the line, and cycle through alternate head/tail symbols*

2.3 Select the far-right section indicator.

- Pick the Properties icon from the Options Bar.

- Pick Edit/New in the Element Properties dialogue.

- Pick Duplicate.

- In the Name field, type **Wall Section**, as shown in Figure 6.12.

- Click OK.

- Change the Section Tag value to Detail View 1 and click OK twice.

  The new section will now have an open section head. The Project Browser now shows the section you changed as a Wall Section.

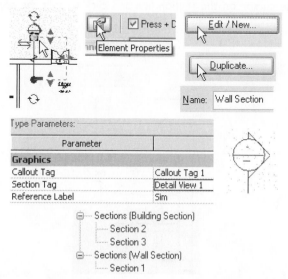

Figure 6.12 *Changing the section head properties for a detail (wall) section*

2.4 You will now have two section types to choose from when adding a section, plus a Detail View, as shown in Figure 6.13.

Detail Views appear in a separate area of the Project Browser.

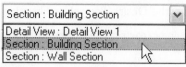

Figure 6.13 *Section type selection*

Many offices differentiate wall sections and building sections with different section heads: thus the solid fill and no fill section heads. In addition, many experienced Revit Architecture users find that using a temporary section to study a model and a different, permanent one for documentation purposes is useful. Some offices develop a temporary section that is graphically different to indicate to other users that it is permissible to move, edit or even delete it without consequences to the documentation set.

2.5 Select the middle section.

- Cycle the head, as shown in Figure 6.14, so that the section shows as a double tail section indicator.
- Select Split Segment from the Options Bar.
- Place the cursor on the section line and click to split the section.
- Drag the lower portion into the restrooms as shown.

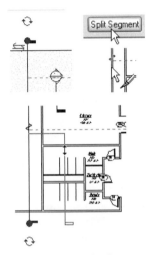

Figure 6.14 *Change a section head symbol and split the section to show a different part of the model*

ELEVATION INDICATORS

2.6 From the View Tab on the Design Bar select Elevation.

- Place an elevation indicator on the bottom of the female toilet room, as shown in Figure 6.15.
- Hit ESC to terminate the tool.

 Tip: Revit Architecture will snap the point of the elevation to the nearest wall. You can also use TAB to cycle through pointer locations before placing the elevation indicator.

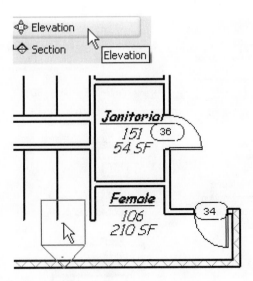

Figure 6.15   *Interior elevation indicator*

You will now create a Schematic Design style elevation marker to go with the Schematic Design room tags you created before.

2.7 From the Menu Bar, pick Settings>View Tags>Elevation Tags.

There are two default types, square and circular.

- Select Duplicate.

- Type **Interior Elevation – SD** in the Name dialogue.

- Click OK.

- Change the Shape and Text Position values, as shown in Figure 6.16.

- Place a check in the Filled value.

- Change the Text Font to Comic Sans.

- Set the Width to **3/8" [10mm]**.

- Click OK twice to close the dialogue box.

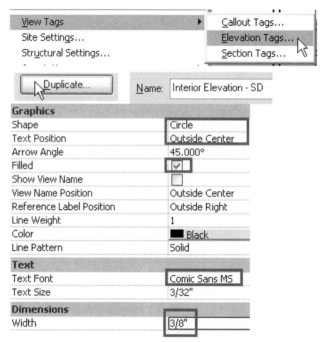

Figure 6.16 *SD elevation tag properties*

Now you apply the new Elevation Tag type to a new Elevation Type.

2.8 Select the square portion of the elevation. Pick the Properties icon from the Options Bar.

- Pick Edit/New from the Element Properties dialogue.

- Choose Duplicate from the Type Properties dialogue.

- Type Interior Elevation – SD in the Name dialogue.

- Click OK.

- Change the Elevation Tag value to Interior Elevation – SD, as shown in Figure 6.17.

- Click OK twice to close the dialogue box.

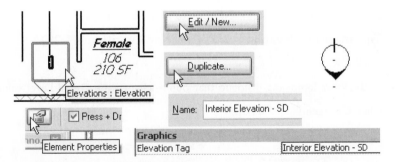

Figure 6.17   *Interior elevation type SD properties*

2.9 Select the elevation marker (now a circle); checkboxes will appear at the quadrants.

- Check the three boxes that are not checked (see Figure 6.18).

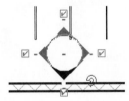

Figure 6.18   *Checkboxes at the quadrants indicate whether an elevation has been generated or not*

Elevation views will be added to your Project Browser (see Figure 6.19).

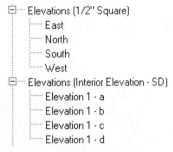

Figure 6.19   *Elevation Views are automatically added when you check the associated checkboxes*

CREATE AN ELEVATION VIEW TEMPLATE

2.10 Double-click on any one of the arrow portions of the interior elevation indicator.

This will open the associated interior elevation.

- Use the View Scale control on the View Control Bar at the bottom of the view window to change the View Scale to 3/8" = 1' – 0" [1:20].

2.11 From the View menu, select View>Create View Template from View.

- Type **Interior Elevation – 3/8" [Interior Elevation – 1:20]**.

- Click OK.

- In the View Templates dialogue that opens, change the Detail Level value to Fine.

- Click OK (see Figure 6.20).

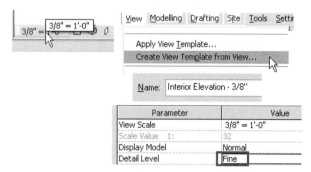

Figure 6.20 *Save view settings for future use*

2.12 CTRL + select the names of the three other new elevations in the Project Browser.

- Right-click and pick Apply View Template.

- In the Select View Template dialogue, select Interior Elevation – 3/8" [Interior Elevation - 1:20].

- Click OK.

  Note that there is an option to apply the template you choose to new views as you create them.

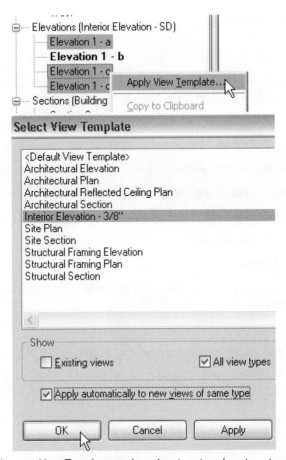

Figure 6.21 *Apply the new View Template to the other interior elevation views*

 Note: In this case, you are using the View Template to simply store the scale and detail level of the Interior Elevations, but this tool can be used to manage any number of visibility settings including linestyles, lineweights and colors. Just about anything you can change regarding view characteristics or visibility can be stored and managed using view templates. It will be well worth your time to explore these settings on this or any other model.

2.13 Save your project file.

EXERCISE 3. CUSTOMIZE AND PLACE OTHER ANNOTATION

TEXT AND NOTES

As you recall, you are hypothetically in Schematic Design, and all of our annotation text so far has used the Comic Sans MS font. You will need to create a note style that matches.

3.1 Open or continue working in the file from the previous exercise.

3.2 Open the 1ST FLOOR – FLOOR PLAN view.

- Zoom to include the washrooms and open area to the right.

3.3 From the Basics Tab of the Design Bar, pick the Text tool.

- Select Text: 3/32" Arial [2.5mm Arial] from the Type Selector pull-down.

- Select the Properties icon.

- Choose Edit/New.

- Click Rename and change the name of the Text Type to **3/32" SD Note [2.5mm SD Note]**.

- Click OK.

- Change the Leader Arrowhead value to Heavy End 1/8" [Heavy End 3mm].

- Change the Text Font to Comic Sans MS and check the Italic checkbox, as shown in Figure 6.22.

- Click OK twice.

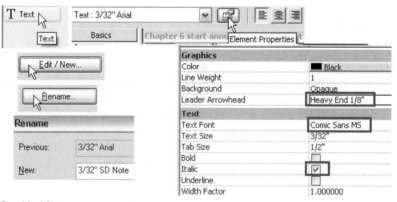

Figure 6.22 *Modifying text properties*

3.4 While still in the text command, pick the Two Segments leader and the Left Justified Text Alignment options from the Options Bar, as shown in Figure 6.23.

Figure 6.23 *Leader type and Text Alignment*

3.5 To start a text leader, pick a spot on the bottom washroom wall, then a point down and to the right of that, and finally select a point directly to the right of the second point.

- Type **8" MASONRY WALL, FULL HEIGHT. PROVIDE FULL FIRE SEALANT**. **[200mm MASONRY WALL, FULL HEIGHT. PROVIDE FULL FIRE SEALANT.]**

- Pick a point outside the text to end the text string.

- Hit ESC twice to get back to the Modify tool.

3.6 Select the text you just entered to bring up the blue grips and drag the right-hand grip to the left to adjust the width of the text box.

There are many ways to move the text you have created to achieve different results.

3.7 First pick the text and put the cursor over the highlight around the text.

- Drag the text box to drag the text and leader as a stationary unit.

- Now select the text and drag the blue four-headed arrow at the top-center of the text box.

- Note that the text moves but the leader arrowhead remains stationary (see Figure 6.24).

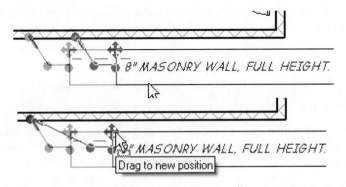

Figure 6.24 *Moving text two ways*

3.8 Continue adding notes, as shown in Figure 6.25.

Revit Architecture automatically snaps the second and endpoints of the leader line to align with the adjacent leaders and notes.

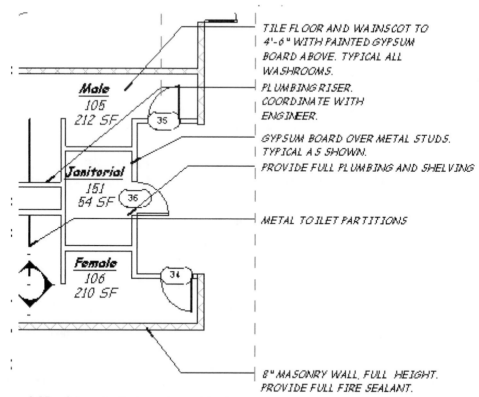

Figure 6.25 *Schematic Design notes and leaders*

PLAN CALLOUTS

For the purpose of linear dimensions, you will use an enlarged plan of the washrooms. You use the Callout tool to create a view of a defined portion of the current view. The outline and number tag of the new view will be visible in the parent view. You can place callouts in plans, elevations and sections.

3.9 From the View Tab on the Design Bar, select the Callout tool.

- Pick a point at the upper-left of the washrooms and pick again at the bottom right.

- Move the callout head to the bottom of the callout lines by selecting the callout and dragging the grip at the intersection of the callout head and the leader to the bottom of the callout, as shown in Figure 6.26.

- Hit ESC once to get back to the Modify command.

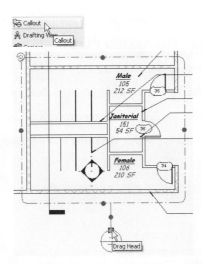

Figure 6.26 *Create a callout and pull the callout head to the bottom of the callout*

3.10 Open the callout by double-clicking the blue callout head.

- In the callout view, type **VP** to open the View Properties dialogue.
- Set the View Name to **ENLARGED WASHROOM PLAN**.
- Clear the Crop Region Visible checkbox, as shown in Figure 6.27.

3.11 Select Edit from the Visibility/Graphic Overrides Parameter (this is the same as typing **VG** from the view).

- Pick the Annotation Categories Tab and clear the Elevations checkbox.
- Click OK twice to see the results.

Identity Data			☆
View Name	ENLARGED WASHROOM PLAN		
Dependency	Independent		
Extents			☆
Crop View	✓		
Crop Region Visible			
Visibility/Graphics Overrides	Edit...		
Model Graphics Style	Hidden Line		

Visibility/Graphic Overrides for Floor Plan: ENLARGED WASHROOM PLAN

Model Categories | Annotation Categories | Imported Categories | Filters

☑ Show annotation categories in this view

Visibility	Projection/Surface Lines	Halftone
☐ Elevations		☐

Figure 6.27 *View Properties settings for the enlarged washroom plan*

3.12 From the Menu Bar select View>Create View Template from View.

- Name the template **ENLARGED PLAN**.

- Choose OK to create the template.

- From the Menu Bar select View>Apply View Template.

- Select the new ENLARGED PLAN template you just created.

- Check the option to apply this template to new views of the same type (see Figure 6.28).

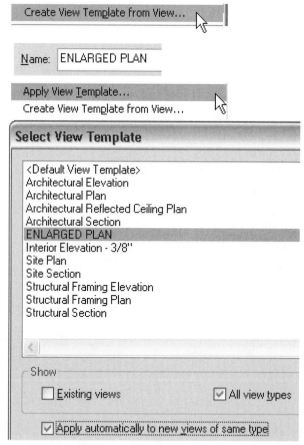

Figure 6.28 *Use the new view settings as a template*

3.13 Use Room Tag from the Drafting Tab of the Design Bar to room tags to the Men's, Women's and Janitor's rooms in this view. See Figure 6.29 for appropriate locations.

- Click Modify to terminate placement.

Notice that the room information is filled in for you. You have previously defined each space as a room; Revit Architecture recognizes it as an object and therefore "reports" the information through the new room tag for this view.

Remember, annotations are view-specific.

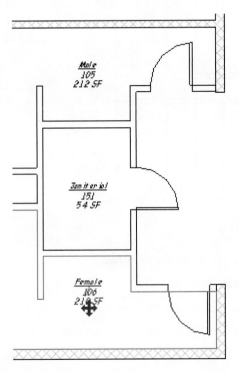

Figure 6.29   *Add room tags*

LINEAR DIMENSIONS

3.14 From the Basics or Drafting Tab on the Design Bar, select the Dimension tool.

- Add dimensions to the plan similar to Figure 6.30.

  You will need to switch between Prefer wall centerlines and Prefer wall faces on the Options Bar during the dimension picks.

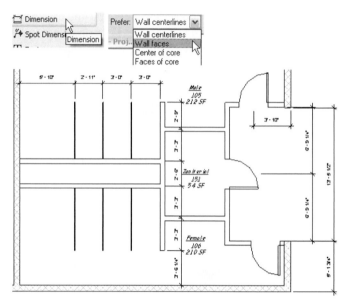

Figure 6.30 *Place dimensions*

3.15 Select Modify to terminate the Dimension tool.

- Select the entire view.

- Pick the Filter tool on the Options Bar.

- Clear all of the categories except Dimensions, as shown in Figure 6.31.

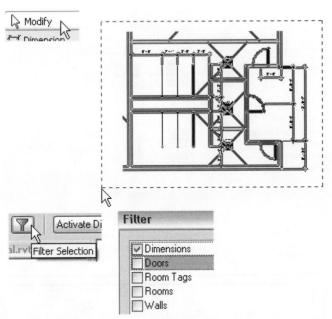

Figure 6.31   *Filter out all categories except the Dimensions*

3.16 From the Options Toolbar, select the Properties icon.

- Select Edit/New.

- Click Rename.

- In the New Name field, name the dimension style **Linear - SD**.

- Click OK.

3.17 Set the Parameters as follows:

- Tick Mark Lineweight: 7

- Centerline Symbol: Centerline [M_Centreline]

- Centerline Pattern: Center

- Text Font Comic Sans MS

  There is no Italic checkbox for dimension text.

- Click OK twice when you are finished to see the results, as shown in Figure 6.32.

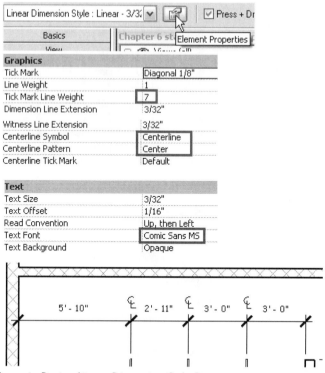

Figure 6.32 *Schematic Design Linear Dimension Style Parameters*

ANGULAR DIMENSIONS

3.18 Open the IST FLOOR – FLOOR PLAN view.

- Zoom to the curtain wall near the doors.

- From the Basics Tab of the Design Bar, select Dimension.

- Pick the Angular icon, as shown in Figure 6.33.

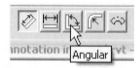

Figure 6.33 *Angular icon*

3.19 Dimension the curtain wall grids, as shown in Figure 6.34.

- Use the Wall Centerlines option to make picking mullions easier.

- Pick Modify from the Basics Design Bar.

- Select the dimension string.
- Click Properties.
- Click Edit/New.
- Click Rename.
- Name the type **Angular – SD**.
- Use all of the same Parameters you used for the SD linear dimensions.
- Click OK twice.

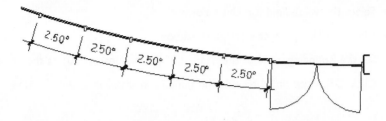

Figure 6.34 *Angular Schematic Design dimensions at curved curtain wall*

Now it is time to dimension the radius of the curtain wall.

3.20 Select the Dimension tool.

- Check the Radial icon.
- Pick the centerline of the curtain wall.

It will pre-highlight with a light dashed line and the Status Bar will read Walls: Curtain Wall: Curtain Wall 1.

- If it does not automatically pre-highlight, tap the TAB key while the cursor is over the center of the wall.
- When the wall centerline does highlight, pick the wall and then pick a location for the dimension line (see Figure 6.35).
- Hit ESC twice to exit the command and put you back in the Modify tool.

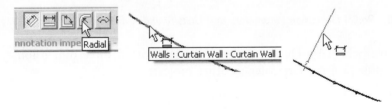

Figure 6.35 *Status Bar feedback and wall centerline highlight*

3.21 Select the radial dimension you just created.

- As before, click Properties.
- Click Edit/New.
- Click Rename.
- Rename the Dimension Type **Radial – SD**.
- Set the Tick Mark to Diagonal 3/64" [Diagonal 3mm].
- Set the Tick Mark Line Weight to 7.
- Set the Text Font to Comic Sans MS.
- Select the value field for Units Format.
- Clear the Use project settings checkbox.
- Set the Rounding value to the nearest 1" [to the nearest 10], as shown in Figure 6.29.
- Click OK three times to exit the dimension dialogue boxes.

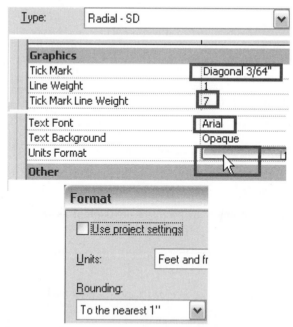

Figure 6.36 *Radial Dimension Parameters and Units/Rounding Format*

You now have Schematic Design dimension styles for all three dimension types, to use anywhere in this project or transfer to other projects.

3.22 Save the project file.

GRIDS

Just like some of these other annotation families you have been modifying, Revit Architecture's grids also hold the parametric controls necessary to change their appearance with a few simple modifications. In this next exercise, you will create a new grid head from scratch.

3.23 Use the File pull-down and select File>New>Family, as shown in Figure 6.37.

The New Family browser will open for you to select a template file.

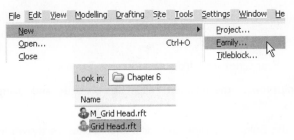

Figure 6.37 *New Family file*

3.24 Navigate to the *Chapter 6/Library* folder.

- Select *Grid Head.rft [M_Grid Head.rft]*.

- Click Open.

  This will open the template file to create any grid head geometry you want. In this case, you will create a Schematic Design head to match our other annotations.

3.25 From the Family Tab (the only one) on the Design Bar, pick the Label tool.

- Make sure that the Center and Middle icons are selected in the Text Alignment section of the Options Bar, as shown in Figure 6.38.

- Place the label centered on the vertical Reference Plane and above the horizontal Reference Plane.

  The label will snap weakly to the Reference Planes. Revit Architecture will ask you for the label name; there is only one Parameter to choose from on a grid, Name.

- Click OK.

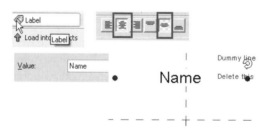

Figure 6.38 *Label Text Alignment*

3.26 Click Modify.

- Choose the Name label.

- Pick the Properties icon from the Options Toolbar.

- Pick Edit/New in the Element Properties dialogue, and change the Font to Comic Sans MS.

- Check the Bold, Underline and Italic checkboxes, as shown in Figure 6.39.

- Click OK twice to see the results.

Figure 6.39 *Label text properties*

The red text and "dummy" line are there for your use in orienting the grid. They are no longer necessary.

3.27 Select and delete them.

- Save the file as ***Grid Head – SD.rfa*** in a location determined by your instructor.

3.28 Select the Load into Projects icon, as shown in Figure 6.40.

If you have more than one project file open, you will see a dialogue with a list of projects from which to choose. The screen will change to the project file.

- Return to the Family file and close it.

- Return to the project file to apply your work on the grid family.

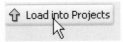

Figure 6.40 *Load grid head into the project*

3.29 Open view IST FLOOR – FLOOR PLAN.

- Zoom to Fit.

- From the Basics Tab on the Design Bar, pick the Grid tool and draw a gridline by selecting a point at the bottom left of the project and another straight above the first.

- The grid head will appear as a circle with the number I in it.

- Click Modify.

3.30 Select the grid.

- Pick the Properties icon from the Options Bar.

- Choose Edit/New.

- Pick Duplicate.

- Name the new type **Grid – SD**.

- Click OK.

- In the Grid Head Value, select Grid Head – SD, as shown in Figure 6.34.

- Click OK twice to see the changes.

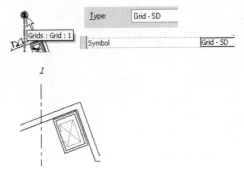

Figure 6.41 *New Grid properties*

3.31 From the Basics Design Bar, select the Grid tool.

- In the Type Options selector, pick Grid – SD, and continue placing vertical grids from left to right on your project.

  Specific locations are not critical. Note that the end points will snap to alignment; in addition, when you drag grid endpoints all of the other aligned grid endpoints will maintain their alignment.

- Choose Modify.

3.32 Select any gridline.

- Pick the Copy tool from the Toolbar.

- Click an arbitrary point in space to establish a start reference, drag the cursor horizontally to the right, type **I [300]**, and hit ENTER.

  You have just copied a gridline I' – 0" [300mm] to the right of the original. The text at the grid head will be overlapped.

- Zoom in Region to an area just around the grid heads.

3.33 Pick the new grid. Click the squiggle grip on the grid line to create an elbow to the right.

The text will move and there will be an angled extension line on the grid line, as shown in Figure 6.42.

- Select the Grid label text to edit it.

- Enter **X.I**, where X is the number of the grid line to the immediate left that you copied.

- Click Modify.

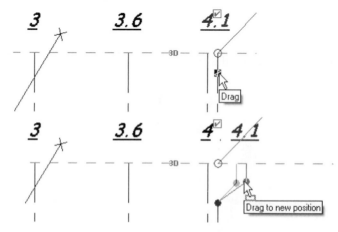

Figure 6.42 *New Schematic Design grids with offset heads*

3.34 Save your file.

EXERCISE 4. TITLEBLOCKS

One of the most common requirements in customizing standard annotation stems from the fact that every firm has an individualized titleblock or sheet frame to incorporate into plan sets. We are going to show you how to create a Schematic Design (horizontal) titleblock easily using an existing titleblock and its labels.

4.1 From the File menu, select File>Open.

- Navigate to the *Chapter 6/Library* folder and open the file *E 34 x 44 Horizontal.rfa* *[A0 metric.rfa]*.

    This is a generic Revit Architecture titleblock that you will use to create our custom Schematic Design titleblock.

4.2 From the File pull-down, select File>Save As and save the file as *34 x 44 Horizontal – SD.rvt [A0 – SD.rfa]*, in a location determined by your instructor.

4.3 Zoom in around the lower-right corner of the sheet.

- While holding down the CTRL key, pick the text and labels circled in Figure 6.43.

Figure 6.43   *Text and labels to be reused in the Schematic Design titleblock*

4.4 Use the Move command to move the selection to the middle of the sheet.

- Repeat this technique to move the Autodesk Revit logo and web address from the top right of the titleblock to the middle of the sheet.

    Your arrangement will look similar to Figure 6.44.

Figure 6.44 *Place all of the necessary labels and text in the middle of the sheet to modify them*

4.5 Delete all of the lines and remaining text from the right side of the sheet.

- Leave the outside perimeter line for size reference.

  You are now going to change the appearance of the remaining text and labels.

4.6 Select the Autodesk Revit logo.

- Pick Properties.

- This is a bitmap image with a height and width.

- Change the Width value to **7" [200]**.

- Click OK (see Figure 6.45).

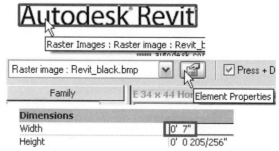

Figure 6.45 *Resize the logo image*

4.7 Select the web address text.

- Change the value in the Type Selector to Text: 1/4" [Text: 5mm].

4.8 Select the Client Name label.

- Click Properties.

- Pick Edit/New.

- Select Duplicate.

- Name the new type **3/4" Comic [18mm Comic]**.

- Click OK.

- Change the Text font to Comic Sans MS.

- Set the Text Size to **3/4" [18mm]**.

- Check Bold, Italic and Underline, as shown in Figure 6.46.

- Click OK twice.

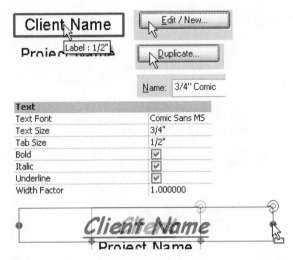

Figure 6.46 *Comic label Parameters for the Client Name, Project Name and Sheet Name labels*

4.9 Drag the right-hand grip well to the right to allow for a long client name.

- Click in open space to clear the selection.

 Note: Remember that the labels represent information that will come from the project and may be longer in some projects than others. If you plan to re-use titleblocks in the future, allow enough space for labels.

4.10 While holding the CTRL key down, select the Project Name and Sheet Name labels.

- Change the value in the Type Selector to Label: 3/4" Comic [Label: 18mm Comic].

- Stretch the labels as in the previous step.

4.11 Edit the Project Number label (not the text) similarly.

- Click Properties.

- Pick Edit/New.

- Select Duplicate.

- Create a **3/8" Comic [10mm Comic]** Label type.

- Set the Text Font to Comic Sans MS.

- Set the Text Size to **3/8" [10mm]**.

- Check the Bold and Italic boxes, as shown in Figure 6.47.

- Click OK Twice.

4.12 Change the type of the Date and Scale labels by selecting them and changing the value in the Type Selector to Label: 3/8" Comic [Label 10mm Comic].

- Stretch the grips on any label that is not long enough to contain its contents in one line.

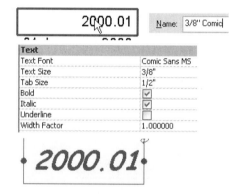

Figure 6.47 *Comic label Parameters used for the Project Number, Sheet Date and Scale labels*

4.13 Repeat the preceding step on the Sheet Number label.

- Click Properties.

- Pick Edit/New.

- Select Duplicate.

- Create a **1" Comic [25mm Comic]** Label Type.

- Set the Text Font to Comic Sans MS.

- Set the Text Size to **1" [25mm]**.

- Check the Italic checkbox. Click OK.

- Set the Horz. Align value to Right and the Vert. Align value to Bottom (see Figure 6.48).

- Click OK.

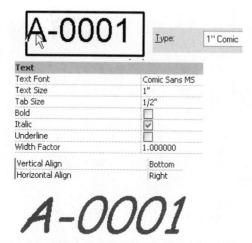

Figure 6.48 *Comic label Parameters used for the sheet number label*

4.14 Edit the vertical Date/Time Stamp label similarly.

- Click Properties.

- Click Edit/New.

- Click Duplicate.

- Create a **1/4" Comic [6mm Comic]** Label Type.

- Set the Font as before.

- Set the Text Size to **1/4" [6mm]**.

- Check the Italic option (see Figure 6.49).

- Click OK Twice.

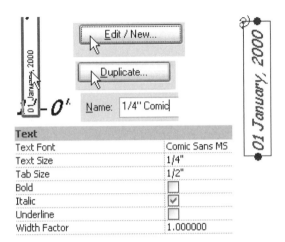

Figure 6.49 *Comic label Parameters used for the vertical Date/Time Stamp date label*

4.15 Select the Project Number text.

- Pick Properties.

- Select Edit/New.

- Pick Rename.

- Rename the type **3/8" Comic [10mm Comic]**.

- Set the Text Font as before.

- Set the Text size to **3/8" [10mm]**.

- Check Italic (see Figure 6.50).

- Click OK twice.

    The Date and Scale text will also display the Type changes you just made.

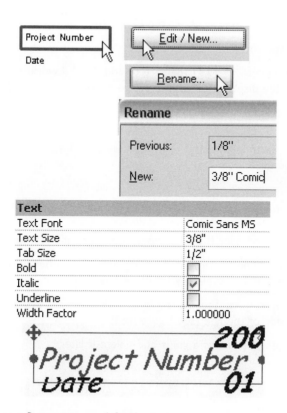

Figure 6.50 *Comic text Parameters used for the text*

4.16 Stretch the three text objects as necessary to make each fit on one line with the text border of an appropriate size.

You now should have a mess in the middle of your sheet—maybe even more congested than what's shown in Figure 6.51. Relax—you will straighten that out right away.

Figure 6.51 *New label and text types at the middle of your sheet—quite a mess!*

4.17 Zoom to Fit (**ZF**) to see your whole sheet.

4.18 From the Options Toolbar, select the Offset tool.

- Set the type to Numerical.

- Set the Offset to **I" [25]** (be sure to add the " if working in imperial units—Revit will alter the display, as shown in Figure 6.52).

- Check the Copy checkbox.

Figure 6.52 *Offset settings*

4.19 Move your cursor over the top of one of the boundary lines.

A green dashed line will appear on one side or the other, indicating where Revit Architecture is going to place your offset line.

- Before selecting, tap the TAB key once to link all four boundary lines and select the interior side of the line (see Figure 6.53).

You will use this line as a guide and delete it later.

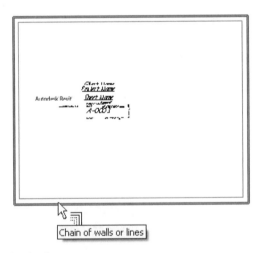

Figure 6.53 *Offset the border lines*

4.20 Change the value of the Offset to **3.5" [80]**.

- Create one line up from the bottom.

This will also be a guideline to be deleted later.

4.21 Click Lines.

- Draw a line from the midpoint of the bottom horizontal boundary, vertically, the full height of the sheet.

- Click Modify.

4.22 Move the text and labels into position, as shown in Figure 6.54.

Location is not too critical. You may notice that Revit Architecture text and labels do not have snap handles per se—eyeballing the text into location is acceptable here. Revit Architecture text and labels do, however, align with each other, so once one is in place the others will align.

Figure 6.54 *Text and label locations for our new titleblock; guidelines are still in place*

Many firms have solid or otherwise wide linework on their titleblocks. The most effective way to re-create this look is by using fills in various shapes and colors.

4.23 From the Design Bar, select the Filled Region tool.

The Design Bar will switch to Sketch mode.

- From the Options Toolbar, select the Rectangles option, as shown in Figure 6.55.

- If you do not want a black border on the edge of your band, make sure you are sketching with <Invisible lines> selected in the Type Selector drop-down list.

Figure 6.55 *Rectangle sketch tool*

You will be creating a long narrow rectangular blue band from the horizontal guideline, the full width of the printable area of the sheet, 3/8" [18mm] high.

4.24 Pick the end point on the guideline above the Autodesk logo.

- Pick another point on the right-hand vertical guideline above the horizontal guideline.

- You will be left with a rectangle that is the width of the guidelines by an ambiguous height.

- Click Modify.

- Select the top line of the rectangle.

- Change the temporary dimension value between the top and bottom of the rectangle to **3/8" [10]**.

4.25 From the Sketch Design Bar, select Region Properties.

- Pick Edit/New.

- Change the Color and Cut Fill Parameters to the values shown in Figure 6.56.

- Click OK twice.

- Click Finish Sketch on the Sketch Design Bar.

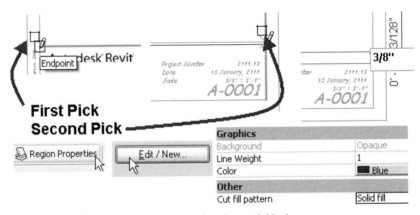

Figure 6.56 *Solid fill Parameters for solid blue band on titleblock*

4.26 Select Label from the Family Design Bar.

- Set the Type to 3/8" Comic [10 mm Comic].

- Select the Center and Bottom alignment icons, as shown in Figure 6.57.

Figure 6.57 *Text Alignment icons for label alignment*

4.27 Pick a point directly below the Client Name label.

Revit Architecture will show you green dashed alignment lines to assist in the placement.

- In the Select Parameter dialogue, select the Project Status Parameter, as shown in Figure 6.58.

- Click OK.

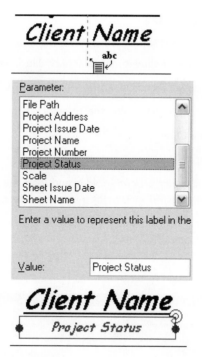

Figure 6.58 *Parameter selection dialogue*

4.28 Hit ESC to terminate the Label tool.

- Select the label you just created.

- Adjust its position if necessary.

4.29 Zoom to Fit.

- Erase all of the guidelines.

- Save the file.

  Congratulations, you have now created a custom titleblock that can be used on any project in the Schematic Design phase (see Figure 6.59).

4.30 Close the file.

Figure 6.59 *Finished Schematic Design titleblock family*

EXERCISE 5. SHEET LAYOUT

Now you'll use that Schematic Design style Titleblock in our project file.

5.1 Open or return to the *Chapter 6 annotations.rvt* file from the previous exercises.

5.2 From the File pull-down menu, select File>Load from Library>Load Family.

- Navigate to the folder that holds the titleblock family you just created.

- Select it and click Open.

5.3 Repeat the last step.

- Navigate to the *Chapter 6/Library* folder and load the generic titleblock family file *34 x 44 Horizontal.rfa [A0 metric.rfa]* into the project.

- You will use both when making the transition from Schematic Design to Design Development annotations.

- See Figure 6.60.

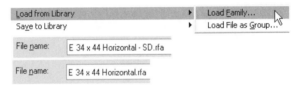

Figure 6.60 *Load two titleblocks*

5.4 Activate the View Tab on the Design Bar.

5.5 Select the Sheet tool.

- Select the 34 x 44 Horizontal – SD [A0 – SD] titleblock from the list.

  Revit Architecture creates and opens a Sheet view named A101 – Unnamed.

5.6 Click on the IST FLOOR – FLOOR PLAN view from the Project Browser.

- Drag it onto the sheet and locate it in the center toward the bottom.

- Drag the ENLARGED WASHROOM PLAN view from the browser to the area in the upper-right side of the sheet.

5.7 Select the Titleblock.

- Pick Properties.

- Change the Parameters, as shown in Figure 6.61.

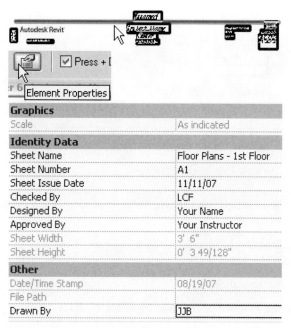

Figure 6.61   *Titleblock Parameters for sheet A1*

 Note: The Drawn By, Checked By, Designed By and Approved By labels are not used in the Schematic Design titleblock but will appear in the next phase.

You will notice that the labels for the floor plan views on this sheet do not use our Schematic Design look with the Comic Sans font. We have preloaded a Schematic Design view title for your use.

5.8 Choose either floor plan view (which will display a red border).

- Select Properties.

- Pick Edit/New.

- Change the Title Parameter to View Title – SD, as shown in Figure 6.62.

- Click OK twice to exit the dialogues.

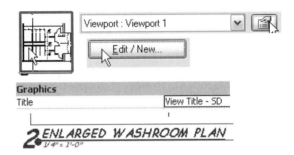

Figure 6.62 *Change the View Title to the SD style*

5.9 From the View Design Bar, use the Sheet tool to create a new sheet using the SD titleblock.

Note that Revit Architecture numbers the new Sheet A2—it picks up the numbering system from the existing sheet.

- Delete the new sheet.

- Change the number of the existing sheet to **A1.1**.

5.10 From the View Design Bar, use the Sheet tool to create six new sheets.

- Modify the sheet names and numbers per Figure 6.63.

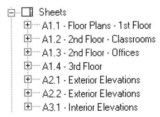

Figure 6.63 *New sheet names and numbers*

5.11 Once the new sheets have been added, rename the four interior elevation views and create an enlarged floor plan view of the elevator area.

5.12 Drag the appropriate views to the new sheets and arrange them as you see fit.

Figure 6.64 shows the sheet names, numbers and views on each of the sheets in our completed set. Save the file.

Tip: Before you move views to sheets, right-click on the name of each floor plan view, right-click and pick Properties, which opens up the View Properties dialogue. Select the Edit button in the Visibility field, and make sure that Elevations is not checked on the Annotation Categories Tab. Do this for each floor plan, and they will take up less space on the sheets. For the Elevation views, clear Topography on the Model Categories Tab.

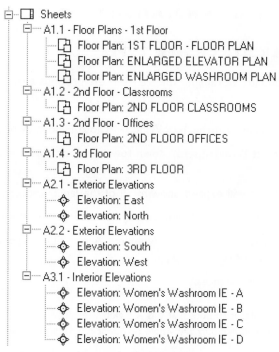

Figure 6.64   *All sheets, names, numbers and subsequent views*

You now have what represents a mini-set of Schematic Design drawings. Take a few minutes to look at each sheet; print them out if possible. This set could be 10 sheets or it could be 100 sheets, but they all look the same and are coordinated with one another. Now it is time for that phase switch mentioned at the beginning of the chapter. This exercise so far has been focused on the Schematic Design look and feel. It is now time to jump to the Design Development phase of our hypothetical project. With that jump comes a completely different set of graphics, tags, notes and sheets—everything needs to look "tighter" (more condensed and composed on the page) and "hard-lined" (not hand-drawn or sketchy in appearance).

It is important to note that you will be using two different techniques to make this change. In exercises in the first part of this chapter, when in the Element Properties dialogue you selected Edit/New and then did one of two things. One option was picking Duplicate (to create a new type) and then modifying Parameters for the new type. The other option was just to modify the Parameters of the existing object. The reason this is important now is that you will be using two different, corresponding techniques to change the appearance of annotation objects. For the items you duplicated (created new annotation sets), you will "swap out" the Schematic Design annotation for the Design Development annotation. For the annotation objects you edited individually, you will re-edit them to change their appearance accordingly.

EXERCISE 6. CHANGE ANNOTATION TYPES

SWAP OUT DUPLICATE OR NEW TYPES

6.1 Open or continue working in the file from the previous exercise.

- Open the view 1ST FLOOR - FLOOR PLAN.

- From the Window menu, pick Window>Close Hidden Windows.

- In the Project Browser, scroll down toward the bottom and pick the + sign next to Families.

  The family list will expand, showing all of the family categories.

- Pick the + sign next to Annotation Symbols to expand this category.

- Expand the Room Tag with the Number - Area – SD category, and right-click on the Room Tag with Area – SD type.

- Pick Select All Instances.

 Note: It is important to note that this is a global selection and will select room tags from all views whether they are visible or not. It is a very powerful tool.

You will notice that all of the room tags in the adjacent view are now selected and highlighted in red.

6.2 From the Type Selector pull-down, change the type to Room Tag, as shown in Figure 6.65.

Notice that the fonts in the room tags are now all Arial, there is a rectangular box around the room number, and the area field is no longer shown.

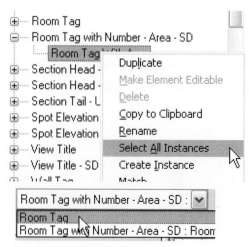

Figure 6.65 *Select all the Room Tags and change the type*

6.3 Repeat the selection step as in Step 5.1 with the Titleblocks, at the top of the list under Families>Annotation Symbols in the Project Browser.

- Select 34 x 44 Horizontal – SD [A0 – SD], right-click to Select All Instances, and use the Type Selector drop-down list to select 34 x 44 Horizontal [A0 metric], as shown in Figure 6.66.

   Revit Architecture will apply this titleblock to all sheets.

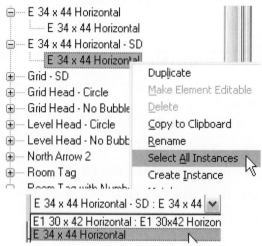

Figure 6.66 *Select All Instances of SD titleblock and change type*

6.4 Open the view ENLARGED WASHROOM PLAN.

- Select one of the dimensions.

- Right-click and choose Select All Instances.

- Change the Type to Linear: 3/32" Arial [Linear: 25mm Arial] (see Figure 6.67).

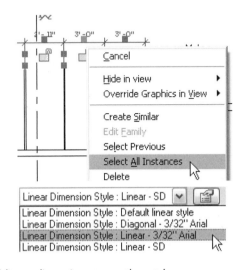

Figure 6.67 *Change all linear dimensions to another style*

6.5 Open the view IST FLOOR – FLOOR PLAN.

- Repeat the changes in Step 5.5 on the angular and radial dimensions.

CHANGING EXISTING PARAMETERS

The second technique you will use to manage the annotation graphics is simply changing the Parameters of the annotation families.

6.6 Select one of the grid lines in the Floor Plan view.

- Pick the Properties icon on the Options Toolbar.

- In the Element Properties dialogue, pick Edit/New.

- Make the Grid Head value Grid Head – Circle, as shown in Figure 6.68.

- Click OK twice.

Figure 6.68   *Change the Grid Head to the original Grid Head – Circle type*

6.7 Select one of the notes you created at the washrooms.

- Pick the Properties icon from the Options Toolbar.

- Select Edit/New.

- Change the Text Font to Arial.

- Clear the Italic checkbox.

- Change the Leader Arrowhead to Arrow Filled 30 Degree, as shown in Figure 6.69.

- Click OK twice.

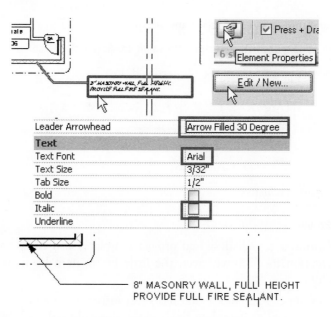

Figure 6.69   *New Parameters for text notes*

6.8 Repeat the previous step on the interior elevation symbol.

- Change the Elevation Tag to 1/2" Square [10mm Square], as shown in Figure 6.70.

    Note that you must select the center of the symbol, as the arrows are separate entities and have properties of their own.

- Click OK twice to exit the dialogue.

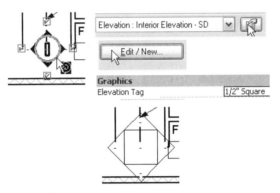

Figure 6.70 *Interior elevation Parameters*

6.9 Open any sheet view and select any viewport.

- Click Properties. Change the Title value to View Title, as shown in Figure 6.71.

- Click OK twice to exit the Properties dialogues.

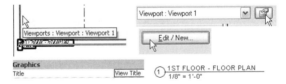

Figure 6.71 *Viewport Parameters*

6.10 Save the file.

In a working environment, well-designed project templates will carry annotation style or appearance information to eliminate even the little bit of item-by-item editing that you just went through.

The replacement of our SD titleblock that has sheet information along its bottom with the generic version that has its project data fields along the right side will make the views on sheets appear a little unbalanced. There is still a little house cleaning to be done to make these sheets completely ready for client presentation, but you have dramatically changed the appearance of an entire set of drawings in a few quick and easy steps.

Keep in mind, these techniques can be used on any annotation family, not just the few you used in this chapter. Imagine that these 7 sheets were a 250-sheet set (not an unusual size for a building of the size of our hypothetical project), and you changed everything about its appearance in 20 minutes. We promised you a few short steps!

SUMMARY

You have just walked through a small portion of Revit Architecture's annotation capabilities. While there is not enough room in this book to cover all of the possibilities and techniques with each annotation tool, what you have just completed will set a solid foundation for creating, editing and managing the other annotation types.

REVIEW QUESTIONS – CHAPTER 5

MULTIPLE CHOICE

1. Changing Type Parameters in a Room Tag family lets you control

   a. The appearance of all instances of that Tag

   b. The size of rooms

   c. The room numbers

   d. The appearance of Interior Elevations for that room

2. Adding a Callout to a Floor Plan

   a. Creates a new sheet in the project

   b. Deletes any dimensions that the callout touches

   c. Creates a separate plan view showing the area enclosed by the callout border

   d. Means the end of civilization as we know it

3. Titleblock Text and Labels

   a. can only be erased from a Titleblock family file, not added

   b. can only appear in one Titleblock family at a time in any single project

   c. can only be left-justified

   d. none of the above

4. Titleblock Labels that are going to read-in values from Project Parameters

    a. have to be created before any text

    b. should be an appropriate size for long names

    c. can be formatted Left, Right or Center

    d. b and c, but not a

5. View Templates

    a. are only for elevation views, not plans or sections

    b. save and apply common view settings

    c. can't be applied after they are changed

    d. can only be used once per file

TRUE/FALSE

6. True _ False _ You can only adjust a grid line header offset when you first create the grid line.

7. True _ False _ Elevation indicators can be square or circular.

8. True _ False _ Dimension Styles for Linear, Angular and Radial dimensions are edited separately.

9. True _ False _ Revit Architecture will align text leaders to model objects, but not to other leaders.

10. True _ False _ You create a room when you place a room tag into a space whose boundary consists of either three or more room-bounding walls or three or more room separation lines.

 Answers will be found on the CD.

CHAPTER 7

Schedules

INTRODUCTION

It has been said that one of the most tedious and unrewarding tasks in an Architectural/ Engineering/Construction firm is compiling, counting and organizing schedules. Whether a schedule lists doors, windows, vents, parking stalls, sheets or anything else, time spent devising and populating the schedule is much better spent elsewhere.

If you are a working designer, does the thought that "today I am going to count, categorize and organize all of the windows on this project" sound depressingly familiar? If you are a student, does the prospect of spending your first years at work preparing lists from plan pages feel like the right preparation for your landmark design?

Computers are unambiguously excellent at these tasks. The developers of Revit Architecture understood this when they set out to develop a software package for the AEC industry. One of the founding principles of Revit Architecture is to let the computer handle the tedious, trivial tasks, and let the designer design. The Scheduling module is the perfect example of this concept. Because Revit Architecture is a central database of building information, scheduling is quick, accurate and simple. The fact that you can create an accurate custom schedule in a matter of minutes means you can spend more time designing and less worrying about the count and sizes of your windows. For that matter, you can create several custom schedules and provide more specific information for better communication, resulting in less confusion in the field. Virtually every object in Revit Architecture can be scheduled.

The following exercises will take you through a variety of schedule types and illustrate several techniques in creating, editing and managing your schedules.

OBJECTIVES

- Understanding the Schedule Properties Dialogues
- Create, modify and manage a single category schedule
- Create, modify and manage a multiple category schedule

REVIT ARCHITECTURE COMMANDS AND SKILLS

Create a Schedule

Add and remove fields

Create new fields

Work with shared and project parameters

Sort Schedule Rows

Create and modify headers and footers

Group Schedule rows and columns

Format Schedule font, alignments and orientations

Manage Schedule appearance

Create and modify a multi-category Schedule

Scheduling is one of the strongest features of Revit Architecture. You can add, modify, remove and change any or all components in the model and Revit Architecture tracks them, no matter what. Revit Architecture's scheduling abilities will report location, size, number or any other parameter associated with virtually any building component.

OVERVIEW OF THE SCHEDULE DIALOGUE BOX

The power of scheduling is controlled by the Schedule Properties dialogue box with its five critical tabs. Because all schedules are controlled by these tabs, it is imperative that you understand what each tab contains and the functions each controls. This chapter will start out by outlining these tabs, and follow with the exercises.

You can examine the Schedule Properties tabs we are about to discuss in any Revit Architecture project file by picking View>New>Schedule/Quantities from the Menu Bar, or picking Schedule/Quantities from the View Tab of the Design Bar. Click OK in the New Schedule dialogue to open the Schedule Properties dialogue, as shown in the following illustrations.

FIELDS TAB

The Fields Tab controls the fields that will be in your schedule (see Figure 7.1).

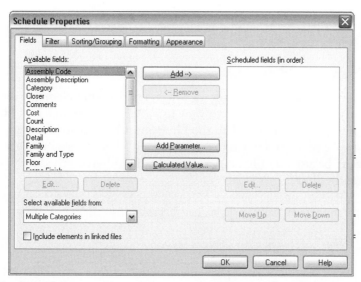

Figure 7.1 *The Fields Tab*

The Available fields area contains a list of the default Family Parameters that can be included in your schedule. This list will be different for each schedule type. The schedule type shown in Figure 7.1 is multi-category. The Schedule Type is selected in the New Schedule dialogue. The Scheduled fields area contains the list of parameters (in order) that will be included in your schedule. The top of the list will be the far-left column, with the next on the list being the next column to the right, and so on.

The Add button adds the highlighted parameter in the Available fields list to the Scheduled fields area, and the Remove button removes the parameter from the Scheduled fields area.

The Add Parameter button allows you to create a custom field, whether it is a project parameter or shared parameter. You will look at these two parameter types later.

The Calculated Value button creates a field whose value is calculated from a formula based on other fields in the schedule. For instance, a Width parameter value could be calculated as a specified fraction or multiple of the Length parameter.

The Edit Field buttons—note that there are two—allow you to edit user-created parameters, and the Delete buttons—also two—allow you to delete a user-created parameter.

The Move Up and Move Down buttons move the highlighted parameter up or down the list and therefore left and right in the schedule.

There is a filter named Select available fields from in the lower-left corner of the dialogue. This is a drop-down list of schedule categories related to the one chosen. When a category is selected in this list, the available fields list changes to show fields from the related category, such as Finishes or Occupancy for Rooms, or From Room and To Room for Doors.

THE FILTER TAB

The Filter Tab shown in Figure 7.2 allows you to restrict what elements display in simple or multi-category schedules (and also to view lists, drawing lists and note blocks). You can set up to four filters on this tab, and all the filters must be satisfied for elements to display. You can use displayed or hidden schedule fields as filters. Certain fields—mostly Type parameters—can't be used as filters. An example of a filtered schedule would be a door schedule filtered by floor.

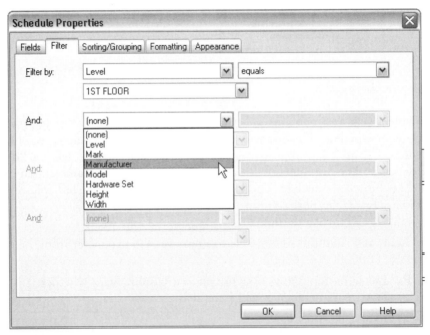

Figure 7.2 *The Filter Tab*

THE SORTING/GROUPING TAB

The Sorting/Grouping Tab shown in Figure 7.3 is a very powerful tab that works well with Revit Architecture's underlying ease of use. There are some concepts that you will have to understand before jumping in. This tab, simply put, controls the order in which the information will be displayed—sorting. An example of this in a door schedule might be sorting by door number. Revit Architecture will display all of the doors in numerical order by door number (which is most often keyed to the room number).

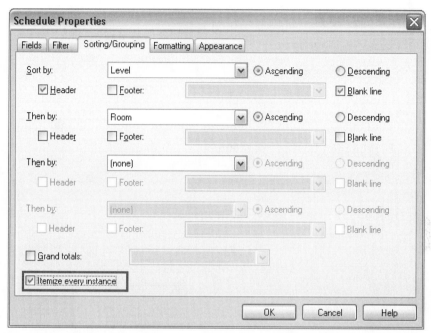

Figure 7.3 *The Sorting/Grouping Tab*

Grouping allows you to group items together. In the door schedule example, you might want to group doors by level or by a particular zone in a building. By virtue of this you could tell Revit Architecture to group the doors by zone and then sort them by door number. To activate grouping, you will need to check the Header or Footer checkbox. The Grand totals checkbox will provide a grand total at the bottom of your schedule, where it applies. The Itemize every instance checkbox allows you to toggle between a list of every item (doors, in our example) and a list that might just show the type and a count of each.

THE FORMATTING TAB

The Formatting Tab shown in Figure 7.4 controls the formatting of the individual columns. There is a field to change characteristics of the displayed heading—Heading, Heading orientation and Alignment. In addition, you can format the units shown for numerical fields using the Field Format button, and have Revit Architecture calculate the totals with a click of the Calculate totals checkbox. The Hidden field checkbox allows you to hide a column. This is useful if you want to group items by field (level, for instance), but don't want to repeat the information in the schedule.

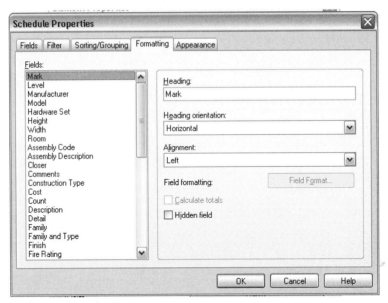

Figure 7.4 *The Formatting Tab*

THE APPEARANCE TAB

The Appearance Tab shown in Figure 7.5 controls the overall graphic appearance of the schedule. The Header font controls manage the font type, its size and whether it is bold, italic or both. This will affect all headers in the schedule. The Body font controls manage the appearance of the remaining fonts. You have the ability to display gridlines, a title and column headers or not, by using the checkboxes at the bottom of the dialogue box. You will use a majority of these settings in the exercises to follow.

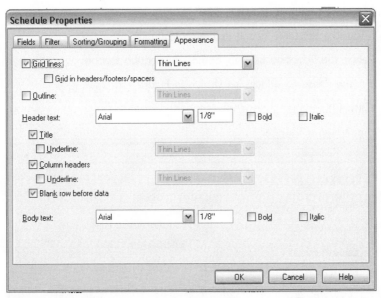

Figure 7.5 *The Appearance Tab*

As stated earlier, Revit Architecture enables you to schedule virtually any object or combination of objects in your project. For the purposes of the next exercise, we will demonstrate the schedule features using a door schedule. You will begin with the basics and add complexity as you go.

EXERCISE I. CREATE A SCHEDULE FOR DOORS

1.1 Launch Revit Architecture and open *Chapter 7 start imperial.rvt [Chapter 7 start metric.rvt]*.

This file is the continuation of the exercise file from Chapter 6 with a few modifications to door tags and numbers.

1.2 Save the file as ***Chapter 7 schedules.rvt*** in a location determined by your instructor.

1.3 From the View Tab on the Design Bar, select the Schedules/Quantities tool, as shown in Figure 7.6.

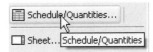

Figure 7.6 *Schedule/Quantities tool*

1.4 In the New Schedule dialogue, select Doors from the Category list.

- Leave the Schedule Name as Door Schedule.

- Select the Schedule building components radio button.

- Set the Phase to Offices, as shown in Figure 7.7.

- Click OK.

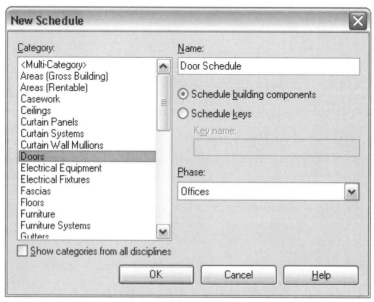

Figure 7.7 *Starting a new Door Schedule for the Offices Phase*

1.5 In the Schedule Properties dialogue, while holding down the CTRL key, select Height, Mark, Thickness and Width, and click the Add button.

1.6 In the Selected fields list, highlight Mark and click Move Up until Mark is at the top of the list.

- Using the Move Up and Move Down buttons, arrange the parameters to match what you see in Figure 7.8.

- Click OK.

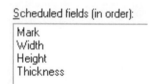

Figure 7.8 *Arranging added fields*

A Schedule view will open. You will now have a list of all of the doors in this building project. Please notice that it is in random order, and that there are some doors labeled using a room number and a sequential letter, and others that are just numbers. This is intentional, to demonstrate the fact that Revit Architecture schedules recognize phases.

1.7 In the Schedule view, right-click and select View Properties to bring up the Element Properties dialogue.

- In the Phase Filter pulldown, select Show Previous + New.

  The Phase parameter was already set to Offices, as shown in Figure 7.9.

- Click OK to see the results.

  Your schedule should now include only doors that are numbered by room number, plus a sequential letter designation, but the numbering is still in random order.

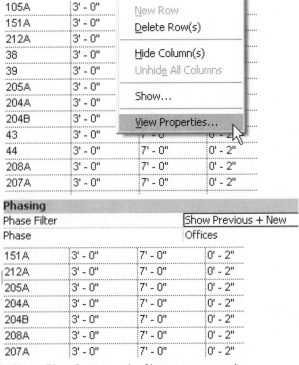

Figure 7.9 *Schedule Phase Filter Settings—the Phase was set earlier*

1.8 In the Schedule view, right-click and select View Properties.

- In the Element Properties dialogue, select the Edit button for the Sorting/Grouping value.

  The Schedule Properties dialogue opens to the Sorting/Grouping Tab.

- In the Sort by pull-down, select Mark and click OK twice.

You now have a basic door schedule that represents all of the doors in the office phase, in ascending numeric order from top to bottom (see Figure 7.10). Let's now really move.

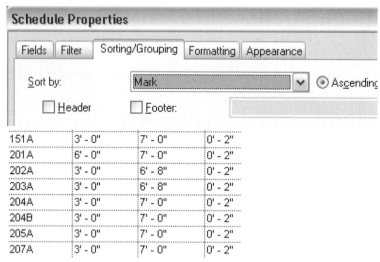

Figure 7.10 *Sort by Mark*

1.9 Right-click in the view again and select View Properties.

- Click Edit for Fields to open the Schedule Properties dialogue.

- Set the properties for Fields, as shown in Figure 7.11.

You will add five Door fields and two Room fields. Room fields can be for the To Room or From Room (i.e., the room the door swings toward or away from).

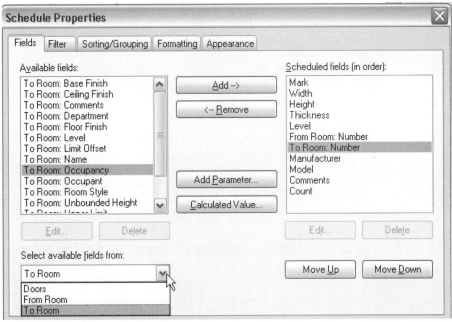

Figure 7.11   *Fields Settings—Add and move the additional fields, and use available Room fields*

- Set the properties for Sorting/Grouping, as shown in Figure 7.12.

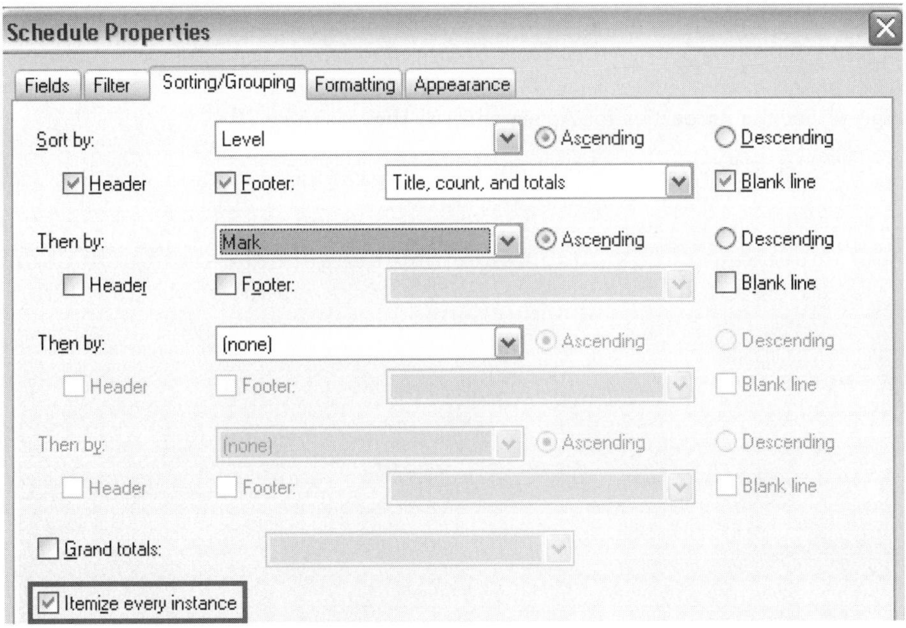

Figure 7.12   *Sorting/Grouping Settings—Sort by Level, check Header and Footer, then Sort by Mark*

- Set the properties for Formatting, as shown in Figure 7.13.

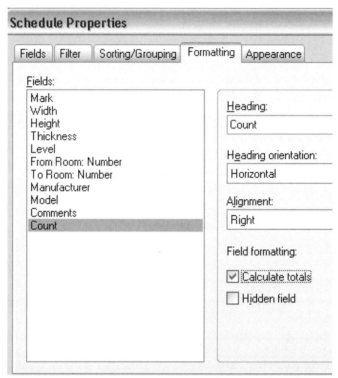

Figure 7.13 *Formatting Settings—set Count to Align at Right, and select Calculate totals*

- Set the properties for Appearance, as shown Figure 7.14.

  Metric Body text will be 1mm.

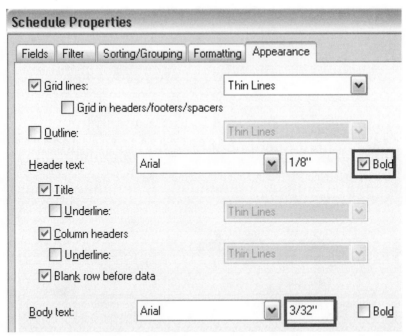

Figure 7.14 *Appearance Settings—set Header font to Bold and Body font to 3/32" [1mm]*

- Click OK twice when the adjustments are complete.

  You will notice that the Appearance Tab does not affect the font appearance in the view. Revit Architecture only displays this when the schedule is placed on a sheet. In addition, you will see that the level information is redundant, as it is already in the header and in each row.

1.10 Right-click and select View Properties.

- Select Edit in the Formatting parameter.

- In the Formatting Tab, select Level.

- Check the Hidden field checkbox.

- Click OK twice (see Figure 7.15).

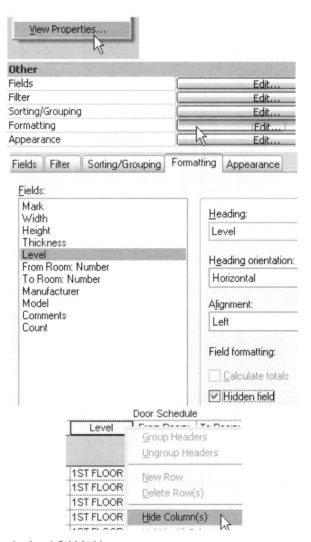

Figure 7.15 *Make the Level field hidden*

 Tip: You may also right-click over the top of any of the cells in the Level column and select Hide Column(s).

1.11 From the View Tab on the Design Bar, select Sheet.

- Select the 36x48 Horizontal – DDCD [A0 metric horizontal – DDCD] titleblock.

- Click OK.

Revit Architecture will create a new sheet view and open it.

- Drag the Door Schedule view from the Project Browser onto the new sheet.

Notice that the columns are all of equal width. There will be triangles at the top and a tilde (squiggle) line at the mid-point of the far right-hand column, as shown in Figure 7.16.

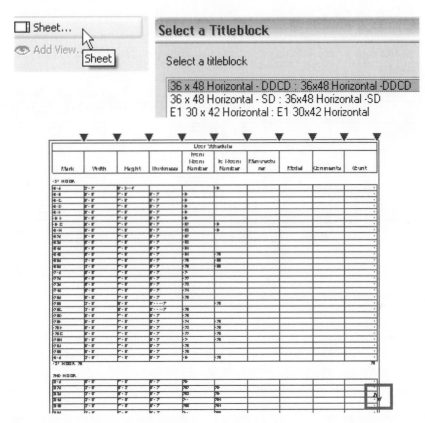

Figure 7.16   *Schedule display control on a sheet—triangles control column width and a squiggle symbol splits the schedule*

1.12 Drag the triangular grips left and right to set the column widths.

  Tip: It is best to set widths starting with the columns at the left and move right.

There will be occasions when you will want to display the same schedule but separate the grouped areas in different locations on the sheet. You will separate level 1 from levels 2 and 3.

1.13 Pick the squiggle symbol at the right of the schedule; this will split the schedule into two individual views.

- Use the Move grip (double-headed arrow) in the middle of the bottom schedule to drag the lower view to the left of the upper portion.

- Drag the grip at the bottom center of the right-hand view up, until it is set at the gap between levels 1 and 2.

  The tops of the schedules will snap to alignment (see Figure 7.17).

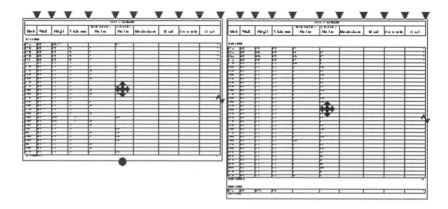

Figure 7.17   *The schedule has been split*

 Note: These adjustments require you to be mindful of doors as they are added and deleted. As the level 1 schedule grows in length it will begin to populate on the level 2 side, and you will need to stretch the grip to adjust it.

 Note: One of the advantages to having a schedule is that information can be viewed and edited in a format conducive to annotative information on a broad brush scale.

1.14 Return to the Door Schedule view.

- Select the cell under Manufacturer in the door 101B row, and type in **Judson Door and Window Co.**

  This change will be applied to all doors of this type—a notice will appear.

- Click OK.

1.15 Repeat the preceding step for doors 105A – **Fox and Sons**, 125B – **Delmar Door Corp.**, and 204A – **Lucas Woodworks**.

- Use the same technique to fill out model information.

- Fill in individual comments as you see fit, as shown in Figure 7.18.

  The column widths may not display all of the information you have input.

1.16 Move the cursor over the top of one of the vertical column lines to get a double-headed arrow (see Figure 7.18).

- Drag the column edge to the appropriate width.

  Note that this does not affect the schedule display on the sheet.

2ND FLOOR								
201A	6' - 0"	7' - 0"	0' - 2"		201	Judson Door and Window Co.	7284-2-ABC	Verify Clearence
202A	3' - 0"	6' - 8"	0' - 2"	202	201	Western Door and Window	1236-KJ-RW	Under cut 1"
203A	3' - 0"	6' - 8"	0' - 2"	203	201	Western Door and Window	1236-KJ-RW	
204A	3' - 0"	7' - 0"	0' - 2"	211	204	Lucas Woodworks	CDW-3672-PSS	Paint Red

Figure 7.18 *Schedule—manufacturer, model and comments*

1.17 Open the sheet view that you placed the schedule on and make sure that the new information is showing correctly—a single line of text per schedule row.

- Use the triangulated grips to adjust the column widths if necessary.

  It is now time to add key card reader information to the schedule, but it is not on the list of parameters for this schedule.

1.18 Right-click on the schedule in the sheet view and select Edit Schedule.

The Door Schedule view will open.

- Right-click and click View Properties.

- Select the Edit button for Fields.

- Select Add Parameter (see Figure 7.19).

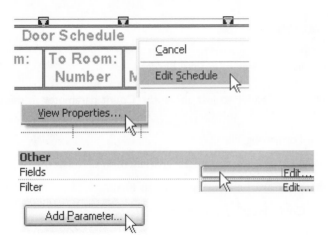

Figure 7.19 *Prepare to customize the schedule fields*

ADD A PARAMETER, CHANGE THE FORMAT

1.19 In the Name field of the Parameter Properties dialogue, type **Key Card Reader**.

- Set the Type to Yes/No.

- Select the Instance radio button, as shown in Figure 7.20.

  Doors of the same type will have different Key Card Reader values.

- Click OK.

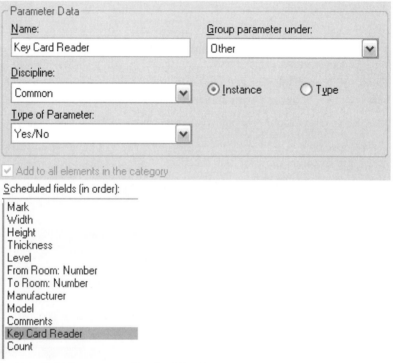

Figure 7.20 *New Field Parameter—Key Card Reader*

1.20 Move Key Card Reader to just above the Count parameter in the order of scheduled fields.

1.21 Go to the Formatting Tab.

- Set the Key Card Reader Heading orientation to Vertical.

- Set the Key Card Reader Alignment to Right.

- Set the Count Heading orientation to Vertical.

- Click OK twice.

- Check the boxes in the new Key Card Reader column for various doors.

1.22 Open the sheet with the Door Schedule on it.

- Adjust the column widths for Key Card and Count to take advantage of the vertical header alignment, as shown in Figure 7.21.

    Note that the doors for which you checked the Key Card Reader Yes/No boxes show Yes in the schedule and others are blank.

Manufacturer	Model	Comments	Key Card Reader	Count
Judson Door and Window	72832-ABC	Verify Clearance	Yes	1
Western Door and Window	1236_KJ_RW	Under cut 1"	Yes	1
Western Door and Window	1236_KJ_RW		Yes	1
Lucas Woodworks	CDW-3672-PSS	Paint Red	Yes	1
Lucas Woodworks	CDW-3672-PSS			1
Lucas Woodworks	CDW-3672-PSS		Yes	1

Figure 7.21 *Vertical heading orientation*

 Note: Traditionally, architects provided a door schedule once, in one format, and it was up to the contractor and door manufacturers to sort through this information to get the information they needed for their particular use. Because Revit Architecture is a centralized database, information is always up-to-date; the architect using Revit Architecture has the ability to group, sort and provide it in a concise manner. These next few steps fall under the "Revit Architecture Magic" category, allowing you to manipulate the same data and provide it in a completely different manner for different uses.

EXERCISE 2. CREATE A SECOND DOOR SCHEDULE—MORE FORMATTING CHANGES

2.1 From the Project Browser, right-click on the Door Schedule and select Duplicate.

This will make a view called Copy of Door Schedule the active view.

2.2 In the new view, right-click, and select View Properties.

- Change the View Name to **Door Schedule by Mfr and Type**.

- Click Edit in the Fields Section.

- Use Add and Remove to set the Fields, as shown in Figure 7.22.

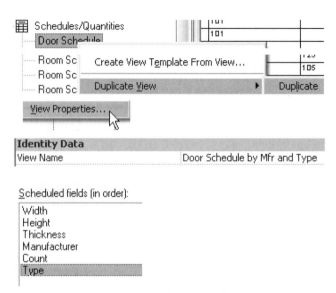

Figure 7.22 *View Properties—name the new schedule and set the fields*

- Set the Sorting and Grouping Tab, as shown in Figure 7.23.

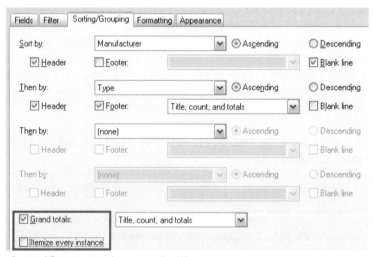

Figure 7.23 *Sorting/Grouping in the new schedule*

2.3 From the Formatting Tab, select Count.

- Check the Hidden field checkbox.

- From the Appearance Tab, clear the Display grid lines checkbox, as shown in Figure 7.24.

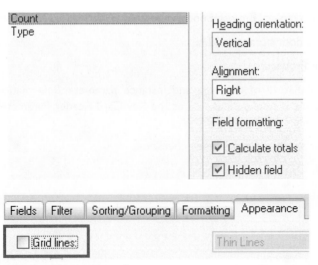

Figure 7.24 *Formatting and Appearance changes*

- Click OK twice.

2.4 Open the sheet view with the first door schedule on it.

- Drag the new Door Schedule by Mfr and Type onto the schedule sheet.

  Note that it appears markedly different from the first schedule.

- Drag the right-hand part of the schedule over the left-hand part to combine them into one, as shown in Figure 7.25.

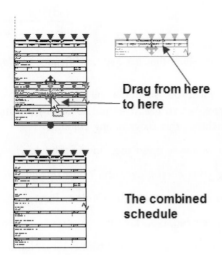

Figure 7.25 *Place and condense the schedule*

2.5 From the Project Browser, open view 2ND FLOOR OFFICES – FLOOR PLAN.

- Select door number 212A.

- Click Properties.

  Note that all of the type and instance parameter information you input on the schedule is displayed, as well as the Key Card Reader Parameter you defined in the schedule (see Figure 7.26).

- Clear the Key Card Reader checkbox.

- Click OK.

Parameter	Value
Head	
Jamb	
Muntin	
Detail	
Hour/Min	
Label	No
Glazing	
Louver	
Closer	No
Panic Hardware	No
Hardware Set	43
Threshhold Material	None
Material	Paint
Frame Finish	
Floor	2nd Floor
Key Card Reader	

Figure 7.26 *Door Parameters—note that the Key Card Reader now appears*

On many occasions, your schedule will have several columns that are related to each other. Many firms like to call attention to this by putting a header above the related columns. You achieve this by grouping columns.

2.6 From the Project Browser, open the Door Schedule view.

- Click and drag your cursor over the Width, Height and Thickness column heads.

- Select the Group button from the Options Bar.

2.7 In the cell that is created, type **Door**.

- Repeat for To Room: Number and From Room: Number, using **Room** for the group header.

2.8 Click in the To Room: Number heading cell and enter **To**.

- Do the same for From Room: Number, entering in **From** (see Figure 7.27).

- Click OK.

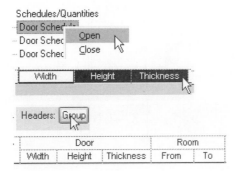

Figure 7.27 *Grouping column headers*

This can also be accomplished in the Schedule (View Properties) dialogue, as shown in Figure 7.28.

Figure 7.28 *Changing the display name on a parameter*

2.9 Open the schedule sheet to see the results.

- Adjust column widths to take advantage of the formatting changes.

2.10 From the Project Browser, open the Door Schedule by Mfr and Type that you just created.

- Modify the width on the 36" x 84" [0915 x 2134] doors supplied by Lucas Woodworks to **3' - 6" [1050]**.

All of the doors of that type change in both door schedules.

2.11 From the Project Browser, open 2ND FLOOR – FLOOR PLAN.

Note that the adjusted door widths of the doors for rooms 212–234 are now represented graphically in the plan as well (see Figure 7.29). The dimension has been added in the illustration.

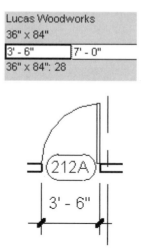

Figure 7.29 *Change one field to edit all doors of that type*

    2.12 Save the file.

MULTI-CATEGORY SCHEDULES

Most schedules you produce will be similar to the schedules you have just completed: door, window, wall, curtain panel, furniture and so on. These types of schedules report information on a single category type. On occasion you will need to schedule components from different categories that might have the same parameter. In addition to single category schedules, Revit Architecture allows you to schedule components from different categories, known as Multi-Category Schedules. One example that comes to mind is the electrical requirements for particular components or building parts. Any number of building components have electrical requirements and you may be asked or, better yet, *paid* to assimilate this information. The following exercise is designed to expose you to the process of creating a multi-category schedule.

To enable the multi-category schedule functionality, the categories of the components must have a common parameter. Revit Architecture calls these parameters "shared parameters." Shared parameters are parameters that you can share with multiple Autodesk Revit Architecture families and projects. You define them once and save them to a shared parameter file. You can use shared parameters in multi- and single-category schedules, and to tag elements.

There are a few key concepts to understand before using multi-category schedules. Multi-category schedules consider *all* components within the model and return the desired field information. Of course, a schedule on all of the components in a model is of little use. When you set up a multi-category schedule, you will make use of the Filter Tab in Schedule Properties. This tab allows you to filter out all of the components that do not contain the filter parameter(s), so that you schedule only the components that do.

EXERCISE 3. CREATING A SHARED PARAMETER FILE

3.1 Open or continue working with the file from the previous exercise.

3.2 From the File pull-down menu, select Shared Parameters.

3.3 In the Edit Shared Parameters dialogue, select Create.

- Enter **Chapter 7 – Shared Parameters** in the Name field.

- Save it in a location as indicated by your instructor.

3.4 From the Groups area, select New.

- Enter **Electrical** in the Name field, as shown in Figure 7.30.

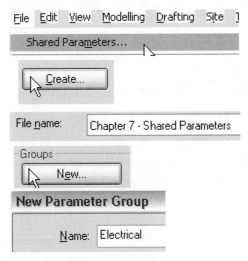

Figure 7.30 *New Shared Parameter Group—Electrical*

3.5 From the Parameters area of the dialogue, select New.

- Enter **Amps** in the Name field.

- From the Type pull-down list, select Number.

- Click OK (see Figure 7.31).

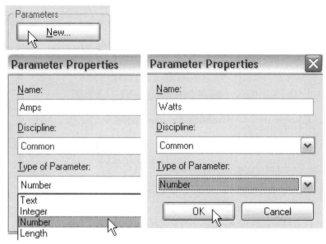

Figure 7.31 *Parameter Properties—it's important to specify the type here, as it is not editable afterward*

3.6 Repeat the last step adding **Watts** (also a Number) to the Electrical group.

- Click OK twice.

 Note: You have now created two new parameters that will be available within this project, as well as other projects you create. This is a great way to create custom parameters for your company that you know will be maintained as standards throughout your projects. You will now need to indicate which categories have the parameters.

3.7 From the Settings menu, select Project Parameters.

- Click Add.

- Select the Shared Parameter radio button.

- Click Select.

- Highlight Amps.

- Click OK.

3.8 In the Parameter Properties dialogue, select the Type radio button.

- Select Electrical in the Group Parameter Under drop-down list.

- Check the following categories: Furniture Systems, Lighting Fixtures, Mechanical Equipment and Specialty Equipment (see Figure 7.32).

- Click OK.

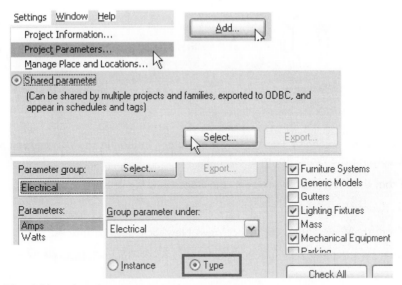

Figure 7.32 *Adding shared paramters to the project*

You have now added a parameter to the categories checked that will be included in all components in those categories.

 Note: Selecting Instance or Type values is only available when you add the parameter to the project.

3.9 Repeat the preceding steps and add Watts to the project parameters.

- Remember to select the Type radio button.

- Click OK twice to exit the dialogues.

3.10 From the Project Browser, open the 2ND FLOOR OFFICES– FLOOR PLAN.

- Pick one of the Workstation Cubicles.

- Click Properties.

- Select Edit/New.

- Enter a numerical value for Amps and Watts.

  The actual numbers don't matter in this case (see Figure 7.33).

- Click OK twice.

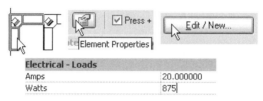

Figure 7.33 *Specify values for the new parameters*

3.11 Repeat the preceding step for the Computer Monitor in the Workstation and the Copier in the Work Room area.

3.12 From the Project Browser, open the Roof view.

- Repeat Step 3.10 on the Air Conditioner and Boiler.

3.13 From the View Design Bar, select View>New>Schedule/Quantities.

- From the New Schedule dialogue box, choose <Multi-Category>, as shown in Figure 7.34.

- Click OK.

Figure 7.34 *Create a new Multi-Category Schedule*

3.14 Use the Fields Tab to add and order fields for the new schedule, as shown in Figure 7.35: Family, Type, Room, Count, Amps and Watts.

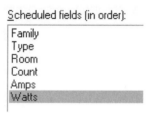

Figure 7.35 *Multi-Category fields*

3.15 On the Filter Tab, select Amps from the drop-down list, as shown in Figure 7.36.

Figure 7.36 *The Filter Tab*

Once a parameter is selected, the right-hand drop-down list becomes active, so that you can further filter against the parameter value: equals/does not equal, greater than/less than and so forth.

- Don't make any changes to that option.

3.16 On the Sorting/Grouping Tab, Sort by Family, as shown in Figure 7.37.

- Clear the Itemize every instance checkbox.

Figure 7.37 *Sorting by Family*

3.17 On the Format Tab, set the Count, Amps and Watts with Right Alignment.

- Check the Calculate totals checkbox (see Figure 7.38).
- Click OK.

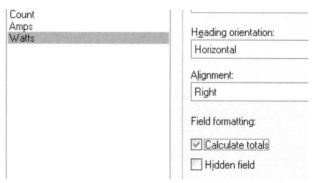

Figure 7.38 *Formatting*

There you have it—a schedule showing all of the Amperage and Watts for specified equipment in the entire building.

3.18 If you do not see the Copier, Monitor and Workstation in your schedule, right-click and select View Properties, then set the Phase for the schedule to Offices (see Figure 7.39).

The view that was active (on the screen) controls the initial Phase setting for any Schedule that you create.

3.19 Save the file.

View Properties...

Phasing	
Phase Filter	Show Complete
Phase	Offices

		Multi-Category Schedule			
Family	Room	Type	Count	Amps	Watts
Air Conditioner-Outside		37" x 28" x 2	24	50	56400
Boiler		21" x 32" x 3	9	65	22500
Copier-Floor		Copier-Floor	5	30	6250
Monitor		17" Monitor	25	20	16250
Work Station Cubicle		96" x 96"	44	20	38500

Figure 7.39 *The new multi-category schedule tracks electrical requirements for different types of equipment*

Note: Keep in mind that this technique can be applied to just about any categories. You can customize the Filter by creating a new Shared Parameter just for this use. While you have covered considerable ground regarding schedules, you have barely scratched the surface of Revit Architecture's potential. Continue to explore new uses and combinations; you will never fail to amaze yourself with the power of schedules.

SUMMARY

Congratulations! You now have the base knowledge to create virtually any schedule type imaginable. These exercises have exposed you to the tools you need to create any schedule type and modify it to suit any customized requirements.

Keep in mind, a schedule is just another view into your project data and there is no need to create schedules with "blinders on"— in other words, like the non-linked, non-smart schedules CAD software creates. You have a new tool, so use it effectively. Remember that different people need the same information in different ways. You might have an overall master door schedule and a separate schedule in which you group and total all the doors by hardware set for the hardware installer. The same data can be shown in different ways for different uses (see *Sheet A9.91* in your exercise file).

In addition, you can display graphical information in coordination with your schedule. For instance, you could create three different room schedules (by Department, Floor Finish and Occupancy Type) and put them on the same sheet with color-fill diagrams for analysis (see *Sheet A9.92* in your exercise file).

We have included several different schedules and their uses for your review. Open the A9.9 series of sheets and check out some of the schedules that Revit Architecture provides.

REVIEW QUESTIONS – CHAPTER 7

MULTIPLE CHOICE

1. The Schedule Properties dialogue controls

    a. Fields and Filters

    b. Sorting, Grouping, Formatting and Appearance

    c. Both a and b

    d. b only

2. To hide a schedule column

    a. Use the Hidden Field checkbox on the Schedule Properties Formatting Tab

    b. Right-click in a field cell in the Schedule View and click Hide Column

    c. Delete the Schedule and start over

    d. a or b

3. Which property of the view that is on-screen when you create a schedule affects the schedule output?

    a. Zoom

    b. Detail level

    c. Phase

    d. Crop region

4. Multi-Category Schedules make use of

    a. Shared Parameters

    b. Schedule Filters

    c. Cost Estimates

    d. a and b, but not c

5. Project or Shared Parameters can include fields of the following types:

    a. Text, Number, Length

    b. Area, Volume, URL

    c. Material, Angle, Yes/No

    d. All of the above

TRUE/FALSE

6. True _ False _ A Schedule Filter can be set to equal or not equal a certain value.

7. True _ False _ Project Parameters can appear in Schedules but not in Tags.

8. True _ False _ Shared Parameters can be shared by multiple projects and appear in Schedules and Tags.

9. True _ False _ Calculated Parameters are text-only values.

10. True _ False _ If you Group Schedule Column Headers, you have to delete the Schedule from any Sheets it appears on and then replace it for the change to appear.

 Answers will be found on the CD.

CHAPTER 8

Beyond the Design—Area and Room Plans

INTRODUCTION

Revit Architecture's intelligent building modeler provides tools that fill a couple of basic needs of the people who pay designers to do their work. Design clients, usually the owners of the building being designed, almost always need to know the relative areas reserved for different spaces and functions within the design. Sometimes this information will be used for taxation purposes, most often for rental or other occupancy classifications. Throughout the life span of a commercial building in particular, accurate occupancy information can be critical. Revit Architecture models can supply this information—as anticipated uses or building configurations change during and after construction—in ways that static 2D drawings or previous 3D models never could. This may have long-range implications for design firms as suppliers of "life-cycle" management (or facilities management) services for buildings long after architectural functions are complete.

One of the easiest ways to collect and display basic area information about a Revit Architecture model is the Color Fill diagram, which can be either an Area plan or a more sophisticated room plan with associated schedules. These plans are displayed in Views of their own. Schedules are also Views, as you have seen in Chapter 7; room and area plans, like schedules, can be located on sheets.

OBJECTIVES

- Create and edit a room schedule
- Work with room tags
- Define room styles
- Apply room styles to rooms
- Create and edit a Color Fill room diagram
- Create multiple Color Fill diagrams from one set of room tags
- Place Color Fills and schedules on a sheet
- Create a Gross Building Area Plan
- Create a Rentable Area Plan
- Create a Custom Area Plan
- Create and edit a Color Fill Area diagram

- Export Revit Architecture information via ODBC
- Export a schedule
- Export to other formats

REVIT ARCHITECTURE COMMANDS AND SKILLS

Room Schedules

Schedule Keys

Room Styles

Color Fills

Area Settings

Area Plans

Export ODBC

EXERCISE 1. CREATE A ROOM SCHEDULE

1.1 Open the file *Chapter 8 start imperial.rvt [Chapter 8 start metric.rvt]*.

1.2 Floor Plan: 1ST FLOOR – FLOOR PLAN should be the active view.

- Save the file as **Chapter 8 area plans.rvt** in a location specified by your instructor.

1.3 Make Floor Plans: 2ND FLOOR OFFICES the active view.

- Make Room and Area the active tab on the Design Bar.

1.4 From the View menu, choose View>New>Schedule/Quantities.

- In the New Schedule dialogue, select Rooms under Category.
- Accept the defaults, as shown in Figure 8.1.
- Click OK.

Figure 8.1 *Make the new schedule category Rooms*

1.5 On the Fields tab of the Schedule Properties dialogue, select Area from the Available fields: list.

- Click Add--> to move it to the Scheduled fields (in order): list.

1.6 Select Department, Name and Number and add them to the Scheduled fields, as shown in Figure 8.2.

Figure 8.2 *Move fields from Available to Scheduled*

1.7 Select Name in the Scheduled fields.

- Click Move Up twice to put Name at the top of the list.

- Place Number in the second position (see Figure 8.3).

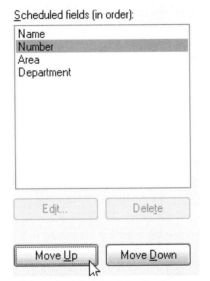

Figure 8.3 *Change the order of the scheduled fields in the list*

1.8 On the Appearance tab, check Bold and Italic for the Header font, as shown in Figure 8.4.

This change will appear when the schedule view is placed on a Sheet.

- Click OK.

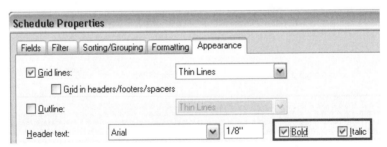

Figure 8.4 *You can adjust the appearance of the schedule on this tab*

Revit Architecture creates and opens a view named Room Schedule under Schedules/ Quantities in the Project Browser and makes it the active view. Eighty rooms are listed.

1.9 Select New on the Option Bar in the Schedule view.

Revit Architecture will place a room with Name **Room**, Number **81** and Area Value **Not Placed** in the last row of the schedule. This blank Room definition has not yet had an area tag placed inside a boundary or room.

1.10 Select New again to create a second unplaced room in the schedule (see Figure 8.5).

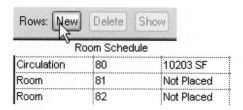

Figure 8.5 *Insert two room definitions in the schedule*

1.11 Open the view 2ND FLOOR OFFICES.

- On the Room and Area tab of the Design Bar, select Room Separation.

- On the Option Bar, check Draw (the pencil icon) and Chain (see Figure 8.6).

Figure 8.6 *The Room Separation tool and drawing options*

1.12 Sketch two lines to the left of the corridor door, as shown in Figure 8.7.

The exact dimensions are not critical.

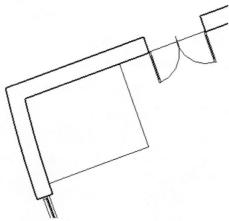

Figure 8.7 *Sketch a room where there are no walls*

1.13 Sketch two more lines, as shown in Figure 8.8.

The exact dimensions are not critical.

- Select Modify.

- Move the Circulation area tag to the right of the new lines.

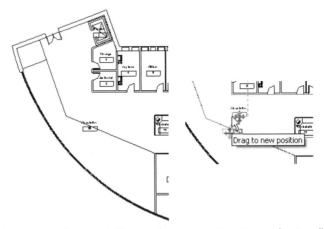

Figure 8.8 *Sketch two more lines to define another room separation without walls*

1.14 Select Room from the Room and Area tab of the Design Bar.

- On the Option Bar in the Room: field, click the drop-down arrow to expose the list. The two new room definitions are available to use.

- Select 81 Room, as shown in Figure 8.9.

- Place the tag in the small rectangular area.

- Select 82 Room from the Room: list and place the tag in the larger boundary area.

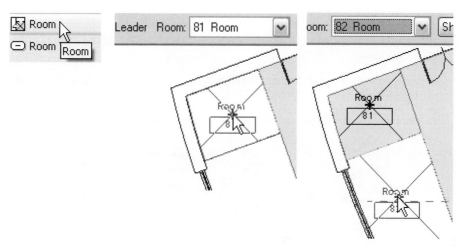

Figure 8.9   *Pick the room values from the Option Bar list*

1.15 Select Modify. Select the Room Tag for room 81. Pick the room name field a second time to make it editable. Change the Name Value to **Reception**.

1.16 Select Room Tag 82. Change the Name Value to **Waiting Area**. Click away from the Tag to exit edit mode.

1.17 Select the following Rooms and change their Department Values according to the following table:

4–10	**Human Resources**
11–22	**Finances**
23, 24	**Student Affairs**
29–40	**Student Affairs**
41–55	**Finances**
56–63	**Human Resources**
64–72	**Fund-raising**
73–79	**Dean's Office**

Tip: Use Crossing or Inside selection windows with Filters as necessary. Once a new Value is typed into a field, Revit Architecture remembers the value and makes it available when that field is selected for editing later in another room.

1.18 Open the Room Schedule view to see the new Department Values.

Parameter values that are not driven by the model can be edited in the Schedule View.

- Select the Name cell for room number 82 and change it to **Lounge**, as shown in Figure 8.10.

Your area value for room 82 may be different than shown.

Office	77	312 SF	Dean's Office
Office	78	461 SF	Dean's Office
Office	79	254 SF	Dean's Office
Circulation	80	9014 SF	
Reception	81	88 SF	
Lounge	82	1101 SF	

Figure 8.10 *Edit a Name value in the Schedule view*

1.19 Save the file.

Room schedules often hold information about physical characteristics to be applied to rooms, such as the type of flooring or wall covering. Rather than enter this data into individual schedule cells, you can create keys for schedules that place the information automatically. Schedule keys can be applied to many types of schedules.

EXERCISE 2. CREATE A SCHEDULE KEY

2.1 Open or continue working in the file from the previous exercise.

2.2 Make the View tab of the Design Bar active.

- Select Schedule/Quantities.

2.3 In the New Schedule dialogue, select Rooms in the Category: field.

- Select the Schedule keys radio button.

- Accept the Schedule name and Key name default values, as shown in Figure 8.11.

- Click OK.

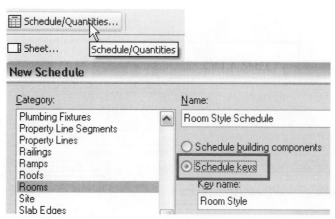

Figure 8.11   *Pick Schedule keys to make the new schedule a Room Style Schedule*

2.4 In the Schedule Properties dialogue, select Available fields Base Finish, Floor Finish and Wall Finish and add them in turn to the Scheduled fields, as shown in Figure 8.12.

- Click OK.

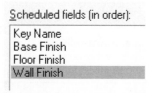

Figure 8.12   *Fields for the Room Style Schedule*

Revit Architecture will create a Room Style Schedule view and make it active.

2.5 Select New three times from the Option Bar to create three rows within the schedule.

2.6 Fill in the row cells, as shown in Figure 8.13.

Figure 8.13 *Values for the Room Style Schedule*

2.7 Open the view 2ND FLOOR OFFICES.

- CTRL + select the Rooms in large Offices 77 and 78, the three Conference Rooms and the Library.

- Select the Properties icon from the Options Bar.

- There will be a Room Style Parameter now available.

- Select VP Office from the drop-down list, as shown in Figure 8.14.

- Click OK.

Other values will fill in automatically.

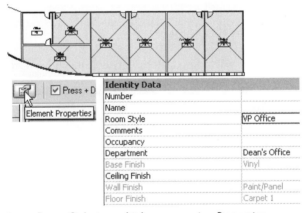

Figure 8.14 *Assigning a Room Style to multiple rooms using Properties*

2.8 Select Office 79 and all the Offices along the upper wall.

- Click the Properties icon.

- Make the Room Style Parameter Std. Office.

- Click OK.

2.9 Select the following rooms: Vending 25, Toilets 26 and 28, Services 27, Storage 2, Janitor 3 and Copiers 4.

- Click the Properties icon.

- Make the Room Style Parameter value Services.

- Click OK.

2.10 Zoom to Fit.

- Make the Drafting tab of the Design Bar active.

- Select Color Scheme Legend.

  This tool also appears on the Room and Area tab.

  When you move the cursor over the Drawing Window, a Color Scheme legend notice will be ghosted under the arrow.

2.11 Click to place the Color Scheme legend to the left of the model.

- Click OK in the notice box that appears about applying a color scheme to the view (see Figure 8.15).

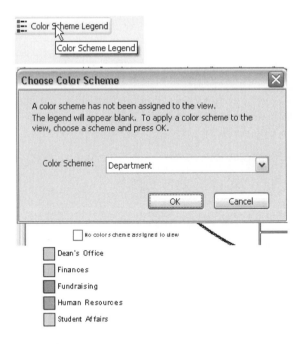

Figure 8.15 *Create a Color Scheme Legend*

2.12 Select the Color Scheme Legend.

- Choose Edit Color Scheme from the Options Bar.
- In the Edit Color Scheme dialogue, select the Color field for Finances.
- Select the Red swatch under Custom colors, as shown in Figure 8.16.
- Click OK.

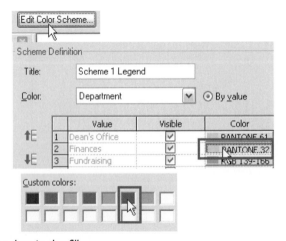

Figure 8.16 *Editing colors in the fill*

2.13 Select the Color field for Human Resources.

- Pick the Cyan swatch under Custom colors.

- Click OK (see Figure 8.17).

	Value	Visible	Color
1	Dean's Office	☑	PANTONE 61
2	Finances	☑	Red
3	Fundraising	☑	RGB 139-166
4	Human Resources	☑	Cyan
5	Student Affairs	☑	PANTONE 62

Figure 8.17   *The Color shows in the Edit Color Scheme panel*

2.14 Click OK to return to the View window.

- Hit ESC to terminate the edits (see Figure 8.18).

- Save the file.

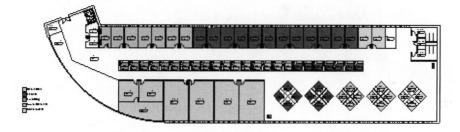

Figure 8.18   *The finished Departmental Color Fill diagram*

EXERCISE 3. WORK WITH COLOR FILLS

CREATE A NEW SCHEDULE FIELD

3.1 Select the Room Schedule view in the Project Browser.

- Right-click and select Properties, as shown in Figure 8.19.

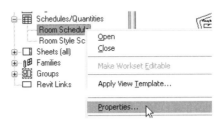

Figure 8.19 *Edit the Room Schedule Properties*

3.2 In the Element Properties dialogue, select Edit in the Fields value.

- Choose Add Parameter (see Figure 8.20).

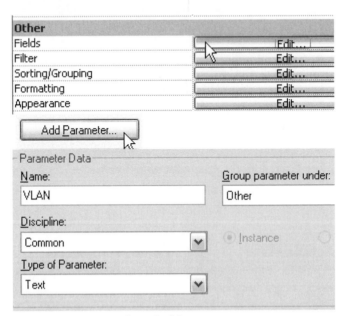

Figure 8.20 *Add a new parameter to this schedule*

3.3 In the Parameter Properties dialogue, type **VLAN** in the Name field.

- Accept the defaults.

- Click OK.

    VLAN stands for Virtual Local Area Network. The new field will identify ethernet computer hookup nodes for offices and workstations.

- Click OK three times to exit the dialogues.

3.4 Open the Room Schedule view.

- Give the rooms the following VLAN values:

Room Tag Number	VLAN Value
4–15	1
16–24	2
29–49	3
50–63	4
64–73	1
74–77	4
78–82	1

Figure 8.21 shows a selection of the Schedule view with the VLAN values being applied.

Room Schedule				
Name	Number	Area	Department	VLAN
Janitorial	3	80 SF		
Copiers	4	145 SF	Human Resources	1
Office	5	165 SF	Human Resources	1
Office	6	169 SF	Human Resources	1
Office	7	169 SF	Human Resources	1
Office	8	169 SF	Human Resources	1
Office	9	169 SF	Human Resources	1
Office	10	169 SF	Human Resources	1
Office	11	169 SF	Finances	1
Office	12	169 SF	Finances	1
Office	13	169 SF	Finances	1
Office	14	169 SF	Finances	1
Office	15	169 SF	Finances	1
Office	16	169 SF	Finances	2
Office	17	169 SF	Finances	2
Office	18	169 SF	Finances	2

Figure 8.21 *Assign VLAN values to rooms*

CREATE NEW COLOR FILLS FROM THE SAME INFORMATION

3.5 Pick the 2ND FLOOR OFFICE view in the Project Browser.

- Right-click and select Duplicate View > Duplicate, as shown in Figure 8.22.

Figure 8.22 *Duplicate the view*

Note: When you duplicate a view, the annotations do not duplicate by default. Use Duplicate with Detailing to keep notes, etc. visible in the new view. In this case, you don't need to see the Room Tags.

3.6 Revit Architecture will open the new view, named Copy of 2ND FLOOR OFFICES.

- Choose the new view name in the Project Browser, right-click and select Properties.
- Change the View Name to **2ND FLOOR OFFICES BY DEPARTMENT**.
- Change the View Scale value to **1" = 20'-0" [1:200]** (see Figure 8.23).
- Click OK.

View Scale	1" = 20'-0"
Scale Value 1:	240
Display Model	Normal
Detail Level	Coarse
Visibility/Graphics Overrides	Edit...
Model Graphics Style	Hidden Line
Advanced Model Graphics	Edit...
Underlay	None
Underlay Orientation	Plan
Orientation	Project North
Wall Join Display	Clean all wall joins
Discipline	Architectural
Color Scheme	Scheme 1
Identity Data	⌃
View Name	2ND FLOOR OFFICES BY DEPARTMEN

Figure 8.23 *Edit the View properties*

3.7 On the Room and Area tab of the Design Bar, click Color Scheme Legend. Place the legend to the left of the model, as shown in Figure 8.24.

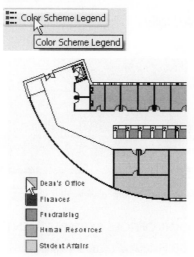

Figure 8.24 *Place a Color Scheme Legend in the new view*

3.8 Select 2ND FLOOR OFFICES in the Project Browser.

- Right-click and select Properties.

- Click Department in Color Scheme.

- Click (none), as shown in Figure 8.25.

- Click OK.

    The Color Scheme field will show none.

- Click OK.

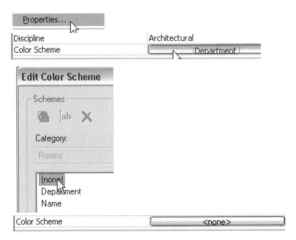

Figure 8.25 *Remove the Color Scheme from a view*

3.9 Select the 2ND FLOOR OFFICES BY DEPARTMENT view in the Project Browser.

• Right-click and pick Duplicate View > Duplicate.

Revit Architecture will open the new view.

3.10 Select the new view name in the Project Browser, right-click and choose Rename.

• Change the view name to **2ND FLOOR OFFICES BY STYLE** (see Figure 8.26).

• Click OK.

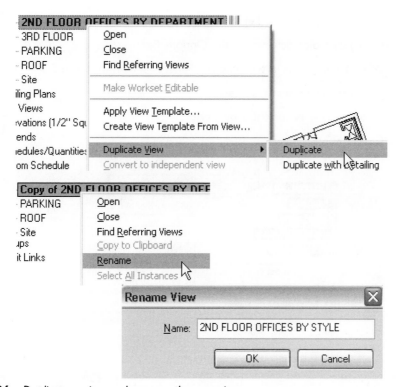

Figure 8.26 *Duplicate a view and rename the new view*

3.11 On the Room and Area tab of the Design Bar, click Color Scheme Legend.

- Place the Legend to the left of the model.

3.12 Select the Color Scheme Legend in the new view.

- Click Edit Color Scheme on the Options Bar.

- Select Department in the Schemes Pane, as shown in Figure 8.27.

- Right-click and select Duplicate.

- Make the new Name value by **Style**.

- Click OK.

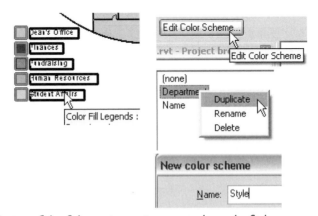

Figure 8.27   *Duplicate a Color Scheme to create a new scheme by Style*

3.13 In the Style Definition Section, set the Title to **Style Legend**.

- Pick Room Style from the drop-down list for Color.

- Click OK in the announcement box.

- Change the (none) Color to Basic Color White, as shown in Figure 8.28.

- Click OK twice to exit the dialogues.

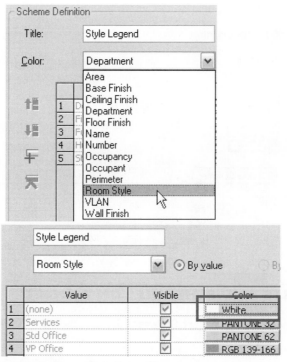

Figure 8.28 *The colors for the Room Style Color Fill*

3.14 Pick the 2ND FLOOR OFFICE BY STYLE view in the Project Browser.

- Right-click and select Duplicate View > Duplicate.

  Revit Architecture will open the new view.

3.15 Pick the new view name in the Project Browser, right-click and pick Rename.

- Change the view Name to **2ND FLOOR VLAN**.

3.16 On the Room and Area tab of the Design Bar, pick Color Scheme Legend.

- Place the Legend to the left of the model.

- Select the new Legend.

- Click Edit Color Scheme on the Options Bar.

- Click Style and right-click.

- Click Duplicate.

- Make the new Name value **VLAN**.

- Click OK.

- Enter **VLAN Legend** in the Title field.

- Select VLAN for the Color.
- Click OK in the warning.
- Click OK.

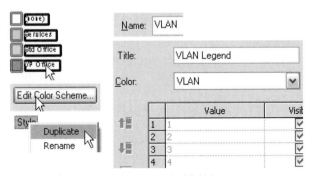

Figure 8.29 *Set up a duplicated view to show color by VLAN*

PUT COLOR FILLS AND SCHEDULES ON A SHEET

3.17 Expand Sheets in the Project Browser.

- Open view A801.

3.18 In the Project Browser, select view 2ND FLOOR OFFICES BY DEPARTMENT.

- Drag the view onto the sheet and place it in the upper-left corner.

3.19 Repeat for the views 2ND FLOOR OFFICES BY STYLE and 2ND Floor VLAN (see Figure 8.30).

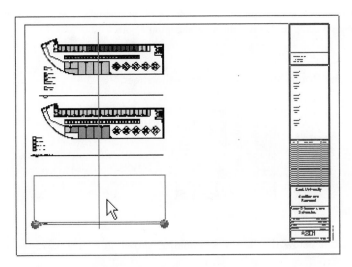

Figure 8.30 *Drag the views onto the page—Revit Architecture will supply alignment planes*

3.20 Pick and Drag the Room Schedule view onto the right half of the sheet.

The schedule is a little too long to fit.

- Zoom in Region to the top rows of the schedule.

- Click the blue arrow-shaped grip that controls the width of the Department field and drag it to the right so that the row text fits on one line per row (see Figure 8.31).

Note: The appearance edits you made earlier to the Room Schedule header appear now that it has been placed on a sheet.

Name	Number	Area	Department	LAN
Elevator	1	51 SF		
Storage	2	101 SF		
Janitorial	3	80 SF		
Coplers	4	145 SF	Human Resources	1
Office	5	165 SF	Human Resources	1
Office	6	169 SF	Human Resources	1
Office	7	169 SF	Human Resources	1
Office	8	169 SF	Human Resources	1
Office	9	169 SF	Human Resources	1

Figure 8.31 *Adjust the column width for better fit*

Tip: Revit Architecture provides a control point on schedule views to allow you to split a long schedule to fit on a sheet, as shown in Figure 8.32.

3.21 Drag the Room Style Schedule onto the sheet.

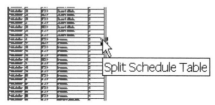

Figure 8.32 *You can split a long schedule*

- Place it below the Room Schedule, as shown in Figure 8.32.

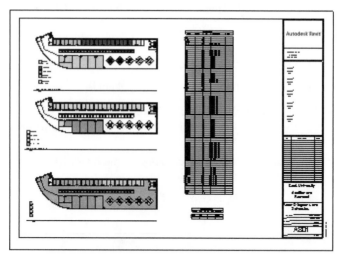

Figure 8.33 *The new page with Color Fills and associated schedules*

3.22 Save the file.

You have just quickly created and displayed five views of the same data. Using the floor plan, room tags and Color Fills, you have provided graphic information that will be valuable to firms working on the building—interior finish contractors and systems electricians, for example. This same information will also be useful to the facilities manager during the life of the building. The room and room style schedules provide other matrix views of the same information shown in the Color Fills.

AREA PLANS AND COLOR FILLS

Revit Architecture's Area Plans are views that will appear in the Project Browser once created. Area plan views show spatial relationships of floor-plan levels based on Area Schemes. Area Schemes are definable. Revit Architecture creates two types of Area Schemes by default: Gross Building, which cannot be deleted or edited, and Rentable, which can be copied and edited to create additional Area Schemes. Area Boundaries and Area Tags are the building blocks of area plans.

EXERCISE 4. CREATE A GROSS BUILDING AREA PLAN

4.1 Open or continue working with *Chapter 8 area plans.rvt*.

- Make the 1ST FLOOR view active.

4.2 From the Room and Area tab of the Design Bar, select Settings.

- In the Room and Area Settings dialogue, open the Area Schemes tab.

- Choose New.

- Click in the Name Field for the new Area Scheme and change the Value to **Classrooms and Labs**.

- Change the Description Value to **Measurements – Temporary Conditions** (see Figure 8.34).

   You will use this new Area Scheme definition later.

- Click OK.

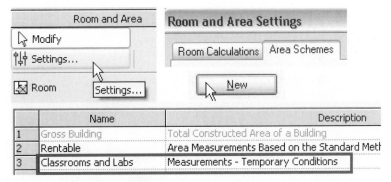

Figure 8.34 *Name and Description for the new Area Scheme*

4.3 Select Area Plan from the Room and Area tab of the Design Bar.

- In the New Area Plan dialogue, select Gross Building for the Type and 1ST FLOOR for the level, as shown in Figure 8.35.

- Click OK.

- Click Yes to create area boundary lines from all external walls.

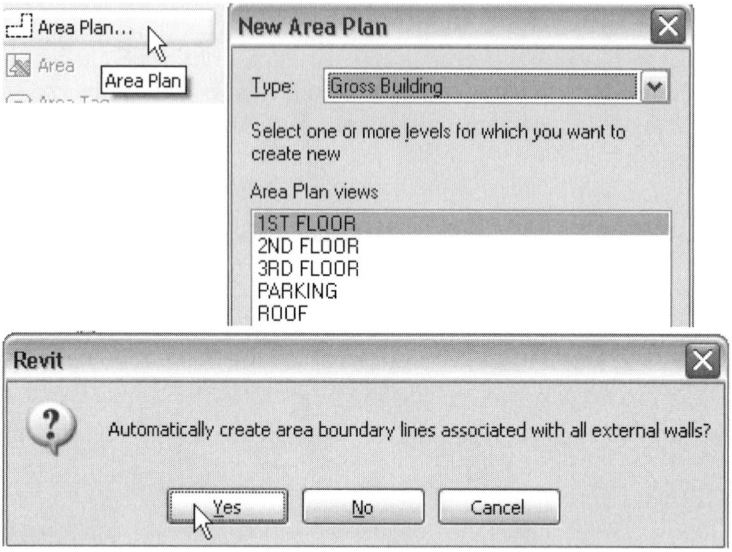

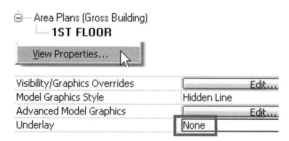

Figure 8.35 *The Area Plan tool. Revit Architecture will read and use the external walls for you.*

Revit Architecture will create a new view named Area Plan (Gross Building): IST FLOOR, as shown in Figure 8.36, and make it active.

4.4 Open the View Properties dialogue.

- Set the Underlay value to None.

This will clean up the screen.

- Zoom to Fit.

Figure 8.36 *The new area plan appears in the Project Browser*

4.5 From the Room and Area tab of the Design Bar, pick Area.

A tag will appear under the cursor.

- Zoom in Region to the left side of the model so you can read the tag text.

4.6 Hold the cursor outside the model.

The second line of the tag will read Not Enclosed—it is not within an area boundary.

- Move the cursor inside the exterior walls of the model.

  The tag will read an area value. This value will be the same whether the cursor is within a room or not, as shown in Figure 8.37.

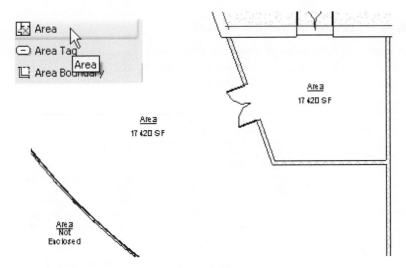

Figure 8.37 *Inside the model, an area tag displays the area of whatever area boundary the tag is within*

4.7 Place the cursor inside the model near the curved curtain wall and left-click to place it.

The new area will highlight.

- Pick Modify.
- Select the new tag.
- Select the Name text to open the field for editing.
- Change the Name Value to **Gross Area 1st Floor**, as shown in Figure 8.38.
- Click OK.
- Zoom to Fit.

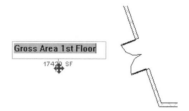

Figure 8.38 *The edited tag name*

CREATE A RENTABLE AREA PLAN

The Gross Building Area scheme contains two simple Area Types: Gross Building Area and Exterior Area. The Rentable Area scheme contains the Exterior Area type, plus five others, defined according to the following table:

Office Area	Contains tenant personnel, furniture or both
Store Area	Suitable for retail occupancy and use
Floor Area	Areas available primarily for use of tenants of the given floor: washrooms, janitorial spaces, utility or mechanical rooms, elevator lobbies, public corridors
Building Common Area	Lobbies, conference rooms, atriums, lounges, vending areas, food service facilities, security desks, concierge areas, mail rooms, health/daycare/locker/shower facilities
Major Vertical Penetration	Vertical ducts, pipe or heating shafts, elevators, stairs, and their enclosing walls

4.8 Select Area Plan from the Room and Area tab of the Design Bar.

- In the New Area Plan dialogue, choose Rentable for the Type and 1ST FLOOR for the level, as shown in Figure 8.39.

- Click OK.

- Answer Yes in the question box that appears to create area boundary lines from all external walls.

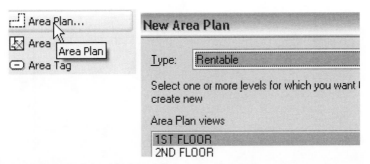

Figure 8.39 *Settings for a Rentable Area plan*

Revit Architecture will create a new view named Area Plan (Rentable): 1ST FLOOR, as shown in Figure 8.40, and make it active.

Figure 8.40 *The new area plan appears in the Project Browser*

4.9 Type **VP** to adjust the Visibility settings of the new view.

- Change the Underlay to None.

- Click Edit in the Visibility field.

- On the Annotations Categories tab, clear Elevations.

- Click OK twice.

- Zoom to Fit.

4.10 From the Room and Area tab of the Design Bar, select Area.

- A tag will appear under the cursor.

- Place it in the same location as on the previous plan (see Figure 8.41).

  The new area will highlight.

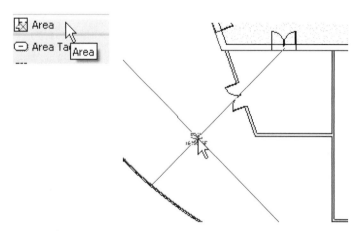

Figure 8.41 *Place the first area tag in open space*

4.11 Select Area Boundary from the Room and Area tab of the Design Bar.

- On the Options Bar, leave the Pick option selected and Apply Area Rules checked.

- Zoom in Region to the left side of the model.

4.12 Select the two walls of the elevator enclosure facing the interior of the building, as shown in Figure 8.42.

Note that once you choose the second wall, the area value of the existing area tag changes.

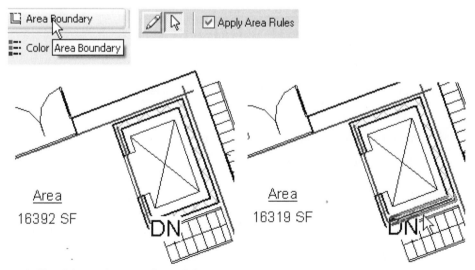

Figure 8.42 *Select only two walls to define the elevator area*

4.13 Continue selecting all the room-defining interior walls (see Figure 8.43).

- Do not pick walls inside the washroom areas.

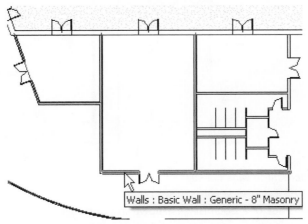

Figure 8.43   *Selecting a wall and Revit Architecture will create the corresponding boundary line*

4.14 Select the Draw option on the Option Bar.

- Draw two lines, as shown in Figure 8.44, to create an area boundary not defined by walls.

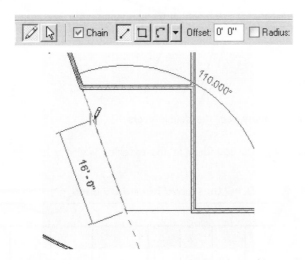

Figure 8.44   *Switch to the Draw option to create boundaries where there are no walls*

4.15 Draw two sides of a boundary, as shown on the right side of Figure 8.45.

- Draw a line across the entrance to the washroom area.

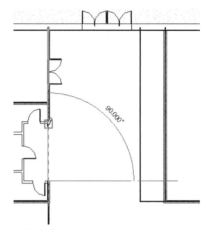

Figure 8.45 *Boundaries in a wide hallway*

4.16 Draw two sides of a boundary, as shown in Figure 8.46. The actual dimensions are not important.

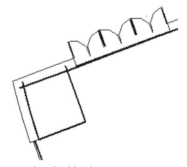

Figure 8.46 *Another boundary near the double doors*

4.17 Draw two sides of a boundary at the exterior walkway, as shown in Figure 8.47.

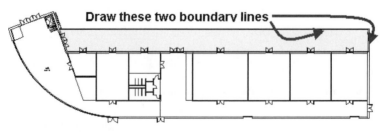

Figure 8.47 *Boundary lines include the walkway outside the walls*

4.18 Select Area from the Room and Area tab of the Design Bar.

- Place areas inside each room and boundary area you have created.

- Include the outside walkway (see Figure 8.48).

- Click Modify.

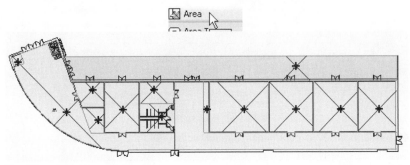

Figure 8.48 *Areas in place—14 in all*

4.19 Zoom to Fit.

- Select everything visible in the view.

- Select the Filter.

- Clear all categories except Areas, Area Tags and <Area Boundary>.

- Click OK.

4.20 Click the Temporary Hide/Isolate control on the View Control Bar.

- Click Isolate Category, as shown in Figure 8.49.

   This will make picking the areas easier in the next few steps.

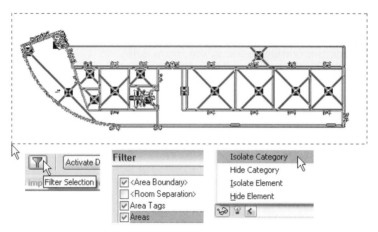

Figure 8.49 *Isolate the area information without walls*

4.21 Edit each area tag Name and Area Type in turn, based on the information contained in Figures 8.50 and 8.51 and the table shown in Step 4.23.

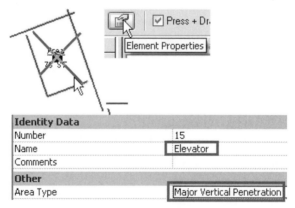

Figure 8.50 *Using Properties to edit the area in the elevator*

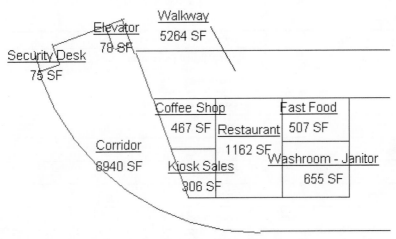

Figure 8.51   *Renamed area tags—they have have been enlarged in this illustration*

4.22   For the area tag in the Vending-Seating area, select Vertical orientation from the Option Bar (see Figure 8.52).

- For the tag inside the walkway, pick the Leader option from the Option Bar, then drag the tag outside the walkway boundary (as shown in Figure 8.49) so that it will be sure to show when the Color Fill is applied in the next steps.

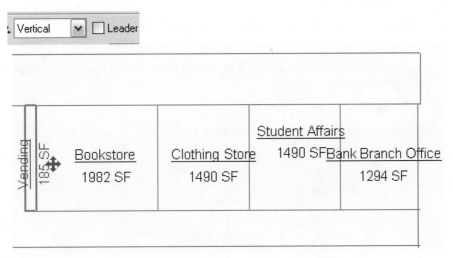

Figure 8.52   *More area tags—note the vertical orientation at Vending*

4.23 Edit the Area Type values for the area tags according to the following table:

Security Desk	Building Common Area
Elevator	Major Vertical Penetration
Corridor	Floor Area
Walkway	Exterior Area
Kiosk Sales	Store Area
Coffee Shop	Store Area
Restaurant	Store Area
Fast Food	Store Area
Washroom—Janitor	Floor Area
Vending—Seating	Building Common Area
Bookstore	Store Area
Clothing Store	Store Area
Student Affairs	Office Area
Bank Branch Office	Store Area

4.24 Zoom to Fit.

- From the View Control Bar, select Temporary Hide/Isolate > Reset Temporary Hide/Isolate.

- Choose Color Scheme Legend from the Room and Area tab of the Design Bar (see Figure 8.53).

- A legend will appear under the cursor.

- Left-click to locate it to the left of the model.

- Click OK.

Figure 8.53 *Restore the view, place a Color Scheme Legend*

4.25 Revit Architecture will supply color fill to the tagged areas, as shown in Figure 8.52. The Color scheme is editable. You have edited a color scheme in a previous exercise.

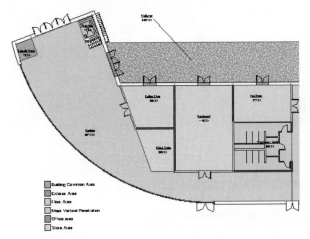

Figure 8.54 *The Color Fill in place*

4.26 Save the file.

EXERCISE 5. CREATE A CUSTOM AREA PLAN

5.1 Open or continue working in the file from the previous exercise.

- Select Area Plan from the Room and Area tab of the Design Bar.

- In the New Area Plan dialogue, choose Classrooms and Labs as the Type, and 2ND FLOOR as the Area Plan view's level (see Figure 8.55).

- Click OK.

- Choose Yes in the query box about creating boundary lines from the external walls.

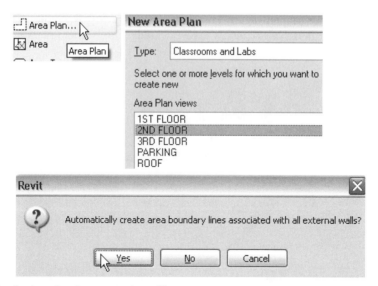

Figure 8.55 *Settings for the custom Area Plan*

5.2 Revit Architecture will create a new view named Area Plans (Classrooms and Labs): 2ND FLOOR, as shown in Figure 8.56, and make it active.

Figure 8.56 *The new Area Plan appears in the Project Browser*

5.3 Type **VP** to open the View Properties dialogue.

- Set the Phase value to Temporary Classrooms.

- Set the Phase Filter to Show Complete.

- Set the Underlay to None (see Figure 8.57).

- Click OK.

Figure 8.57 *Set the View Phase and Underlay*

5.4 On the Room and Area tab of the Design Bar, select Area Boundary.

- Pick walls, as shown in Figures 8.58 and 8.59.

- Do not select the walls of the washroom in the upper-right corner of the model or the elevator shaft walls.

- Do not select partition walls between similarly sized spaces.

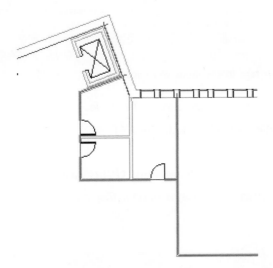

Figure 8.58 *The first wall picks do not include the elevator shaft walls or walls inside comparable spaces*

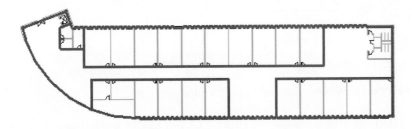

Figure 8.59 *The interior walls selected for boundaries*

5.5 On the Room and Area tab of the Design Bar, select Area.

- Place a tag inside each major area whose walls you previously picked (four in total). See Figure 8.60.

- Select Modify to terminate tab placement.

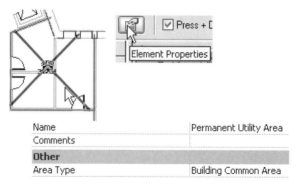

Figure 8.60 *Place four tags in the areas you enclosed*

5.6 Zoom in Region to the left side of the model.

- Choose the area shown in Figure 8.61.
- Pick the Properties icon.
- Make the Name Value **Permanent Utility Area**.
- Leave the Area Type Value as Building Common Area.
- Click OK.

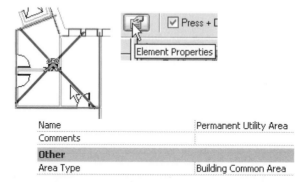

Figure 8.61 *Name and Area Type values for the area tag*

5.7 Select the area tag shown in Figure 8.62 and edit its properties.

- Make the Name Value **Temporary Classroom**.
- Make the Area Type Value **Floor Area**.
- Click OK.

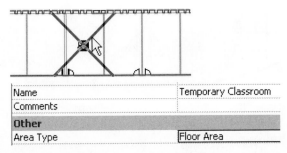

Figure 8.62 *Edit the tag in the room areas on the north wall*

5.8 Select the area tag shown in Figure 8.63 and edit its properties.

- Make the Name Value **Temporary Labs**.

- Make the Area Type Value **Store Area**; these spaces will be considered rentable during their existence.

- Click OK.

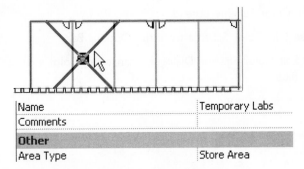

Figure 8.63 *Edit this tag in the rooms at the southeast corner*

5.9 Select the remaining area tag and edit its properties.

- Make the Name Value **Permanent Offices**.

- Make the Area Type Value **Office Area** (see Figure 8.64).

- Click OK.

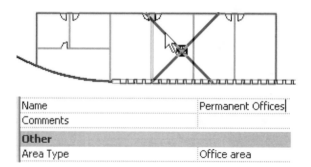

Figure 8.64 *Edit the final tag*

5.10 Zoom to Fit.

- Choose Color Scheme Legend from the Room and Area tab of the Design Bar.

- Locate the Color Scheme Legend to the left of the model.

- Click OK in the notice.

5.11 Select the Color Scheme Legend.

- Choose Edit Color Scheme from the Options Bar.

- In the Edit Color Scheme Dialogue, change the Color value to **Name**, as shown in Figure 8.65.

- Click OK in the notice.

- Click OK.

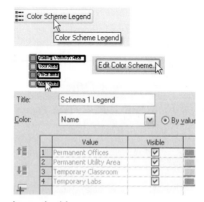

Figure 8.65 *Color the new scheme by Name*

The Color Scheme Legend will change to show the names you just assigned to the area types in this area plan, as shown in Figure 8.66.

Figure 8.66 *The legend now shows the assigned Name values*

5.12 Zoom to Fit (see Figure 8.67).

- Save the file.

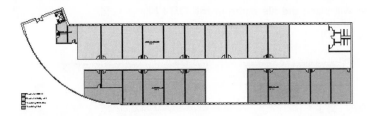

Figure 8.67 *The custom Area Plan with color fill*

EXERCISE 6. EXPORTING SCHEDULE INFORMATION

Schedule information is valuable inside Revit Architecture, but even more useful to many firms when it's taken out to other applications. Revit Architecture will export data to any ODBC-compatible (Open Database Connectivity—the Microsoft standard for accessing data from a variety of sources) database management application. This requires setting up the ODBC interface by specifying drivers to use. For this exercise, you will use the driver supplied by Revit Architecture to export to Microsoft Access. The process would be similar for any ODBC-compliant application.

EXPORTING MODEL INFORMATION VIA ODBC

6.1 Open or continue working in the file from the previous exercise.

- From the File menu, pick File > Export > ODBC Database, as shown in Figure 8.68.

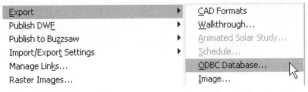

Figure 8.68 *Export database information to an external application*

The Select Data Source dialogue opens. It contains two tabs and a file browser window.

ODBC requires Data Sources with Data Source Names (DSN) for connection. Data Sources can be located on the current workstation (Machine Data) or anywhere on the network (File Data). Machine Data Sources cannot be shared, and can be limited to individual users on any given machine. File Data Sources can be shared.

Revit Architecture supplies a DSN file that appears on the File Data Source tab of the dialogue (see Figure 8.69).

6.2 Select the file *revitdsn.dsn* that appears in the dialogue window.

This will place the file name in the DSN Name field.

- Click OK.

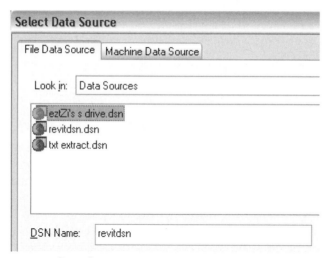

Figure 8.69 *Create a new Data Source*

6.3 The ODBC Microsoft Access Setup dialogue will appear, as shown in Figure 8.70.

- Leave the System Database option on None.

- Select Create.

6.4 In the New Database dialogue, give the new database the name **chapter 8**, in a location as directed by your instructor.

- Accept the other defaults in the dialogue.

- Click OK.

The Locale list provides a list of available languages. The Format selection refers to the Jet engine version inside Access (to allow you to convert to older formats if necessary). System Database and Encryption options allow you to define the database as a System file, which then requires system Administrator status to manage or to password-protect the file.

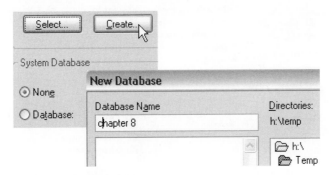

Figure 8.70 *Next, create a database file to load*

6.5 Revit Architecture will display a confirmation box, as shown in Figure 8.71.

- Click OK twice to exit the setup.

Figure 8.71 *Click OK at the notification*

6.6 If Microsoft Access is installed on your computer, launch that application and open the *chapter 8.mdb* file you just created (or use your File Manager/Computer Browser to find the new file name and double-click it to open Access).

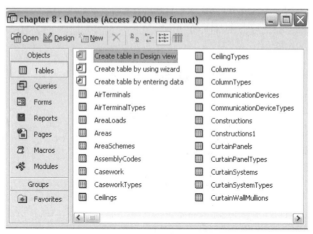

Figure 8.72 *The list of tables in the new Access file*

6.7 Access will open to a window listing the tables created by Revit Architecture's export, as shown in Figure 8.72.

6.8 Select Rooms in the Table list and right-click.

- Choose Open.

  Revit Architecture's table assigns an ID to each room and shows the contents of each cell value as you saw in the schedule, except for Level and Room Style, which are identified by a numerical ID rather than a name (see Figure 8.73).

WallFinish	FloorFinish	Area	Number	Name
			121	Bookstore
			122	Clothing Store
			123	Student Affair
			124	Bank Branch
			125	Corridor
		4.78427911601	1	Elevator
Vinyl WC	Tile	9.42460177760	2	Storage
Vinyl WC	Tile	7.47126156088	3	Janitorial
Vinyl WC	Tile	13.4637620215	4	Copiers
Paint	Carpet 2	15.3209802926	5	Office
Paint	Carpet 2	15.6870637058	6	Office
Paint	Carpet 2	15.6870637058	7	Office
Paint	Carpet 2	15.6870637058	8	Office
Paint	Carpet 2	15.6870637058	9	Office

Identity Data	
Number	8
Name	Office
Room Style	Std Office
Comments	
Occupancy	
Department	Human Resources
Base Finish	Vinyl
Ceiling Finish	
Wall Finish	Paint
Floor Finish	Carpet 2

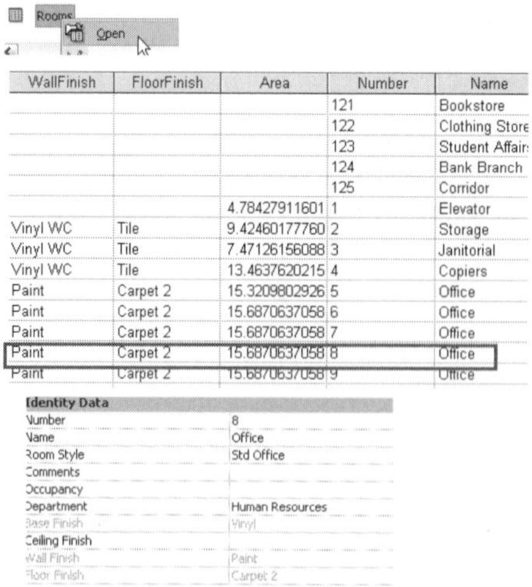

Figure 8.73 *Revit Architecture's room schedule information in the Access database*

6.9 Revit Architecture exports the Room Style Schedule you created as a separate table that uses the Room Style key ID shown in the Rooms table (see Figure 8.74).

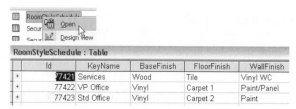

Figure 8.74 *The Room Style Schedule is translated as well*

6.10 Exit Access.

Revit Architecture does not maintain a live link to this external file, so design changes must be re-exported.

6.11 To update a previously created database with new design information, select File > Export > ODBC Database from the File menu, as before.

6.12 In the Select Data Source dialogue, select the *revitdsn.dsn* file as before.

- Click OK.

6.13 In the ODBC Microsoft Access Setup dialogue, choose Select.

- In the Select Database dialogue, pick the name of the file you previously created (see Figure 8.75).

- Click OK twice to update that file.

At this point, you could also select another database file to load the Revit Architecture information into, or create a new one.

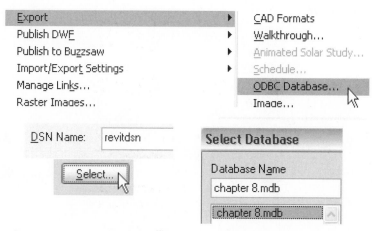

Figure 8.75 *Click OK to update the listed file, pick another or create a new one*

You do not have to export the entire model contents in order to use the project information in an external application. You can export a single schedule with a quick, easily repeatable process.

EXPORTING A SINGLE SCHEDULE

6.14 Open the Room Schedule view.

- From the File menu, select File > Export > Schedule.

    This option is only available in a Schedule view. The Schedule export creates an ASCII text file, a very simple format that is read by many applications.

6.15 Locate the output file in a location you can find (see Figure 8.76).

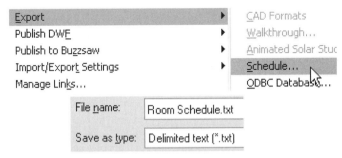

Figure 8.76 *Export a text file from the schedule*

6.16 In the Export dialogue, accept the defaults (see Figure 8.77).

There are options for what to include and how to format the text file indicators for fields and their contents. These options will affect how applications such as Excel import the file.

- Choose OK.

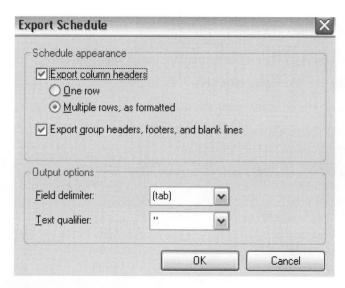

Figure 8.77   *Output options*

6.17 Revit will create the file.

If you have Microsoft Excel on your system, you can use that application to open the text file and use the schedule data. This is a quick and efficient way to export information to spreadsheets (see Figure 8.78).

	A	B	C	D	E
1	Room Schedule				
2	Name	Number	Area	Department	VLAN
3					
4	Elevator	1	51 SF		
5	Storage	2	101 SF		
6	Janitorial	3	80 SF		
7	Copiers	4	145 SF	Human Resources	1
8	Office	5	165 SF	Human Resources	1
9	Office	6	169 SF	Human Resources	1
10	Office	7	169 SF	Human Resources	1
11	Office	8	169 SF	Human Resources	1
12	Office	9	169 SF	Human Resources	1
13	Office	10	169 SF	Human Resources	1

Figure 8.78   *The text file opened in Excel*

Like the ODBC, this is export is not a live link. The schedule must be re-exported after changes so that the spreadsheet file can be updated.

EXPORTING TO OTHER FORMATS

6.18 Open view Area Plans (Classrooms and Labs) 2ND FLOOR.

- From the File menu, select File > Export > Room/Area Report.

- Accept the supplied name and locate the file in a location you can find later.

- Select Current View (see Figure 8.79).

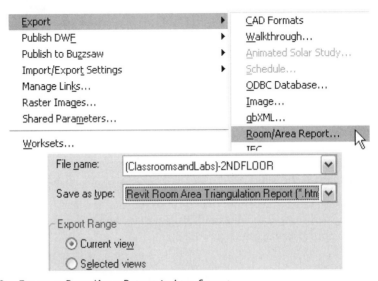

Figure 8.79 *Export a Room/Area Report in htm format*

6.19 Select Settings to view format possibilities for the report, as shown in Figure 8.80.

- Choose Cancel.

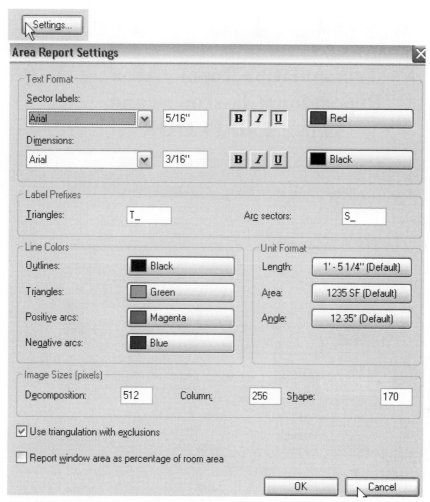

Figure 8.80   *Settings for the area report export*

6.20 Choose Save in the Export dialogue.

Revit Architecture will work for a while.

Revit will not display the export.

6.21 Use your File browser to navigate to the folder where you saved the file, and double-click the file name to open it in Microsoft Internet Explorer (see Figure 8.81).

Triangulation Room - Area Report for (Classrooms and Labs): 2ND FLOOR

23/08/2007 8:51:33 PM

Area calculation for Room __ Circulation #80 __

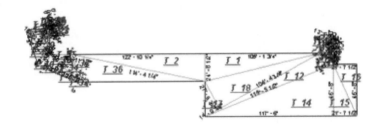

#	Calculation	Area
	Triangles	
1	½ * 24' - 8 1/2" * 105' - 1 3/4"	1299 SF
2	½ * 24' - 8 1/2" * 122' - 10 1/4"	1518 SF
3	½ * 12' - 4 1/4" * 5' - 10"	36 SF
4	½ * 5' - 1" * 13' - 6 1/2"	34 SF
5	½ * 5' - 1" * 13' - 6 1/2"	34 SF

Figure 8.81   *The area report opened in a browser*

6.22 Microsoft Excel can also open *htm* files, as shown in Figure 8.82.

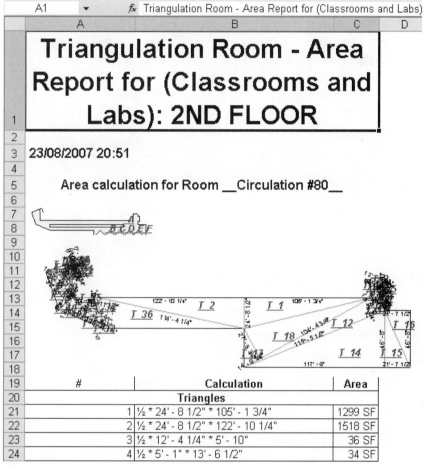

Figure 8.82 *The area report opened in Excel*

6.23 Room Area reports are primarily for the European market.

6.24 Revit Architecture will export and import Industry Foundation Classes (IFC) files, as shown in Figure 8.83.

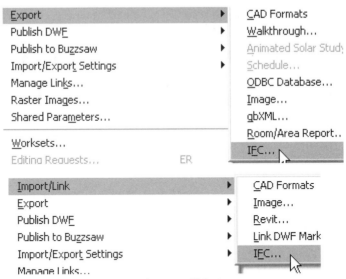

Figure 8.83 *The menu picks to export and import IFC*

IFC export and import will be covered in a later chapter.

6.25 Save and close the Revit Architecture file.

6.26 Open a new project file.

- From the Basics tab of the Design Bar, pick Wall.

- Create four walls using the rectangle option.

- Pick Room, and place a single room inside the new walls, as shown in Figure 8.84.

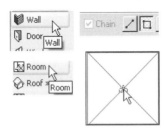

Figure 8.84 *Walls and a Room in a new file*

6.27 As preparation for the next steps, select Settings > Project Information from the Menu Bar.

- Choose the Edit button for Project Address.

- Provide this project file with a hypothetical address, complete with Zip code, as shown in Figure 8.85.

- Post Code information is used in the export you will perform next.

- Click OK.

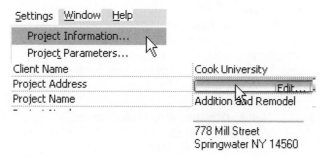

Figure 8.85 *Modify the Project Address*

6.28 From the File menu, choose File > Export > gbXML (see Figure 8.86).

This creates a report in XML (extended markup language) format. There are no options; this is an automatic process.

- Export the file to a convenient location.

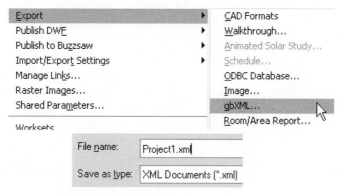

Figure 8.86 *Generate an XML file*

6.29 Use your file browser to navigate to the folder where you created the XML file.

- Double-click the file name, and it will open in Microsoft Internet Explorer, showing a list of field data, as shown in Figure 8.87.

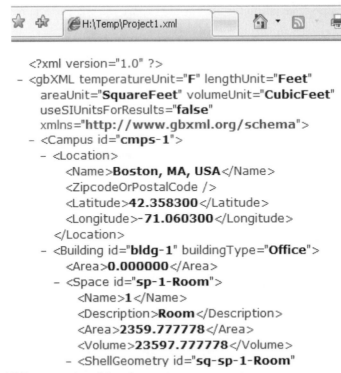

```
<?xml version="1.0" ?>
- <gbXML temperatureUnit="F" lengthUnit="Feet"
    areaUnit="SquareFeet" volumeUnit="CubicFeet"
    useSIUnitsForResults="false"
    xmlns="http://www.gbxml.org/schema">
  - <Campus id="cmps-1">
    - <Location>
        <Name>Boston, MA, USA</Name>
        <ZipcodeOrPostalCode />
        <Latitude>42.358300</Latitude>
        <Longitude>-71.060300</Longitude>
      </Location>
    - <Building id="bldg-1" buildingType="Office">
        <Area>0.000000</Area>
      - <Space id="sp-1-Room">
          <Name>1</Name>
          <Description>Room</Description>
          <Area>2359.777778</Area>
          <Volume>23597.777778</Volume>
        - <ShellGeometry id="sq-sp-1-Room"
```

Figure 8.87 *XML export viewed in a browser*

XML is a developing format for sharing data across platforms. It is recognized by a growing number of applications. Figure 8.88 shows the same file when opened in Excel 2003.

A1 fx /gbXML

	A	B	C	D	E
1	/gbXML				
2	/@areaUnit	/@lengthUnit	/@temp	/@useSIUnitsForResults	/@volumeUni
3	SquareFeet	Feet	F	FALSE	CubicFeet
4	SquareFeet	Feet	F	FALSE	CubicFeet
5	SquareFeet	Feet	F	FALSE	CubicFeet
6	SquareFeet	Feet	F	FALSE	CubicFeet
7	SquareFeet	Feet	F	FALSE	CubicFeet
8	SquareFeet	Feet	F	FALSE	CubicFeet
9	SquareFeet	Feet	F	FALSE	CubicFeet
10	SquareFeet	Feet	F	FALSE	CubicFeet
11	SquareFeet	Feet	F	FALSE	CubicFeet
12	SquareFeet	Feet	F	FALSE	CubicFeet
13	SquareFeet	Feet	F	FALSE	CubicFeet
14	SquareFeet	Feet	F	FALSE	CubicFeet
15	SquareFeet	Feet	F	FALSE	CubicFeet
16	SquareFeet	Feet	F	FALSE	CubicFeet
17	SquareFeet	Feet	F	FALSE	CubicFeet

Figure 8.88 *XML export viewed in Excel*

6.30 Close the project file without saving.

SUMMARY

In this chapter, you used Revit Architecture's tools for cataloguing areas and rooms. Area Plans and Room Schedules provide quick graphic representations in Color Fill diagrams, which are customizable and update automatically (always in Revit Architecture) as the model changes. Revit Architecture will export component and schedule information to database applications using a variety of formats.

REVIEW QUESTIONS – CHAPTER 8

MULTIPLE CHOICE

1. Room definitions can be created

- **a.** before any Room Schedules are created, not after
- **b.** after all Room Schedules are created, not before
- **c.** only while a Room Schedule is the active view
- **d.** anytime

2. Schedules and Color Fills are created from Plan views, therefore you need to be aware of

    a. the Underlay of the original view

    b. the Detail level of the original view

    c. the Phase of the original view

    d. the Area of the original view

3. Area Plans are based on Area Schemes. Revit Architecture supplies these Area Schemes by default:

    a. Gross Building (not editable) and Rentable (editable)

    b. Custom (editable) and Standard (not editable)

    c. Exterior (not editable) and Interior (editable)

    d. all of the above

4. The Rentable Area Scheme contains the following Area Types:

    a. Exterior, Office

    b. Store, Floor

    c. Building Common, Major Vertical Penetration

    d. all of the above

5. Color Fill color schemes can be keyed by

    a. default schedule field parameters only

    b. text schedule field parameters only

    c. numeric schedule field parameters only

    d. any schedule field parameter

TRUE/FALSE

6. **True** _ **False** _ You can select walls to make Area Boundary lines, but not Room Separation lines.

7. **True** _ **False** _ Area Plans, unlike Room Plans, can't be placed on Sheets.

8. **True** _ **False** _ Color Fills are not editable—once you place a fill, you have to delete it to change a color.

9. **True** _ **False** _ Revit Architecture will create Room definitions in Schedules without having Room Tags placed in Floor Plan views.

10. **True** _ **False** _ Revit Architecture can export schedule information into external database files in all recognized formats except Microsoft Access.

 Answers will be found on the CD.

CHAPTER 9

Illustrating the Design—Graphic Output from Revit Architecture

INTRODUCTION

By now you have worked with a number of building models using Revit Architecture and have built pages for construction document sets. Revit Architecture's ability to produce informative views of building models is hardly restricted to the plans, elevations and sections used for standard construction documents. All Revit Architecture models are built in three dimensions, with information about materials and surfaces as integral parts of the model objects. Three dimensional orthographic (parallel projection) views are a basic part of the designer's toolkit.

Three dimensional perspective views are also quick and easy. Revit Architecture contains a rendering module so that colored, lit (day or night, interior or exterior) pictures can be captured showing any part of the model at any point during design development. Revit Architecture incorporates a walkthrough tool that allows the user to create an animation based on a camera moving along a path. For users familiar with stand-alone rendering software applications, Revit Architecture will export files—images, models or animations—that these other products can receive and use.

OBJECTIVES

- Create and change perspective views, interior and exterior
- Apply and edit material textures to building objects
- Work with render settings—environment, lighting, location
- Create an exterior daylight rendering
- Capture a rendering image
- Export Revit Architecture views to external files
- Use the View Section Box to restrict the scope of a view and rendering
- Work with the Radiosity tool for interior lighting
- Create an interior nighttime rendering
- Create and edit a walkthrough path/camera combination
- Export a walkthrough animation file

- Plot Revit Architecture views using the Revit Architecture PDF writer
- Export 2D and 3D DWF
- Prepare Revit Architecture views for plotting using standard printers
- Export the Revit Architecture model for use in Autodesk VIZ

REVIT ARCHITECTURE COMMANDS AND SKILLS

Camera properties

New camera

View grip editing

View Properties Eye/Target editing

Export Image

Dynamic View Walkthrough controls

Region Raytrace

Raytrace exterior settings: environment, sun, lighting

Site components

Materials: create and assign

Section Box

Entourage components

Raytrace interior settings: limit model geometry

Light Groups, Daylights

Radiosity

Walkthrough: key frames, camera edit, path edit

Export walkthrough

Print/Plot

Export PDF

Export DWG for rendering

VIEWING THE MODEL

EXERCISE 1. THE CAMERA TOOL

1.1 Launch Revit Architecture.

- Open the file *Chapter 9 start imperial.rvt [Chapter 9 start metric.rvt]*.

  The file will open to show the 3D View: {3D} at the default southeast orientation.

- Pick File>Save As from the Menu Bar.

- Save the file as **Chapter 9 render.rvt** in a location determined by your instructor.

1.2 Open Floor Plan: Level 1.

- Make a right-to-left selection of the Toposurface and a tree, as shown in Figure 9.1.

- Type **VH** at the keyboard to turn off the visibility of Planting and Topography categories in the view.

  This will simplify the screen.

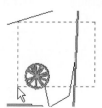

Figure 9.1   *Select and then turn off topography and a tree in this view for clarity*

3D Views have already been created in this file.

1.3 Choose 3D View 1 in the Project Browser tree.

- Right-click and pick Show Camera, as shown in Figure 9.2.

Figure 9.2   *Show the view camera*

1.4 Place the cursor over the camera/field of view symbol (see Figure 9.3).

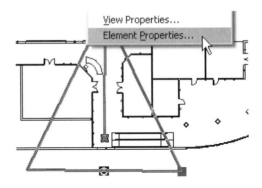

Figure 9.3 *The camera displayed and selected*

- Right-click and pick Element Properties.
- Study the properties of this view tool for a moment (see Figure 9.4).
- Click Cancel to exit the dialogue without making any changes.

Parameter	Value
Extents	
Crop Region Visible	☑
Far Clip Active	☑
Far Clip Offset	85' 0 57/64"
Crop View	☑
Section Box	☐
Camera	
Perspective	☑
Eye Elevation	5' 6"
Target Elevation	5' 6"
Camera Position	Explicit
Render Scene	None
Render Image Size	Edit...

Figure 9.4 *Properties of the camera that controls 3D View 1*

1.5 3D View 1 in the Project Browser tree is pre-selected.

- Right-click and pick Open.

The view will open with its border highlighted red and blue control dots (grips) displayed, as shown in Figure 9.5. Clicking on any of the grips will allow you to change the field of view.

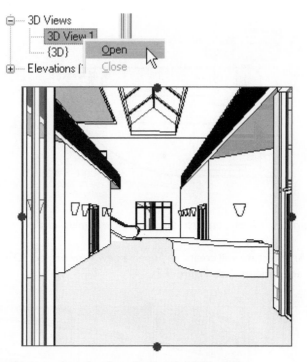

Figure 9.5   *3D View 1—note the grips for adjusting the field of view*

1.6 Open the Level 1 Floor Plan.

- Zoom to Fit.

- From the View tab of the Design Bar, choose Camera, as shown in Figure 9.6.

- Click to place the camera eye position at the lower-right side of the model.

- Leave plenty of space between the camera and the exterior wall of the building, so that perspective sight lines will not be extreme.

- Click inside the model near the curved desk in the main entry.

508

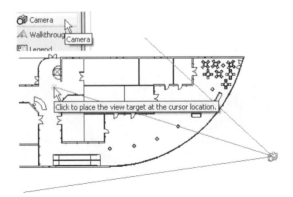

Figure 9.6 *Create a new camera; the first click places the camera at the eye of the viewer.*

Revit Architecture will create and open a new view named 3D View 2, which will need some adjustments (see Figure 9.7).

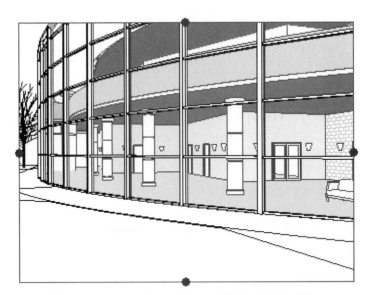

Figure 9.7 *The new 3D view in its rough state. Yours will look slightly different.*

1.7 Use the grips on the view to display the model completely, as shown in Figures 9.8 and 9.9.

As you change the shape of the view, you may find it useful to Zoom Out (2X) to give yourself room to work, then Zoom to Fit when the adjustments are complete. The appearance of your view may be slightly different from the illustrations shown here, since your camera will not be in exactly the same place.

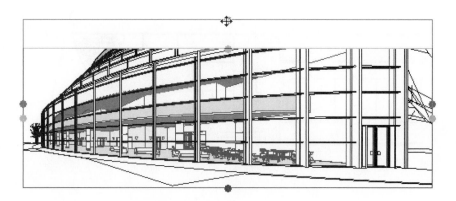

Figure 9.8   *Drag the grips to alter the field of view*

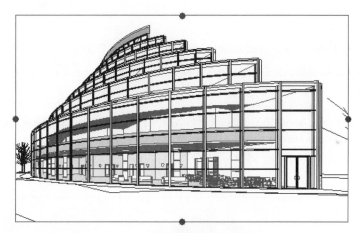

Figure 9.9   *The view now shows the entire model from this angle*

1.8 With the View grips still active, open the Level 1 Floor Plan.

Note the changed appearance of the camera field of view indicator, as shown in Figure 9.10.

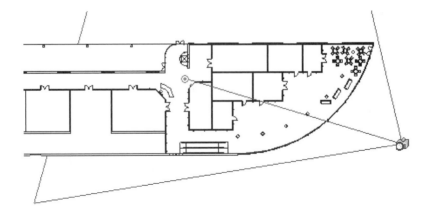

Figure 9.10 *Note how the field of view has altered*

1.9 Return to 3D View 2.

- From the View menu, select View>Shading with Edges.

   Revit Architecture will take a few seconds to process the display shown in Figure 9.11.

 Tip: The two-stroke typed shortcut for the Shading with Edges display is SD. HL sets the display to Hidden Line.

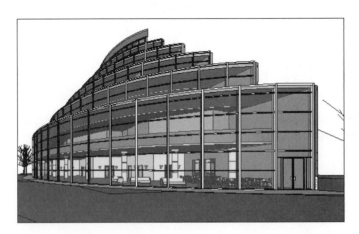

Figure 9.11 *The perspective view, shaded*

1.10 From the File menu, select File>Export>Image, as shown in Figure 9.12.

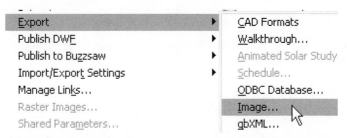

Figure 9.12 *The Export Image menu pick*

1.11 In the Export Image dialogue, give the file the name **Chapter 9 east exterior** and place it in a folder that you can find again easily.

- Select Visible portion of current window in the Export Range section.

- Leave the other settings.

- Note the output choices for file format, as shown in Figure 9.13.

- Click OK to create the file.

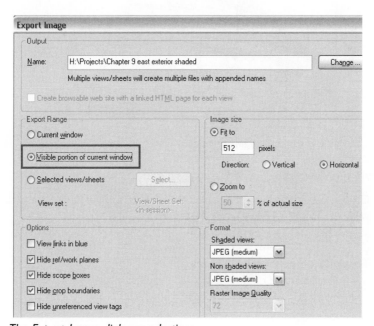

Figure 9.13 *The Export Image dialogue selections*

1.12 Type **HL** at the keyboard or use the View Control bar to return to Hidden Line mode in the 3D View.

- Right-click in the view window and select View Properties.

- In the Element Properties dialogue, change the Target Elevation value to **25' 6" [7650]**, as shown in Figure 9.14.

- Click OK.

The angle of the view will change.

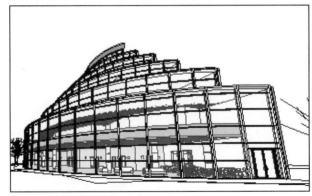

Figure 9.14 *Change the camera target elevation*

1.13 Open the View Properties dialogue again.

- Change the Target Elevation value back to **5' 6" [1350]** and change the Eye Elevation value to **55' 6" [16650]** (see Figure 9.15).

- Click OK.

The angle of the view will change.

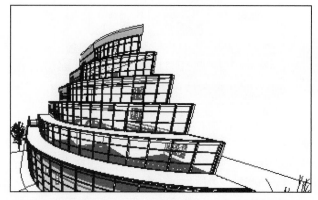

Figure 9.15 *Now change the height of the camera while looking at the original target*

1.14 Select the Undo tool on the Toolbar to undo the last view edit and lower the camera to eye level.

- Pick the Dynamically Modify View tool on the Toolbar, as shown in Figure 9.20, to open the Dynamic View controls.

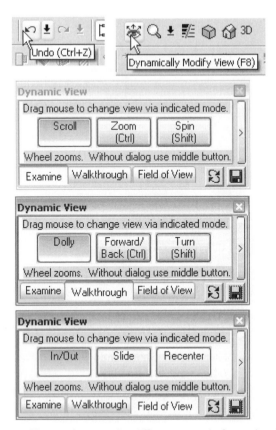

Figure 9.16   *The Dynamic View tool—note the different controls for each tab*

The Dynamic View control dialogue contains three tabs, with three different types of camera motion controls.

The dialogue also contains a control to undo a view change, a control to save the current state of the view, and an expandable panel with controls to orient the view using directions, views or planes, as shown in Figure 9.17. The control to orient the view to a plane opens a dialogue you have used before when sketching profiles in elevation views.

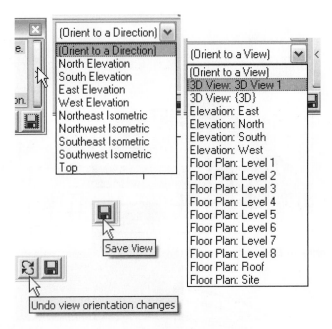

Figure 9.17 *Additional controls on the Dynamic View dialogue*

1.15 On the Dynamic View dialogue, select the Walkthrough tab.

Make the Dolly control active, as shown in Figure 9.16.

Note the instructions for the controls.

1.16 Hold down the left mouse button and Dolly (move laterally) the camera up/down and left/right to adjust the view.

Moving the cursor up has the effect of moving the camera position down, and vice versa (see Figures 9.18 and 9.19).

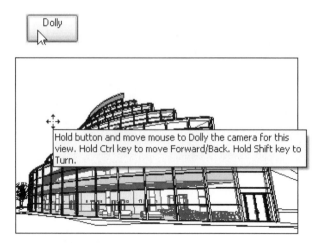

Figure 9.18 *Preparing to Dolly the camera*

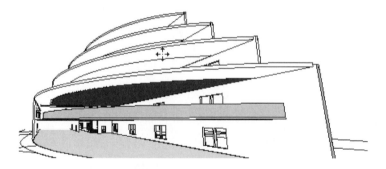

Figure 9.19 *Revit Architecture simplifies the display during motion*

1.17 Hold down the CTRL key to make the Forward/Back control active and repeat—alter the camera position in and out (see Figures 9.20 and 9.21).

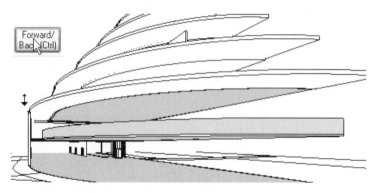

Figure 9.20 *The Forward/Back cursor has a different appearance from the Dolly cursor*

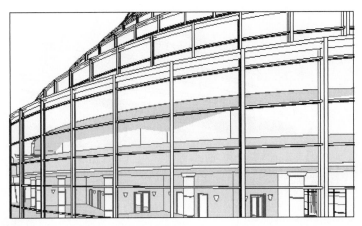

Figure 9.21 *Results of moving the camera Forward*

1.18 Make the Turn control (SHIFT) active and repeat the previous motions (see Figures 9.22 and 9.23).

Note that the camera holds position and turns right as the cursor moves left.

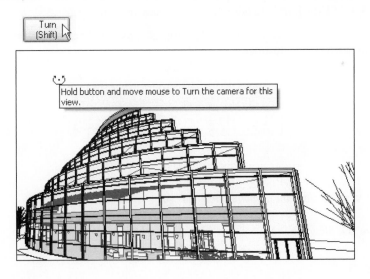

Figure 9.22 *Preparing to Turn the camera*

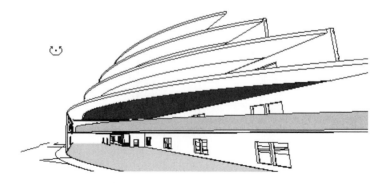

Figure 9.23 *Turning the camera left*

1.19 Hold the cursor over the title bar of the Dynamic View control.

- Pick the X controls (or right-click and select **Close**) to close the dialogue (see Figure 9.24).

- If necessary, use View Properties to return the Target and Eye Elevation values to **5' 6" [1350]**, and readjust the field of view using the grips to view the entire model.

Your camera may not be in its original position because of your left/right and forward/back adjustments.

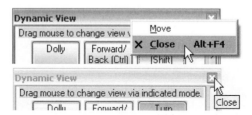

Figure 9.24 *Ways to close the Dynamic View dialogue*

1.20 Save the file.

RENDERING

Now that you have experimented with camera placement and have changed a perspective view to show the model from the angle or angles that suit your purposes, you can work on making the model come alive with color rendering.

EXERCISE 2. EXTERIOR SETTINGS

2.1 Open or continue working in the file from the previous exercise.

2.2 Make the Rendering tab of the Design Bar active.

- Pick the Region Raytrace tool, as shown in Figure 9.25.

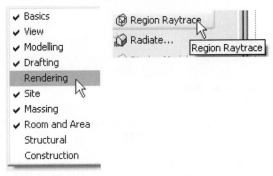

Figure 9.25   *The Region Raytrace tool*

2.3 Pick two points around the double door, as shown in Figure 9.26.

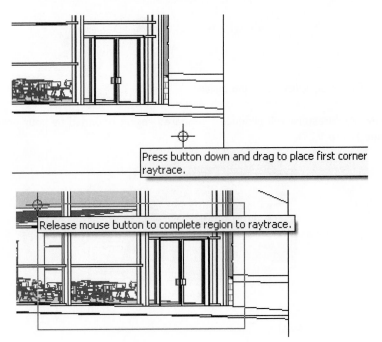

Figure 9.26   *The picks to define a region raytrace*

Revit Architecture will display the Scene Selection dialogue.

2.4 Select Exterior, as shown in Figure 9.27.

- Click OK.

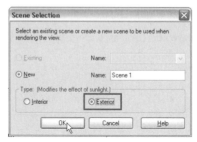

Figure 9.27 *Select Exterior Scene, then click OK*

2.5 In the query box that appears, click Yes to turn off lights inside the building for the render process, as shown in Figure 9.28.

The model already contains light fixtures—since this is an exterior scene, you have the choice to turn off the lights to save rendering time or leave them on for realism (as in a night scene).

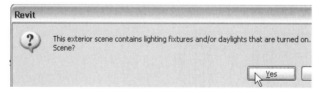

Figure 9.28 *Turning off lights inside the building*

Revit Architecture will produce a rendered region bounded by your previous picks (see Figure 9.29).

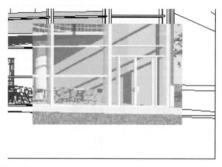

Figure 9.29 *The first rendered test area*

2.6 Choose Settings from the Rendering tab of the Design Bar, as shown in Figure 9.30.

- In the Render Scene Settings dialogue, select Rename.

- In the Rename dialogue, enter **Exterior – Gallery Side**. Click OK.

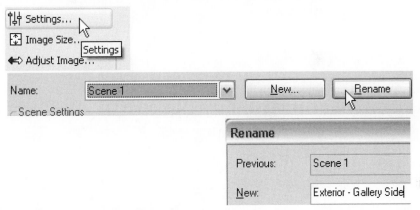

Figure 9.30 *Open Settings and rename the Scene*

2.7 Select Environment.

- In the Environment dialogue, accept the default Automatic Sky as the Background Color value.

- Pick Clouds in the Advanced section, as shown in Figure 9.31.

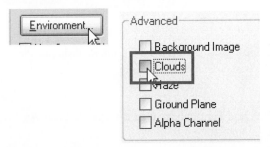

Figure 9.31 *Turn on Clouds*

Revit Architecture activates the Clouds tab, as shown in Figure 9.32.

2.8 Click OK to accept the defaults.

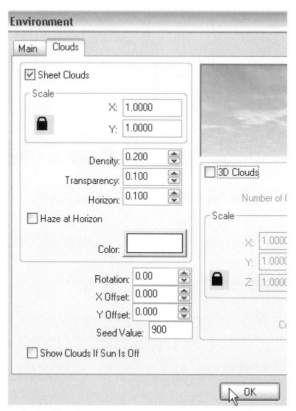

Figure 9.32 *The controls for Clouds*

2.9 In the Scene Settings area of the Environment dialogue, clear Use Sun and Shadow Settings from View.

- Click Sun.

- On the Date and Time tab, change the Specify Solar Angles value to By Date, Time, and Place.

- On the Place tab, use the Cities drop-down list to find **Nashville, TN, USA** and select it, as shown in Figure 9.33.

  You can also select your own location.

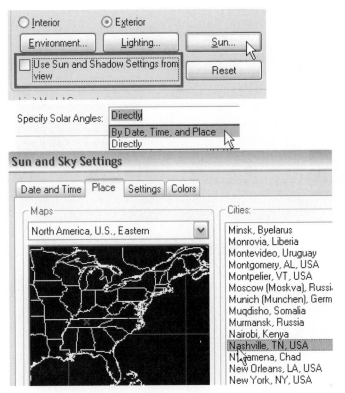

Figure 9.33   *Assign a location on the Place tab*

2.10 Select Date and Time.

- Set the Date to June 15 (December 15 if your location is in the Southern Hemisphere) and the time to mid-morning (see Figure 9.34).

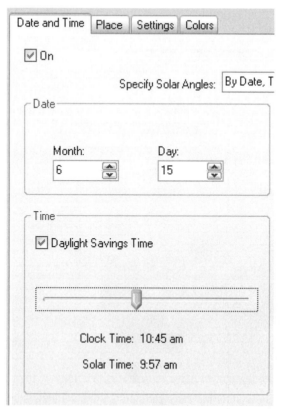

Figure 9.34 *Set the Date and Time for a summer morning*

2.11 Select the Settings tab and Colors tab in turn to view the controls, as shown in Figures 9.35 and 9.36.

- Accept the defaults.

- Click OK.

 Note: The Settings tab allows you to save settings that govern cloudiness, sun and sky intensity, and the direction of solar north, or load in previously saved setting files.

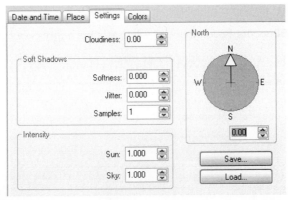

Figure 9.35   *The Settings tab*

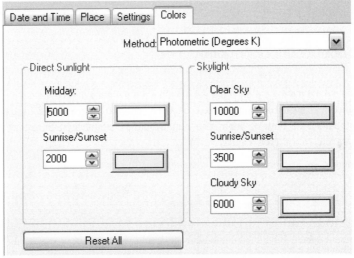

Figure 9.36   *The Colors of the sky can be adjusted here*

> **2.12** Change the Plant Season value on the main Render Scene Settings dialogue page to **Summer**, to match the sun settings (see Figure 9.37).
>
> - Click OK to accept the remaining settings.

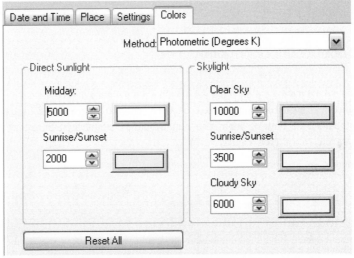

Figure 9.37   *Match the foliage to the sun angle*

2.13 Select Image Size on the Rendering tab of the Design Bar.

- Enter **300** as the Resolution dpi value, as shown in Figure 9.38.

- Hit TAB to view the resulting image size.

- Click Cancel.

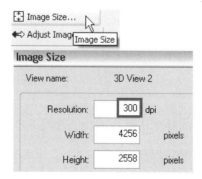

Figure 9.38 *The Image Size can be controlled here*

2.14 Choose Adjust Image on the Rendering tab of the Design Bar.

- Note the available Image Controls: Brightness, Contrast and Indirect (ambient light intensity not coming from the sun or other point light sources) and a toggle for dynamic range (see Figure 9.39).

- Click Cancel.

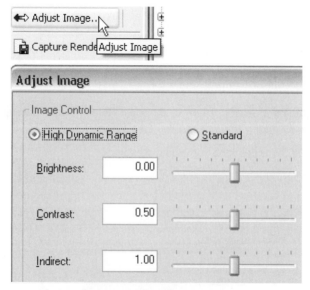

Figure 9.39 *You can tweak overall picture settings here*

MATERIALS AND PLANTINGS

2.15 From the Settings menu on the Menu Bar, choose Settings>Materials, as shown in Figure 9.40.

- In the Materials dialogue, use the Name drop-down list to find **Finishes - Exterior - Curtain Wall Mullions**.

There is no Texture assigned in the AccuRender area.

- Choose the select icon to the right of the Texture field.

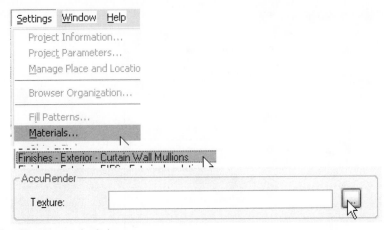

Figure 9.40 *The Materials dialogue*

2.16 Once you're inside the Material Library dialogue, navigate to the *_accurender\Metals\Aluminum, Anodized* folder in the tree pane.

- Select **Bronze, Medium** in the Name pane, as shown in Figure 9.41.

- Click OK.

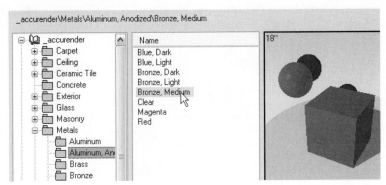

Figure 9.41 *Navigate among the many selections for materials*

2.17 Use the Name drop-down list to find **Glazing - Curtain Wall Glazing**.

- Pick the select arrow next to the Texture field to change the default value.

2.18 Back inside the extensive Material Library dialogue, navigate to the _accurender\Glass\Tinted_ folder in the tree pane.

- Select **Bronze, Medium, Smooth** in the Name pane, as shown in Figure 9.42.

- Click OK.

- Click OK again to exit the Settings dialogue.

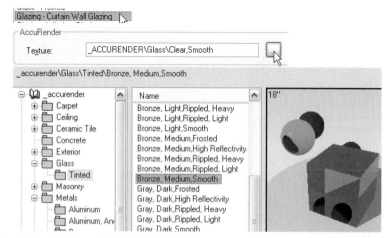

Figure 9.42 *Assign a Material for Curtain Wall glazing*

2.19 On the Rendering tab of the Design Bar, select Display Model to remove the partial rendering.

- Select one of the curtain wall mullions, as shown in Figure 9.43.

    You may have to use the TAB key to cycle through the possible picks—the Type Selector should read **Rectangular Mullion: 2.5" x 5" rectangular [Rectangular Mullion: 62mm x 125 mm rectangular]**.

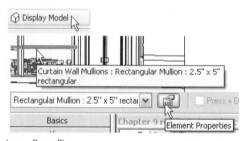

Figure 9.43 *Pick a curtain wall mullion*

2.20 Click the Properties icon.

- Click Edit/New.

- Change the Material parameter Value to **Finishes - Exterior - Curtain Wall Mullions**, as shown in Figure 9.44.

- Click OK three times to exit the dialogues.

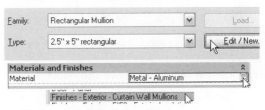

Figure 9.44 *Assign the mullion material via Properties*

2.21 Carefully choose one of the vertical 5" x 10" [125 x 250] mullions.

- Repeat the Materials assignment as in Step 2.19 to **Finishes - Exterior - Curtain Wall Mullions** (see Figure 9.45).

- Click OK three times to exit the Element Properties dialogue.

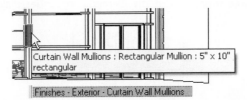

Figure 9.45 *Assign the Curtain Wall Mullion finish to the vertical mullions*

2.22 Select the double door component.

You may have to use TAB to cycle among the possible selections.

- Click Properties.

- Click Edit/New.

- Change the Materials parameter Value to **Finishes - Exterior - Curtain Wall Mullions**, as shown in Figure 9.46.

- Click OK three times to exit the Element Properties dialogue.

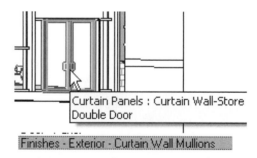

Figure 9.46 *Assign the same material to the Double Door frame*

2.23 Carefully select a curtain wall panel above the door.

You may have to TAB through the possible selections.

- The Type Selector will read System Panel when your selection is correct.

- Click Properties.

- Click Edit/New.

- Change the Materials parameter Value to **Glazing - Curtain Wall Glazing**, as shown in Figure 9.47.

- Click OK three times to exit the Element Properties dialogue.

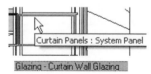

Figure 9.47 *Assign the tinted glazing material to the curtain wall panels*

2.24 Choose Region Raytrace from the Rendering tab.

- Pick two points around the double door as before.

- Note the difference in tints now that bronze finishes and glazing have been selected, as shown in Figure 9.48.

Figure 9.48 *The Region Raytrace shows different colors and shadows after materials are changed and sun angle specified*

2.25 Choose Display Model to remove the rendered image from the view.

2.26 Open the Floor Plan Level 1 view.

- Type **VG** to open the Visibility/Graphics Overrides dialogue.

- Check Plantings and Topography.

- Click OK.

2.27 Make the Site tab active in the Design Bar.

- Choose Site Component, as shown in Figure 9.49.

- Pick **Shrub: Rhodo 24" [Shrub: Rhodo 600mm]** from the Type Selector on the Options Bar.

- Pick the Properties icon.

Figure 9.49   *The Site Component tool on the Site tab*

2.28 Pick Edit/New.

- Click Rename.

- Change the Type name to **Rhodo 16" [Rhodo 400mm]**.

- Change the Plant Height Value to **1' 4" [400]**, as shown in Figure 9.49.

- Click OK.

- Click OK to exit the Element Properties dialogue.

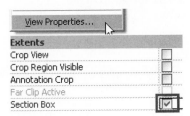

Figure 9.50   *Values for the new Type*

2.29 Place a number of plants beside the concrete walkway, as shown in Figure 9.51.

The exact number and placement are not important.

- Click Modify to terminate the component placement.

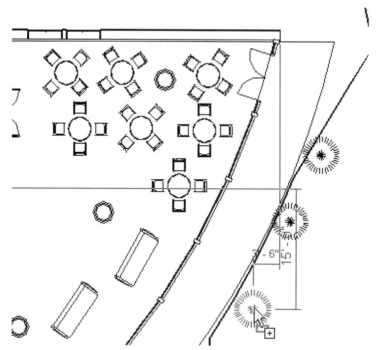

Figure 9.51 *Place shrub instances near the exterior walkway on the east side*

2.30 Open 3D View 2.

- Make the Rendering tab active in the Design Bar.

- Select Raytrace on the Rendering tab, then pick GO on the Options Bar.

  Revit Architecture will generate a rendering using the sunlight, environment and materials settings you have previously defined. This may take a few minutes, so be patient. See Figure 9.52 for an example.

Figure 9.52   *The rendered image*

2.31 When the rendering is complete, select Capture Rendering on the Rendering tab of the Design Bar.

Revit Architecture will add a view named 3D View 2 under Renderings in the Project Browser (see Figure 9.53). You can rename, duplicate and copy this view to the Clipboard.

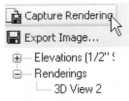

Figure 9.53   *The Capture Image tool*

2.32 Select Export Image on the Rendering tab of the Design Bar.

- Place a copy of the image in a location you will be able to find later.

- Note that you can specify the name and file type of the image using this export routine (see Figure 9.54).

Figure 9.54 *Export the new image to an external file*

2.33 Select Display Model so the 3D view no longer shows the Raytraced image.

- From the File menu, select File>Export>Image.

- Note the additional options in this dialogue, as shown in Figure 9.55. Click Cancel.

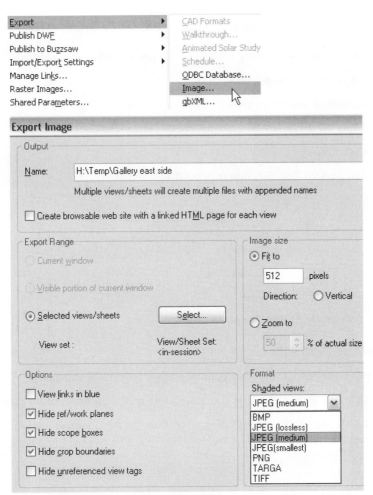

Figure 9.55 *Export Image dialogue from the File menu*

2.34 Save the file.

Interior renderings depend on lights inside the model for their illumination. Revit's standard light components all have lighting values, and so can provide enough light for renderings. Raytracing with multiple light sources is considerably more resource-intensive than with only one source, the sun. Nearly all the interior renderings you create will need to be planned and cropped carefully, so that you are not asking the raytracing engine to calculate light reflections outside the area of the rendering. You can create a 3D crop, called a Section Box, around an area of the model and apply that when rendering. You can create more than one Section Box.

EXERCISE 3. INTERIOR SETTINGS—SECTION BOX CROP

3.1 Open or continue working with the file from the previous exercise.

- Open the default 3D View {3D}.

- Right-click and select View Properties, as shown in Figure 9.56.

- In the View Properties dialogue, check Section Box.

- Click OK.

- The Section Box will appear around the model.

Figure 9.56 *Activate the Section Box for this view*

3.2 Select the Section Box.

It will highlight red, and blue control arrows (grips) will appear.

- Drag the grips and adjust the boundaries of the Section Box to crop the model.

- Isolate the west (left) half of the top floor of the tower, as shown in Figures 9.57 and 9.58.

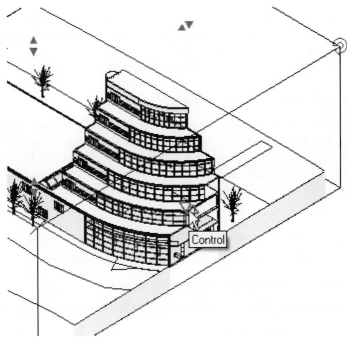

Figure 9.57 *Pick the grip to drag it*

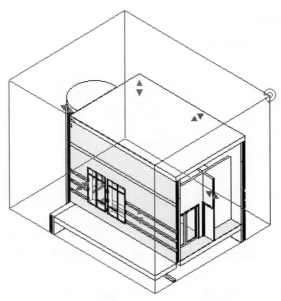

Figure 9.58 *Isolate a portion of the top floor*

Caution: This is an important step. If you do not correctly set and use the Section Box, the Rendering and Radiosity calculations in future steps will run extremely slowly, or not at all.

COMPONENTS—IMAGES OF PEOPLE

Architectural models can look sweeping and grand, or intimate and cozy, when viewed from the proper angle, but even the best spaces tend to appear sterile without human beings or their usual "stuff" in them. Many third-party developers have created quickly rendered models of people, vehicles and accessory items to place in rendered scenes to make them come alive for the viewer, and to provide a sense of scale.

3.3 Open the Floor Plan View Level 7.

- Zoom In Region around the left side of the model, a small lobby or waiting room.

- From the Rendering tab of the Design Bar, select Component.

- Choose Load from Library on the Options Bar.

3.4 Navigate to the *Imperial Library\Entourage* [*Metric Library/Entourage*] folder.

- Select **RPC Female.rfa**.

- Hold down the CTRL key and select **RPC Male.rfa**.

- Click Open.

- This will load both families into the current project file (see Figure 9.59).

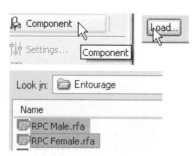

Figure 9.59 *Load in components to represent people*

3.5 Use the Type Selector on the Options Bar to make **RPC Male: LaRon** the current component.

- Check Rotate after placement on the Options Bar.

- Place the instance in the lower-left corner of the room, in front of the elevator doors.

- Rotate it 135° from the default position, which will point the front of the model to the upper right of the plan.

RPC person components display in the plan as circles with a radius line representing the front. You can rotate the component after placement if necessary.

3.6 Use the Type Selector to place an instance of **RPC Female: Cathy** to the right of the desk as shown.

- Click Modify.

- Select and rotate the component approximately -45° so it faces the other RPC person (see Figure 9.60).

Figure 9.60   *Placing the new components, facing each other*

3.7 From the View tab of the Design Bar, select Camera.

- Place the camera to the right of the room, in front of the double doors.

- Place the target behind the desk, as shown in Figure 9.66.

- Relocate the Cathy component if necessary, so it will be in the camera view.

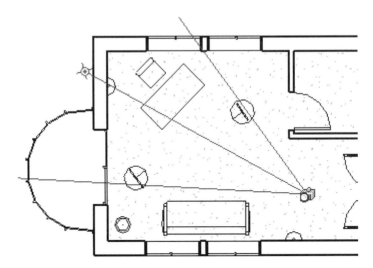

Figure 9.61 *Place and point a camera in the room*

Revit Architecture will open the new camera view. There is a size control on the Options Bar when the view is selected.

3.8 Select the camera field of view.

- Click the button on the Options Bar that shows the current image size.
- Change the width value to **9" [225]** and the height value to **5" [125]**.
- Leave the Field of view radio button selected, as shown in Figure 9.62.
- Click Apply and note the results.
- Click OK.
- Use the field of view grips to adjust the view further if necessary, so that it does not display unnecessary amounts of floor or ceiling. See Figure 9.63 for an example.

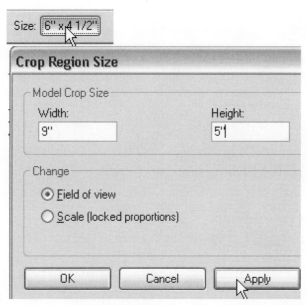

Figure 9.62   *The view size control on the Options Bar and the Crop Region Size dialogue*

Figure 9.63   *The field of view, adjusted—note the simplified appearance of the person components*

    3.9 Choose Settings from the Render tab of the Design Bar.

- In the Scene Selection dialogue, select New and Interior Scene.

- Enter the name Seventh **Floor Lobby**, as shown in Figure 9.64.

- Click OK.

Figure 9.64 *Name the new Scene*

The Render Scene Settings dialogue will appear.

3.10 Click OK to close it for now.

LIGHTING, DAYLIGHTS, RADIOSITY

3.11 Select Lighting from the Rendering tab of the Design Bar, as shown in Figure 9.65.

- Note that a Light Group has been defined in this file—sconce lights on the first floor are all grayed out.

- Clear the On box for light group **1st floor sconces**.

- Click OK.

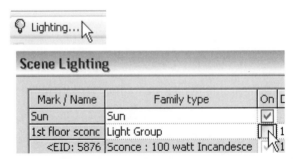

Figure 9.65 *The Lighting tool on the Rendering tab*

Light Groups function as component groups do in Revit Architecture, and are useful for managing groups of lights when you want to study or display alternate lighting conditions. You will not explore Light Groups in any detail in this exercise.

3.12 Choose Daylights from the Rendering tab of the Design Bar, as shown in Figure 9.66.

There is no dialogue associated with this tool.

- Pick the two windows at the right side of the view.

- Pick the double door to the elevator tower.

- Click Modify to terminate the selections.

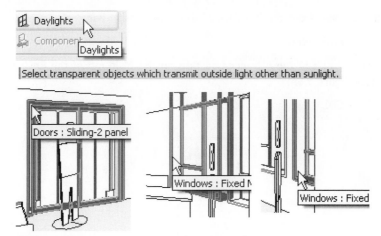

Figure 9.66   *The Daylights tool*

Daylights are sources of exterior light. Glass daylights both admit light and reflect according to the properties of their glass (clear, frosted or tinted).

3.13 Choose Lighting from the Rendering tab of the Design Bar.

- Scroll to the end of the light list.

- Note that the new Daylights are now included (see Figure 9.67).

- Click Cancel to exit the Lighting dialogue.

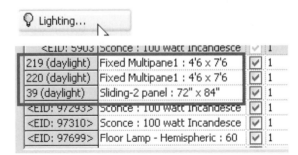

Figure 9.67 *The daylights now show in the lighting list, where they can be turned on and off*

3.14 Click Settings from the Rendering Tab of the Design Bar.

- Click Environment.

- Pick Clouds in the Advanced section on the Main tab of the Environment dialogue (see Figure 9.68).

- Click OK.

Figure 9.68 *Turn on clouds in the sky outside the windows*

3.15 Clear the option to Use Sun and Shadow Settings from view.

- Select Sun.

- Set Specify Solar Angles to Date, Time, and Place.

- Set the Place to **Nashville, TN, USA**, as before.

  You can also use your own location.

- Use the time slider to set the time to **8:15 pm** (just after sunset) on **June 30**.

- If your location is in the Southern Hemisphere, use **December 30**.

- Accept all the other defaults (see Figure 9.69).

- Click OK.

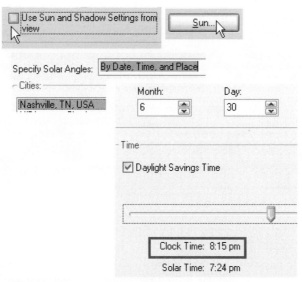

Figure 9.69 *Set the day and time*

3.16 On the main page of the Render Scene Settings dialogue, set the Use View's Section Box drop-down list to display **{3D}**.

This applies the Section Box crop you performed earlier in the exterior view to this render setup.

- Check Back Face Culling and View Culling to save some time in the rendering process (see Figure 9.70).

- Click OK.

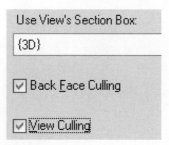

Figure 9.70 *Limit the render geometrically*

3.17 Select Radiate from the Rendering tab of the Design Bar, as shown in Figure 9.76.

Revit Architecture will display a Radiosity Information pane.

- Click OK.

Radiosity is a powerful technique for calculating INDIRECT LIGHTING and performing LIGHTING ANALYSIS. When used properly, in conjunction with ray tracing, this technique can produce images which contain more subtle illumination effects than those produced using traditional ray tracing only.

Radiosity is generally most useful for SMALL ARCHITECTURAL INTERIORS. Use of this option WILL require more memory and processing time. These additional requirements will be sensitive to the size and complexity of the model.

Radiosity is usually NOT appropriate for the following situations:

* Non-architectural models
* Large or complex models
* Exterior renderings

Figure 9.71 *The Radiate tool—useful, but not all-purpose*

Revit Architecture will do a Radiosity calculation on the space and change the view gradually to a completely lighted rendering. This will take some time, so be patient. Note that the RPC persons will not display in the Radiosity (see Figure 9.72). They will display in the Raytrace.

Figure 9.72 *The Radiosity solution displayed graphically*

3.18 When the Radiosity is complete, pick Save Radiosity, as shown in Figure 9.73.

- Save the file in a convenient location.

  This creates a file with a .rad extension to hold the settings for this calculated Radiosity solution. This solution can be loaded later, to save creating another Radiosity solution for this view.

  You can click the Continue button on the Options Bar to cycle the Radiosity through more steps (25 is the default cycle). This will refine the calculations, but will not change the appearance of the radiated view.

Figure 9.73   *The Save Radiosity tool and Continue option*

3.19 Pick Raytrace from the Rendering tab of the Design Bar.

- If a warning appears, click OK.

  This warning appears because you have saved the Radiosity, so Revit Architecture considers that it may now be out of date.

3.20 Set the Resolution field on the Options Bar to Medium (150 dpi).

- Note that the Image Size (grayed out to indicate that you cannot change it independently) changes as you change the Resolution value.

- Click GO! from the Options Bar (see Figure 9.74).

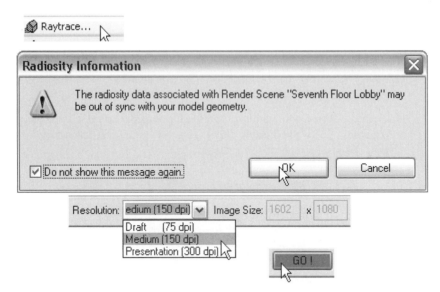

Figure 9.74 *Options Bar controls for the Raytrace rendering*

Revit Architecture will take a few minutes to produce an image of the space (see Figure 9.75). Note that the window glass is both reflective and transparent—a few clouds are visible outside.

Figure 9.75 *The raytrace image with RPC people*

3.21 Capture or export the image as your instructor directs.

- Save the project file.

EXERCISE 4. CREATE A WALKTHROUGH

Still images of the model are not the only graphic output possible. The rendering module in Revit Architecture contains a tool to create a walkthrough, or series of images recorded by a camera that travels along a path. You can view any of these images and export a file that collects the images so they can be played in media players. These files can be sent to clients or located on Websites for playing/download.

Media-savvy designers can, with very little effort and no additional software, provide clients with completely editable, 3D motion pictures of designs at any stage of progress. This capability enables far better decision making by people with strong design sense but limited ability to visualize from standard orthographic plans, elevations and sections. The walkthrough tool will soon elevate client expectations of the quality of the visual information they can expect from building designers. Besides, this is fun.

4.1 Open or continue working with the file *Chapter 9 render.rvt.*

- Open the Floor Plan Level 1.

- Select the Walkthrough tool on the View tab of the Design Bar.

  The cursor will change to a pencil shape and the Tooltip will display the message Click to place Walkthrough key frame.

  The Options Bar will display Walkthrough controls, as shown in Figure 9.76.

Figure 9.76 *Walkthrough controls on the Options Bar*

4.2 Zoom in Region to the right side of the building.

- Place the first key frame inside the double door that opens north.

- Place key frames down the corridor past the curved desk, turn right down the hall past the escalator, and continue placing key frames between the columns and interior walls, moving toward the cafeteria seating (see Figures 9.77 and 9.78).

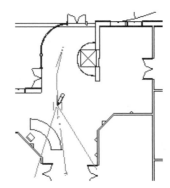

Figure 9.77   *Start in the lobby*

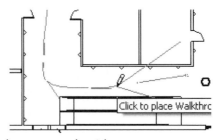

Figure 9.78   *Move around the corner to the right*

4.3   Place 10–12 key frames in all.

- When the field of view indicator reaches the seating area, as shown in Figure 9.79, click Finish from the Options Bar.

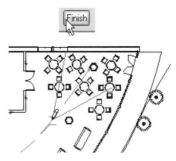

Figure 9.79   *Finish in the seating area*

4.4   Select Edit Walkthrough from the Options Bar, as shown in Figure 9.80.

A second set of controls appears on the Options Bar. Note that the Controls drop-down list lets you choose the type of walkthrough component you can edit.

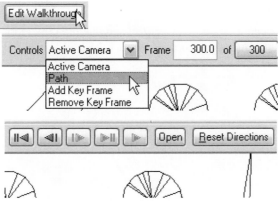

Figure 9.80 *Edit the walkthrough*

4.5 Set the control drop-down list to Path.

- Select the Frame settings button (it shows the number 300 as its caption).

The Walkthrough Frames dialogue opens, as shown in Figure 9.81. This allows you to set the number of frames and the distance between key frames if you do not want a uniform travel speed.

- Change the Total Frames value to **100**. Click OK.

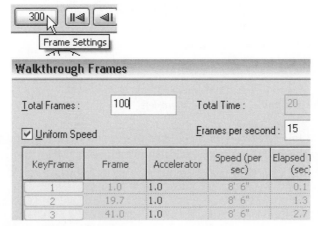

Figure 9.81 *The Frame Settings dialogue*

The camera is located at the last key frame. The Frame step controls on the Options Bar change the position of the camera along the path. You can step per Frame or key frame.

4.6 Select the Previous key frame control one time.

The Camera will move back one dot (key frame point) on the walkthrough path line.

4.7 Set control to Active Camera.

- Pick the camera Target Point control and pull the camera target to the right, pointed toward the couches, as shown in Figure 9.82.

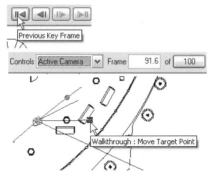

Figure 9.82   *Change the camera direction by moving the target control point*

4.8 Step back two more key frames and point the camera target to the left, as shown in Figure 9.83.

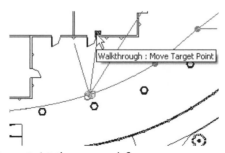

Figure 9.83   *At this key frame, point the camera left*

4.9 Step back to a key frame in the middle of the entry corridor and point the camera target to the left of the path line, as shown in Figure 9.84.

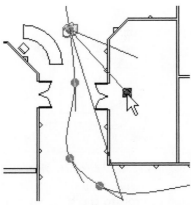

Figure 9.84 *Point the camera left at this key frame*

4.10 Change the Controls value to **Path** and shift the grip on the fifth key frame from the beginning to the left of the original path line, slightly.

- Don't make too big a distortion of the path, or your animation will wobble (see Figure 9.85).

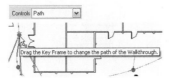

Figure 9.85 *Change the path slightly*

If you accidentally pick a point off of the walkthrough path, Revit Architecture will ask if you are finished editing the Walkthrough.

 Note: The Reset Directions button at the right of the Options Bar will undo all the camera target edits, so that the same path can be set to different view directions and thus create different walkthroughs quickly.

4.11 With the Walkthrough highlighted, choose Open on the Options Bar, as shown in Figure 9.86.

The current camera view will display. Its border will be highlighted and the grips available. You can edit the field of view—this setting will then hold for the entire walkthrough, not just the frame or key frame.

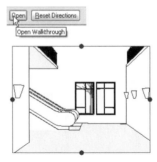

Figure 9.86 *Open the Walkthrough to check the camera view*

4.12 Use the key frame step controls to check the view at each key frame in your walkthrough.

You can type in the number of the frame you want to adjust (0 is the first). You can also play the walkthrough using the key frame controls (see Figure 9.87).

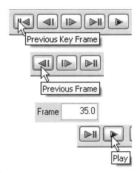

Figure 9.87 *Specific frame controls*

- If you need to change the camera location or direction for clarity, open the Level 1 Floor Plan and select Edit Walkthrough.

- If the Walkthrough path is not visible in the Level 1 Floor Plan, expand the Walkthroughs section of the Project Browser, select Walkthrough 1, right-click and pick Show Camera (see Figure 9.88).

Figure 9.88 *You can show or open the Walkthrough from the Browser*

4.13 When you have finished editing the camera location and direction, choose Walk-through 1 in the Project Browser, right-click and pick Open (see Figure 9.88).

You must do this to make the next step possible.

EXPORT THE WALKTHROUGH TO AN AVI FILE

4.14 From the File menu, select File>Export>Walkthrough.

- Locate the new file in a suitable folder.

- Change the Frames per second value to **5**.

- This will make your walkthrough run for 20 seconds (see Figure 9.89).

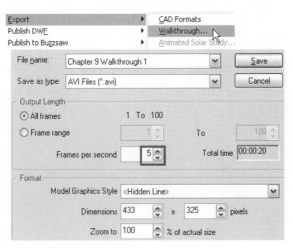

Figure 9.89 *The Export AVI dialogue*

Accept the Hidden Line display mode, but note that Wireframe, Shading, Shading with Edges and AccuRender displays are available.

A rendered walkthrough in a complex model can take a very long time to create. Try experimenting with some settings and render your walkthrough overnight!

4.15 Pick Save in the Export AVI dialogue to create the file.

- In the Video Compression dialogue, choose a compression method, as shown in Figure 9.95.

  Some compression routines are configurable. Some are not supported by certain media players. For our purposes, any one will do (see Figure 9.95). Using Full Frames (Uncompressed) creates a very big file.

- Click OK.

  Revit Architecture will create the walkthrough frame by frame.

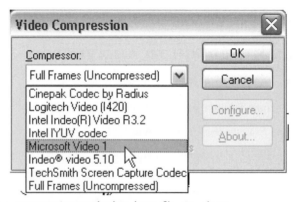

Figure 9.90 *Select a compression method to keep file size down*

- If the Export AVI dialogue appears a second time, click Close.

4.16 When the walkthrough has been created, find the new AVI file in your file manager and open it in your media player software application (see Figure 9.91).

- Play it through a few times; use the player's controls to move from view to view.

Figure 9.91 *The AVI plays in a media player*

4.17 Save the Revit Architecture file.

EXERCISE 5. OUTPUT FORMATS

EXPORT GRAPHICS VIA PLOTTING

Exciting graphics aside, Revit Architecture provides standard plotting and printing controls for putting images or sheets on paper. Revit Architecture's installation CD currently includes a PDF writer option so you can generate images or pages in that format, a standard for exchanging electronic documents. Revit Architecture exports DWF in 2D and 3D. DWF is Autodesk's format for exchanging engineering documents, and allows more control over display of the complex contents of design files than does PDF.

5.1 Open or continue working in the file *Chapter 9 render.rvt.*

- Open view 3D View 2 (the exterior perspective of the curtain wall).

- Zoom to Fit.

- From the Window menu, select Window>Close Hidden Windows, as shown in Figure 9.92.

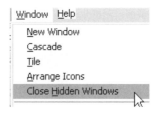

Figure 9.92 *Close other windows*

5.2 Select the Print icon on the Toolbar.

Revit Architecture will open the Print dialogue.

- Select an available network printer, which will be different from what is shown in Figure 9.93.

- Check the Visible Portion of Current Window radio button under Print Range.

- Choose Setup under Settings.

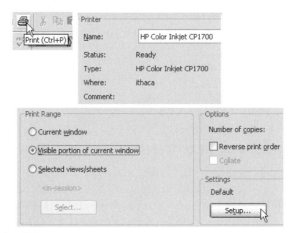

Figure 9.93 *The Print dialogue*

5.3 The Print Setup dialogue opens.

- Note the available print settings, as shown in Figure 9.94.

  You can create different named setups for each printer and save them.

- Check the Fit to page radio button.

- Select Landscape, as this view is wider than it is tall.

- Click OK.

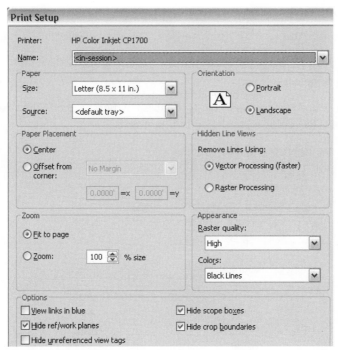

Figure 9.94   *The Print Setup dialogue*

5.4 Select Preview in the Print dialogue.

Revit Architecture will provide a print preview page of the active view based on the print settings (see Figure 9.95).

- Click Close to exit the print dialogue without generating a printed copy.

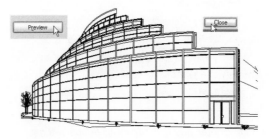

Figure 9.95   *The Print Preview of the View 3D 2*

EXPORT VIA PDF

Previous versions of Revit shipped with a module to create Adobe PDF output. Revit Architecture does not include this option. There are third-party PDF writers available. Adobe Acrobat creates a PDF system printer when installed.

5.5 Click the Print icon on the Toolbar.

- Choose **Adobe PDF** from the Name list, if it is available.

- If a PDF Writer is not in the selection list, move to Step 5.10.

5.6 Select Properties.

- There are two tabs available to establish settings in the PDF Writer 4.2 Document Properties dialogue, as shown in Figure 9.96.

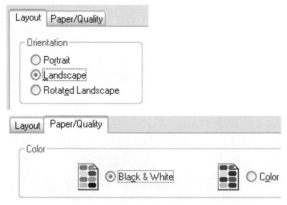

Figure 9.96 *The Revit PDF Writer Properties dialogue, with two tabs*

- The Adobe PDF printer has a third tab, as shown in Figure 9.97.

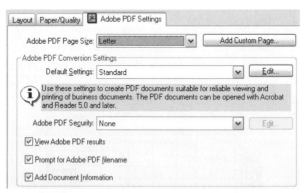

Figure 9.97 *The Adobe PDF printer has a third tab of Adobe settings*

- Choose Cancel.

5.7 In the Print dialogue, select Setup.

- Check that the plot will be landscape orientation on a letter-size page, centered and fitted to the page, as shown in Figure 9.98.

- Choose OK.

- If a question box asks about saving settings, choose No.

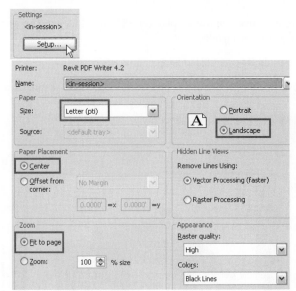

Figure 9.98 *Print settings for the PDF writer*

5.8 Set the Print Range in the Print dialogue to Current Window.

- Choose Preview.

  The image should fill the page.

- Select Print to return to the Print dialogue.

- Click Browse.

- Give the output file a name and location you can find again.

- Click OK to generate the PDF (see Figure 9.99).

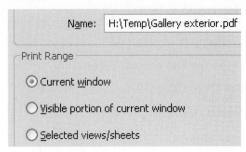

Figure 9.99 *Prepare to print the PDF file*

Adobe will force you to repeat this step. Revit Architecture will generate the new PDF.

Revit Architecture will not display the PDF file. But you can open it in Adobe Reader, as you can see in Figure 9.100.

5.9 Find the new file in your file manager and open it.

- If you are unable to open the file, Adobe Acrobat Reader is a free download.

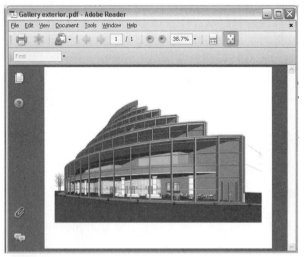

Figure 9.100 *The PDF file in Acrobat*

EXPORT VIA DWF

Autodesk is promoting the use of the DWF (Drawing Web Format) file output from its design products as a file format made for electronic sharing. DWF files are smaller than native (dwg, rvt) files, and can be sent as email attachments or embedded in Web pages for viewing with Autodesk's DWF Viewer. The DWF Viewer is a free download from Autodesk that, once installed, acts both as a stand-alone DWF viewer and a plug-in for browsers. Autodesk markets Design Review, a markup tool for the DWF format.

Revit Architecture will publish 2D or 3D DWF.

5.10 Type **SD** to turn on the Shaded with Edges display in the current 3D view.

5.11 Select File>Publish DWF>3D DWF from the Menu Bar.

There are no options to configure in this process.

5.12 Give the output file a name and location you can find again.

- Click Save in the Export dialogue (see Figure 9.101).

- Click OK to generate the DWF.

It may take a couple of moments.

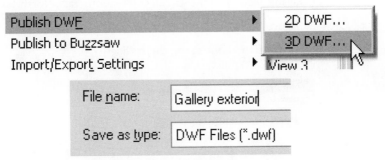

Figure 9.101   *Specify a location for the output file*

Revit Architecture will not display the DWF.

5.13 Launch the Autodesk DWF Viewer, if it is installed on your machine.

If it is not installed, it is a free download from Autodesk.

- Navigate to find the DWF file you just created and open it (or navigate to the file location in your file browser and double-click on the file name).

The Viewer will allow you to move through standard views, zoom, pan and orbit in the viewer pane, turn object categories on and off and print (see Figure 9.102).

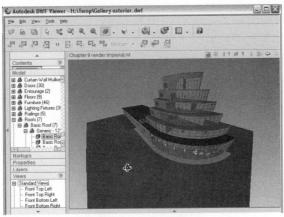

Figure 9.102   *The DWF shows color from the Shaded View*

5.14 Select File>Export DWF>2D DWF from the Menu bar if it is available.

- If the option is grayed out, move to Step 5.20.

  A Print dialogue opens.

5.15 In the Print Range section, choose Selected Views/Sheets, then Select.

- In the list of available views and sheets, select Elevations East, North, South and West.

- Pick OK (see Figure 9.103).

- Click No in the question about saving the settings.

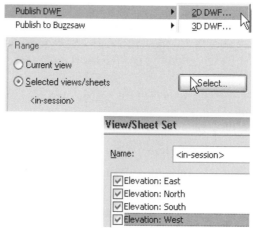

Figure 9.103 *Select views to print to DWF*

5.16 Choose Print Setup.

- In the Print Setup dialogue, make the output fit landscape orientation on a letter size sheet, centered and fit to the page, as shown in Figure 9.104.

- Click OK.

- Choose No in the question about saving settings.

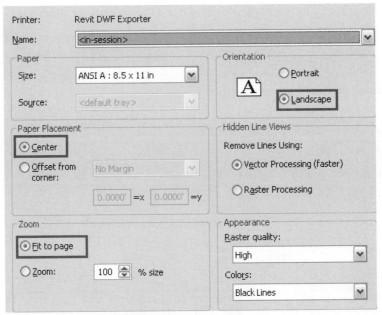

Figure 9.104   *Print Setup for the DWF*

5.17   Make sure Export each view or sheet as a single file is cleared.

- Locate the file in a place you can remember (see Figure 9.105).

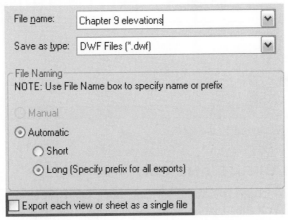

Figure 9.105   *Create a multi-sheet file*

5.18   Choose Save to create the DWF.

Revit Architecture will not display the DWF.

5.19 Launch the Autodesk DWF Viewer, if it is installed on your machine.

- Navigate to find the DWF file you just created and open it (or navigate to the file location in your file browser and double-click on the file name).

The Viewer will allow you to step through the four views, pan and zoom in the viewer pane, and print (see Figure 9.106).

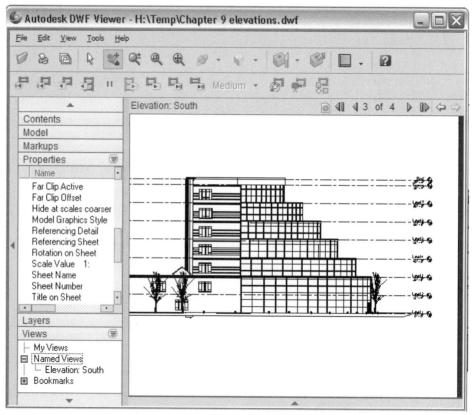

Figure 9.106 *2D DWF output in the viewer*

IMPORT DWF MARKUPS AND EDIT THE MODEL

Revit will import 2D DWF markup files that have been processed by Autodesk DWF Composer, but not DWF files that contain no markups. The imports link to the views that were used to create the original 2D DWF file or files, so long as the views have been placed on sheets.

The links can be loaded, unloaded, removed, turned on/off and saved back to the original DWF. This provides round-trip markup capability for dispersed design teams and their reviewers, who then do not need to have Revit Architecture installed.

5.20 Open view Floor Plan Level 1.

- Zoom to Fit.

- Choose File>Import/Link/Link DWF Markup Set from the Menu Bar.

- Navigate to the folder that holds *Chapter 9 elevations imperial.dwf [Chapter 9 elevations metric.dwf]*.

This file was published from the file Chapter 9 project file and marked up using Autodesk Design Review (see Figure 9.107).

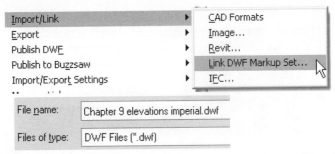

Figure 9.107 *Import a DWF markup*

A dialogue will open listing views for linking (see Figure 9.108). The two sheets in this file will be referenced.

5.21 Click OK.

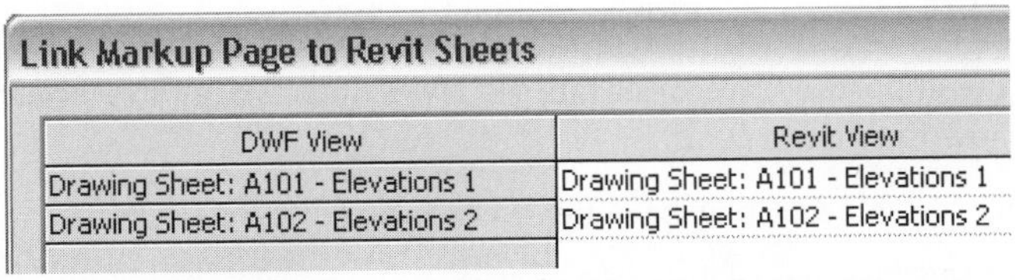

Link Markup Page to Revit Sheets

DWF View	Revit View
Drawing Sheet: A101 - Elevations 1	Drawing Sheet: A101 - Elevations 1
Drawing Sheet: A102 - Elevations 2	Drawing Sheet: A102 - Elevations 2

Figure 9.108 *The DWF link dialogue*

5.22 Open the two sheet views in this file in turn.

They will now display red markup symbols from the referenced DWF (see Figure 9.109).

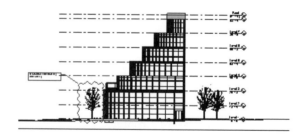

Figure 9.109 *The DWF markups appear in the sheet views*

5.23 Select one of the red DWF symbols.

- Type **VH** at the keyboard.

  The DWF objects will disappear.

- Type **VG** to open the Visibility Graphics dialogue.

  On the Imported Categories tab, the DWF objects are listed according to the sheets, as shown in Figure 9.110.

- Check the objects that have been cleared and choose OK.

  The DWF symbols will be restored to the sheet view. You can turn off DWF objects individually in each view.

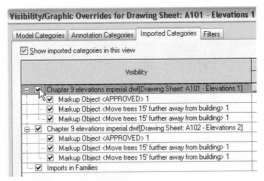

Figure 9.110   *Turn on the DWF in the sheet view*

5.24 Open sheet view A101.

- Zoom in on the markup at the tree.

- Right-click and select Activate View so that you can edit the model from the sheet view (see Figure 9.111).

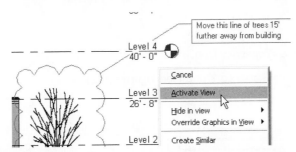

Figure 9.111   *Activate the view*

5.25 Select the trees (there are two, one behind the other).

- Choose Move from the Toolbar.

- Move the trees 15' [4500] to the right, as shown in Figure 9.112.

- Right-click and deactivate the view.

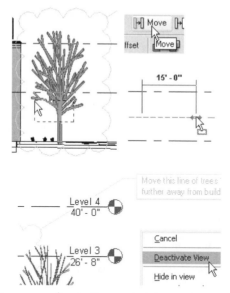

Figure 9.112 *Move the trees according to the markup instruction*

5.26 Select the markup note.

- Select the Properties icon.

- Change the Status to Done (see Figure 9.113).

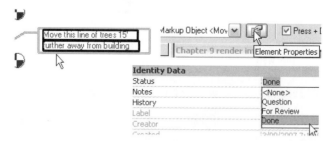

Figure 9.113 *Change the status of the markup once its instruction has been completed*

5.27 Open sheet view A102.

- Select the corresponding markup at the tree and change its Status property to Done, as you did in Step 5.26.

5.28 From the Menu Bar, select File>Manage Links.

- On the DWF Markups tab, select each of the two linked markups (there are two instances, one for each view in which the multi-page markup appears) and choose Save Markups.

This will update the status of the markup objects in the original file (see Figure 9.114).

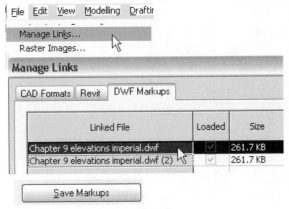

Figure 9.114 *Save the markups back to the linked file*

Saving the Revit Architecture file also updates DWF links. Graphics in the DWF will not update until a new DWF is printed from the affected sheets. If completed DWF notes are turned off and a new DWF printed using the same file name, designers and reviews can have an efficient electronic conversation about project progress.

EXPORT THE MODEL TO AN EXTERNAL RENDERER

For users who prefer to work with an external rendering application such as Autodesk VIZ 2006 or Architectural Desktop's VIZ Render, export to DWG will create files that can be imported into VIZ with a significant amount of object and material information retained. A 2D Revit Architecture view (floor or ceiling plan, section elevation) will create a 2D DWG export. Make a 3D view current in Revit Architecture to generate a 3D DWG file.

When exporting to DWG files for import into VIZ, all the model entities should be designated by either layer or color so that materials can be assigned. This preparatory work is done in Revit Architecture's Export Layer Settings. We have looked at layer export settings before in another context.

5.29 From the File menu, select File>Import/Export Settings>Export Layers DWG/DXF.

The Export Layers dialogue will open. It uses a txt file as its source. Revit Architecture has a number of default files to use as a basis for your own custom settings. The file shown in Figure 9.115 is *exportlayers-dwg-AIA.txt*. Note the color IDs that have been assigned.

5.30 Click Load.

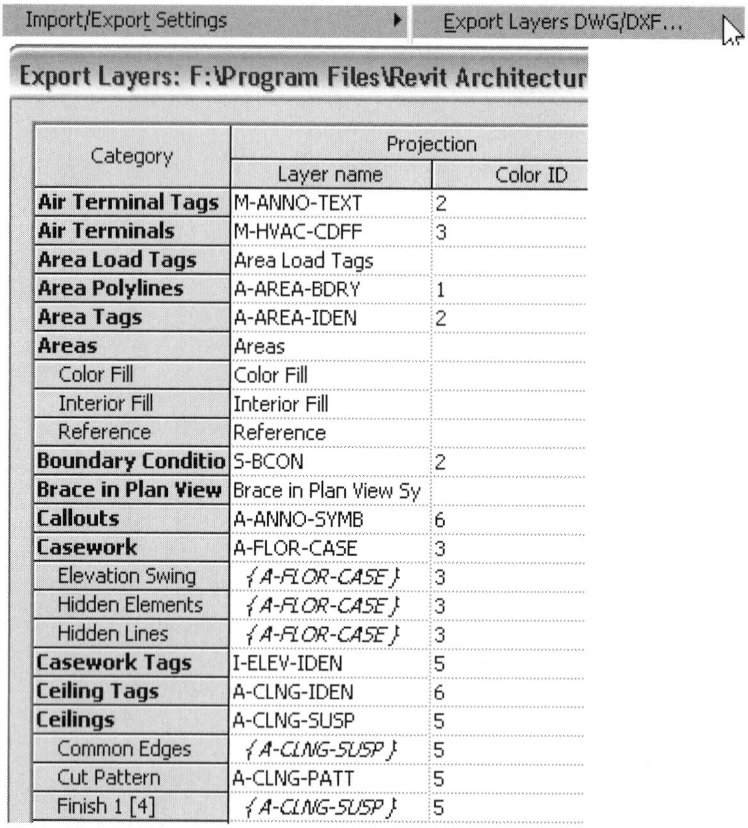

Category	Projection	
	Layer name	Color ID
Air Terminal Tags	M-ANNO-TEXT	2
Air Terminals	M-HVAC-CDFF	3
Area Load Tags	Area Load Tags	
Area Polylines	A-AREA-BDRY	1
Area Tags	A-AREA-IDEN	2
Areas	Areas	
Color Fill	Color Fill	
Interior Fill	Interior Fill	
Reference	Reference	
Boundary Conditio	S-BCON	2
Brace in Plan View	Brace in Plan View Sy	
Callouts	A-ANNO-SYMB	6
Casework	A-FLOR-CASE	3
Elevation Swing	{A-FLOR-CASE}	3
Hidden Elements	{A-FLOR-CASE}	3
Hidden Lines	{A-FLOR-CASE}	3
Casework Tags	I-ELEV-IDEN	5
Ceiling Tags	A-CLNG-IDEN	6
Ceilings	A-CLNG-SUSP	5
Common Edges	{A-CLNG-SUSP}	5
Cut Pattern	A-CLNG-PATT	5
Finish 1 [4]	{A-CLNG-SUSP}	5

Figure 9.115 *The exportlayers-dwg-AIA.txt export settings*

5.31 Click Standard.

- Choose ISO13567, as shown in Figure 9.116.
- Click OK.

Layer names and Color ID values in the Export Layers will change.

Figure 9.116   *You can choose a standard setup export file*

Note that this file contains layer names based on the ISO CAD standards and color IDs.

For export into VIZ, layers and Color IDs should be unique for each object that will receive a different material.

5.32 Click Standard.

- Choose the Standard file appropriate for your country.

- Click OK.

5.33 Enter new Door component Layer name and Color ID properties under Projection and Cut, based on the information shown in Figure 9.117.

Doors					
	A-DOOR	1		A-DOOR	1
Elevation Swing	{A-DOOR}	1		{A-DOOR}	1
Frame/Mullion	A-DOOR-FRAM	12		A-DOOR-FRAM	12
Glass	A-DOOR-GLAZ	13		A-DOOR-GLAZ	13
Hidden Lines	{A-DOOR}	1		{A-DOOR}	1
Opening	A-DOOR-OTLN	6		A-DOOR-OTLN	6
Panel	A-DOOR-PANL	14		A-DOOR-PANL	14
Plan Swing	{A-DOOR}	1		{A-DOOR}	1

Figure 9.117   *Change object layers and colors to provide unique identifiers*

5.34 Click Save As to save your newly edited file for repeated use later.

- Give it a name to distinguish it from the file you modified, as shown in Figure 9.118.

The new file and any others you create can be accessed via the Load button in the import-export settings dialogue.

Figure 9.118 *Save the settings changes in a named file*

5.35 To export a file as DWG, select File>Export>DWG, DXF, DGN, SAT.

The Export dialogue allows you to select a location and name for the new file (see Figure 9.119).

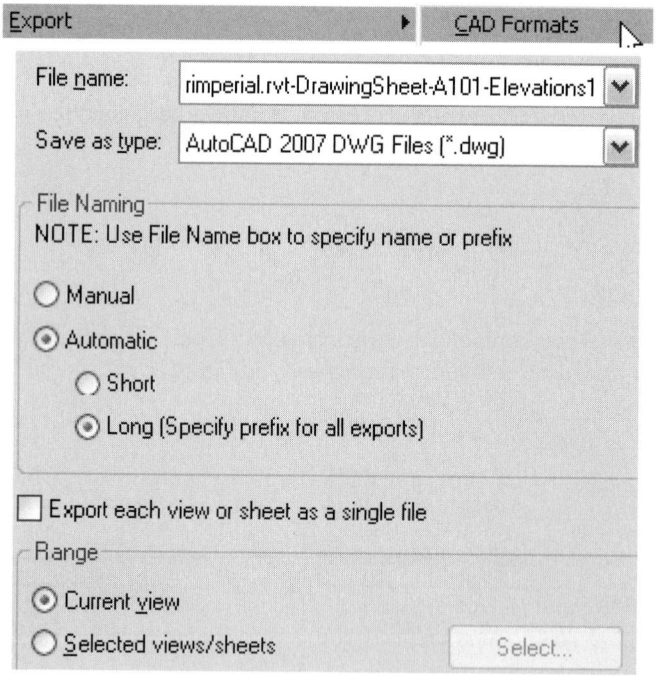

Figure 9.119 *Choose a name, location and layer setting file for your exported DWG*

Note that Revit Architecture adds the name of the current view to the end of the file name. This is important to remember for file exchange purposes. As a Revit Architecture model develops, you may export it numerous times. Be sure to keep the many files you can create from one model organized!

5.36 Select Options.

You can specify output properties of entities, linetype scaling, coordinates and units for the file you are going to generate, and there are other options for solid/face rendering and area analysis (see Figure 9.120).

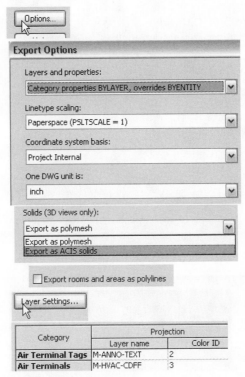

Figure 9.120   *Options for DWG export*

Inside this dialogue you can also switch from one layer output file to another.

5.37 If you have AutoCAD, Architectural Desktop or VIZ on your system, save the export to a DWG file and open it in one of those applications.

- Otherwise, click Cancel to exit the dialogue.

Figures 9.121 and 9.122 show exported DWG information as received in VIZ and AutoCAD.

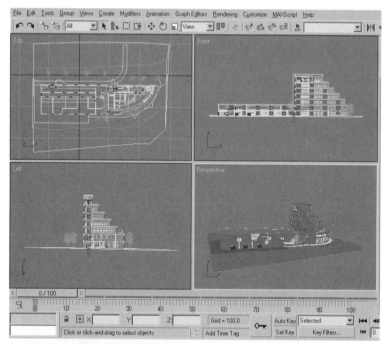

Figure 9.121 *The DWG export linked into VIZ*

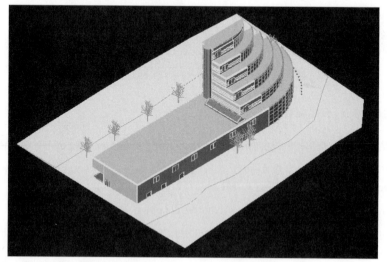

Figure 9.122 *The export DWG opened in AutoCAD*

5.38 Close the Revit Architecture file.

SUMMARY

The lessons in this chapter have shown you how to make use of Revit Architecture's many capabilities for viewing your design within Revit Architecture itself, and for making your standard or enhanced views available as images, printed pages or raw material for external graphics applications.

As with all the exercises in this book, you have seen only a few of the many possible alternative settings and combinations available. Particularly with rendered images, time spent working on the details and nuances of small portions of a building model can make a big difference in how people will see and understand the design intent behind the model. Practice with the render and walkthrough routines in Revit Architecture—you may surprise yourself and delight your clients!

REVIEW QUESTIONS

MULTIPLE CHOICE

1. The Edit Walkthrough control allows you to

   a. Change the Camera orientation at any frame point on the walk-through path

   b. Change the path location at any frame

   c. Add or remove key frames at any point on the path

   d. all of the above

2. To make building components appear as desired in a rendering, you

   a. Apply a Texture definition to a Material

   b. Apply a Material definition to the building components

   c. a and b in that order only, or the definitions won't apply correctly

   d. a and b in any order

3. A Section Box is applied to a view to

   a. Define custom spaces

   b. Provide a restricted volume for rendering calculations

   c. Make a link between floor plans and section views

   d. all of the above

4. You can change properties of a Camera

    a. Using its Properties

    b. Using View Properties of the Camera View

    c. Using the Dynamic View control in the Camera View

    d. all of the above

5. To prepare a Revit Architecture file for export to rendering software, use the following menu selection:

    a. Settings>Materials

    b. View>New>Camera

    c. File>Import/Export Settings>Export Layers DWG/DXF (or Export Layers DGN)

    d. Render>Raytrace

TRUE/FALSE

6. True _ False _ You must adjust the size of a camera view before you place any RPC people components in the model.

7. True _ False _ The Print Setup dialogue lets you make adjustments to printer settings, but not PDF or DWF output.

8. True _ False _ You use Date, Time, and Place settings for an exterior rendering to adjust the angle of the sunlight.

9. True _ False _ DWF output from Revit Architecture can be 2D or 3D.

10. True _ False _ Revit Architecture can export a Walkthrough to an animation (AVI) file in Hidden Line, Shaded or Rendered view mode.

 Answers will be found on the CD.

CHAPTER 10

Augmenting the Design—Design Options and Logical Formulas

INTRODUCTION

Building construction happens over time, as a result of many directed and realized choices. Revit Architecture provides design tools that take care of both space and time (phasing), and also allows explicitly for the process of choice. You can create, order and display alternatives using Design Options. At any stage of the process, you can resolve an option into the model and discard alternatives.

Revit Architecture uses parameters to drive any relationship the user wants to define. Parameters can hold formulas, which include logical operators such as yes/no and if/then conditions. Families can hold features or arrays that can be called or suppressed by family type or instance—cabinets with or without end panels, for example, or brackets that appear under a length of counter. Families can nest within other families, and behave according to parameters in the host. This sounds complicated, but it actually makes three dimensional modeling simpler because you can set up rules to drive shapes and sizes rather than drawing every possibility beforehand.

OBJECTIVES

- Create, manage and resolve Design Options
- Create two different stair styles with custom rails
- Create a custom wall style with a vertical structure
- Create and place a nested component family with family types based on size, utilizing parametric formulas

REVIT ARCHITECTURE COMMANDS AND SKILLS

Design Options

Option sets

Options

Stairs

Railings

Wall structure

Family creation

Family types

Parametric formulas

Conditional operators in formulas

Grid lines

DESIGN OPTIONS

In the previous exercises, you developed various parts of building models. As most design projects progress, you will want, at some point, to explore multiple design schemes. Revit Architecture's Design Options allow you to develop alternate schemes, either simple concepts or detailed engineering solutions. Design Options coexist in the project file with the main model (all building elements that have not been assigned to a named option). You can study and develop each option independently of others, and easily create views to display option combinations. At any point you can resolve an option or option set into the main model to remove the alternates.

EXERCISE 1. DESIGN OPTION SETUP

1.1 Launch Revit Architecture. Open the file *Chapter 10 start imperial.rvt [Chapter 10 start metric.rvt]*. The file will open to show the 3D View: Exterior Iso at the default southeast orientation.

This model is of the arts center for our hypothetical campus project. It contains performance halls, an art display area and cafe on the main floor, shop/storage space on the lower level, and classrooms/offices/rehearsal rooms on the upper levels. No seating has been created in the performance spaces, whose floors slope down to stages on the lower level.

The building owner wants two designs for a staircase from the main level down to the lower level: a simple version with brick sides to echo the walls inside the atrium, and a decorated version if fundraising exceeds expectations.

- Save the file as **Chapter 10 options.rvt** in a location determined by your instructor.

1.2 Open the view Floor Plan: Lower Level.

- Zoom in to the upper half of the central hallway at the lower middle of the plan.

- Open the View Properties dialogue for the view.

- Change the Underlay value to Main Level.

- Choose OK (see Figure 10.1).

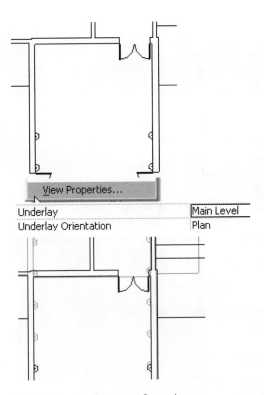

Figure 10.1 *Adjust the Underlay to see the main floor above*

1.3 From the Tools menu, choose Tools>Design Options>Design Options (see Figure 10.2).

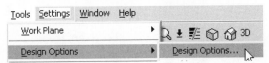

Figure 10.2 *Open the Design Options dialogue*

There are no Design Option Sets (the containers for Design Options) defined in this file.

1.4 Select New under Option Set, as shown in Figure 10.3.

Figure 10.3 *Create an Option Set*

Revit Architecture will open the Design Options dialogue and create a new Option Set named Option Set 1, with Option 1 (primary) as its default subset. You can create as many Option Sets as you desire, each with a tree of Options within it.

1.5 Select Option Set 1 in the tree pane.

- Select Rename under Option Set.

- In the Rename dialogue, type **Stairs**.

- Choose OK.

- Select Option 1 (primary).

- Select Rename under Option.

- In the Rename dialogue, enter **Enclosed**.

- Choose OK (see Figure 10.4).

Rename	
Previous:	Option Set 1
New:	Stairs

Rename	
Previous:	Option 1
New:	Enclosed

Figure 10.4 *Name the new Option Set and Option as shown*

1.6 Choose New under Option, then Rename.

- Name the second option **Open**. Select OK (see Figure 10.5).

Figure 10.5   *Make and name a second option*

1.7 In the Design Options dialogue, select Enclosed (primary), and then choose Edit Selected.

The field text under Now Editing will change from Main Model to Option Set 1: Enclosed (primary) Stairs.

- Choose Close to leave the dialogue and work on the Design Option (see Figure 10.6).

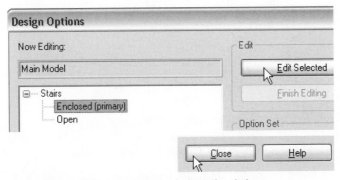

Figure 10.6   *Begin the Option Editing process and close the dialogue*

CREATE AN ENCLOSED STAIR AND LANDING

In the next steps, you will create a stairway "out in space" away from walls, and then align it to the building. In preparation for that, you will first create a couple of Reference Planes as drawing aids.

> 1.8 From the Basics tab of the Design Bar, select Ref Plane.
>
> - Draw a horizontal reference plane across the hallway space between the facing doors.
>
> - Click Modify on the Design Bar.
>
> - Select the new Reference Plane.
>
> - Copy the Reference Plane up **7'-0"** **[2100]** at 90° (see Figure 10.7).

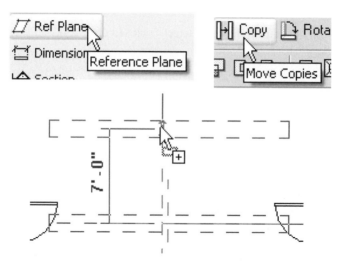

Figure 10.7 *Create Reference Planes for stair creation*

> 1.9 From the Modeling tab of the Design Bar, select Stairs.
>
> - Select Stairs Properties from the Design Bar.
>
> - In the Instance Parameters Section of the Element Properties dialogue, change the Top Level to Main Level.
>
> - Change the Width to **6'** **[1800]**.
>
> - Set the Desired Number or Risers to **23**, as shown in Figure 10.8.
>
> - Click OK.

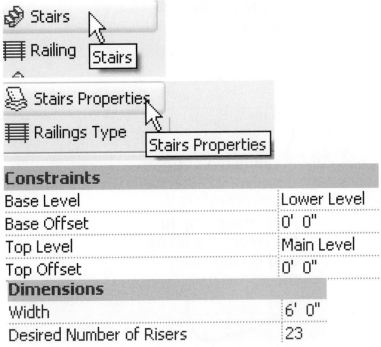

Figure 10.8 *Adjust Top Level, Width and Riser count for the new stairs*

The cursor will show a pencil for Sketch mode.

1.10 Click on a starting point near the right end of the lower Reference Plane.

- Pull the cursor left, until the temporary dimension reads 10'-1" [2750] and the light gray riser counter reads 12 RISERS CREATED, 11 REMAINING, and click to establish the first run of stairs.

1.11 Pull the cursor directly up to the upper Reference Plane.

An alignment line will appear.

- Click and pull the mouse to the right until the sketch outline does not grow any more (9'-2" [2500] minimum) and the riser counter indicates that the layout is complete.

- Click to establish the sketch (see Figure 10.9).

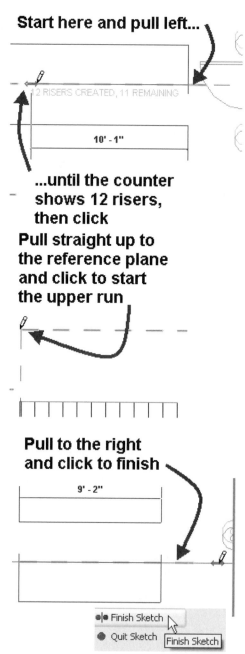

Start here and pull left...

12 RISERS CREATED, 11 REMAINING

10' - 1"

**...until the counter
shows 12 risers,
then click**

**Pull straight up to
the reference plane
and click to start
the upper run**

**Pull to the right
and click to finish**

9' - 2"

Finish Sketch

Quit Sketch   Finish Sketch

Figure 10.9 *Sketch the stairs, with space for a landing—read from top to bottom*

1.12 Click Finish Sketch.

- CTRL + select the Reference Planes and delete them.

- Select the stair's outer railing, not the stair itself. (Select both and use the filter, if necessary.)

- Select the Properties icon.

1.13 In the Element Properties dialogue, choose Edit/New.

- Choose Duplicate.

- In the Name box, type **Single Wall Brick Cap**.

- Choose OK.

- Choose Edit in the Rail Structure value field (see Figure 10.10).

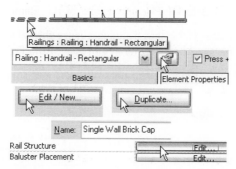

Figure 10.10   *Select the rails and start to define a new type*

In the Edit Rails dialogue, there is a single rail defined for this style.

1.14 For the Profile value, assign Brick Cap: Brick Cap, a custom profile that has been loaded into this project.

- For the Material value, click in the cell, then click the dialogue arrow that appears in the cell.

1.15 In the Materials dialogue, find Concrete-Precast Concrete in the Name pane, and select it.

- Click the arrow to the right of the AccuRender Texture field.

- In the Material Library dialogue, navigate to folder *_accurender\Concrete* and select *Exposed Aggregate, Tan*, as shown in Figure 10.11.

- Click OK three times to exit the Rail editor.

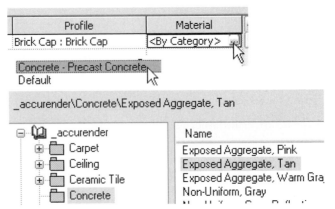

Figure 10.11   *Apply a material and rendering texture to the rail element in the railing*

1.16 Choose Edit in the Baluster Placement field of the Element Properties dialogue.

- In the Edit Baluster Placement dialogue, change the value for Baluster Family to None in row 2 of the Main pattern and in all rows of the Posts, as shown in Figure 10.12.

- Choose OK three times to terminate the rail and stair editing.

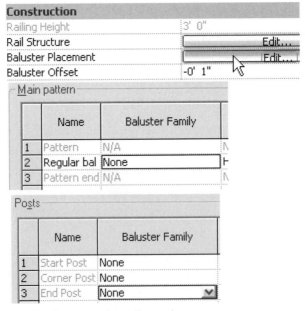

Figure 10.12   *Remove all balusters in this railing style*

1.17 Select the Railing and Stair together.

- Choose the Move tool from the Toolbar.

- Select the upper-left corner of the railing as the starting point for the Move command, and select the intersection of the left side wall and the edge of the Main Level floor in the halftone underlay, as shown in Figure 10.13.

- Zoom out until you can see the whole stair to check the alignment.

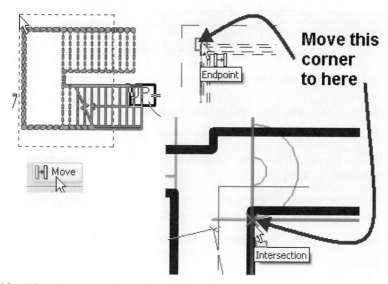

Figure 10.13 *Move the new railing and stair together*

1.18 Open the Main Level view.

- Zoom in around the new stairs.

- Select the Match Type icon from the Toolbar.

  The cursor will change to an empty eyedropper shape.

- Select the outer stair rail to fill the eyedropper, and then select the inner rail to change its type from Rectangular to Single Wall Brick Cap (see Figure 10.14).

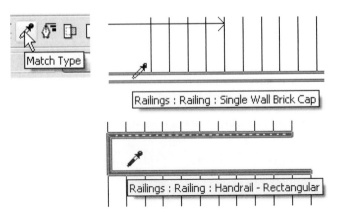

Figure 10.14 *Match properties from one railing to the next*

1.19 Select the Floor tool from the Modeling tab of the Design Bar.

- Choose Floor Properties from the Design Bar.

- In the Element Properties dialogue, change the Type to **Generic – 12" – Filled [Generic – 300]**.

- In the Instance Parameters section, make sure the Level value is Main Level, and set the Height Offset From Level value to **0**.

  Revit Architecture will supply the units, as shown in Figure 10.15.

- Click OK.

Figure 10.15 *Set the Floor Properties for the upper stair landing*

1.20 On the Sketch Bar, choose Lines, and check the Rectangle tool on the Toolbar.

- Select two points, as shown in Figure 10.16: the lower end of the inner rail, and the intersection of the right-hand wall and the floor edge.

- Choose Finish Sketch to draw the upper landing.

- Choose No in the question box about attaching walls.

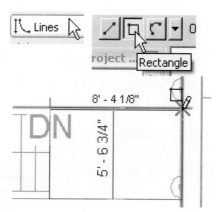

Figure 10.16 *Draw the floor sketch*

CREATE A BALCONY WALL TYPE

1.21 Select Wall from the Modeling tab of the Design Bar.

- Select the Properties icon from the Options Bar. In the Element Properties dialogue, make the current type **Generic – 4" brick [Generic – 90mm brick]**.

- Select Edit/New.

1.22 Select Duplicate.

- In the Name dialogue, type **Double Wall Brick with Cap**.

- Choose OK.

- Select Edit in the Structure Value field.

- In the Edit Assembly dialogue, select Preview in the lower-left corner to open that pane.

- Change the view type to Section, as shown in Figure 10.17.

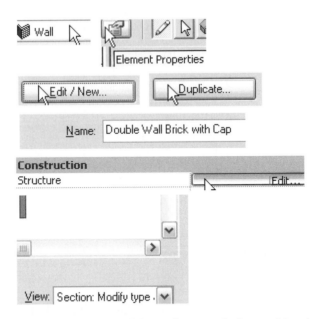

Figure 10.17 *Select the Section preview of this wall type to facilitate editing the new wall type*

1.23 Choose Row 2 in the Layers area.

- Choose Insert twice to create two new layers in the wall.

- Define the layers, as shown in Figure 10.18.

- Set the Material for Row 2 Masonry – Brick with Thickness to **3 5/8" [90]**.

- Set Row 3 Function to Thermal/Air Layer, Material to Misc. Air Layers – Air Space and Thickness to **1 3/4" [45]**.

- Set the Thickness for Row 4 to **3 5/8" [90]**.

  Revit Architecture will show the Total thickness as 9" [225].

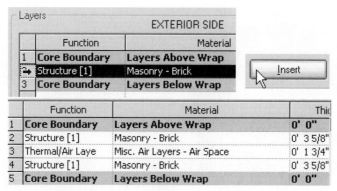

Figure 10.18 *Layers defined in the new wall type*

Revit Architecture allows you to create walls of many layers. The Core Boundary layer defines the point at which finish meets structure, and determines how far floors penetrate into walls, as shown in drafting sections. Since this wall type is for decoration only, you will not alter the placement of the Core Boundary layers.

1.24 Select the Sweeps button at the lower right of the dialogue.

- In the Wall Sweeps dialogue, select Add.

- Assign the Profile Stone Cap: Stone Cap.

  This is a custom profile loaded into the project file.

- Set its Material to Concrete: Precast Concrete.

- Set the AccuRender material value for Precast Concrete to *_accurender\Concrete\Exposed Aggregate, Tan*, as you did before.

1.25 Click OK twice to return to the Wall Sweeps dialogue.

- Set the From Value to Top and the Offset to **–0 4.5 [-112.5]**.

  The offset value was determined by trial and error, to center the profile on the wall.

- Move the dialogue box off the preview if necessary and choose Apply to see the new sweep in place (see Figure 10.19).

- Click OK.

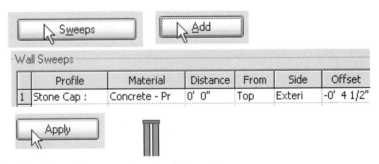

Figure 10.19 *Creating a Sweep at the top of the wall*

1.26 Select the Reveals button at the lower right of the Edit Assembly dialogue.

- In the Reveals dialogue, click Add twice.

1.27 For both rows, make the Profile value Reveal-Brick Course: 1 Brick.

- Make the distance value **–0' 8" [-200]** for both rows.

- Make the From value read Top.

- Make the Side value Exterior for Row 1 and Interior for Row 2.

- Move the dialogue box off the preview if necessary and choose Apply to see the new Reveals in place, as shown in Figure 10.20.

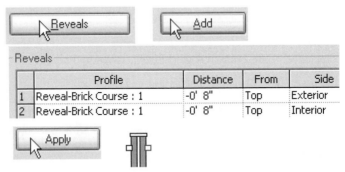

Figure 10.20 *New Reveals added to the wall type*

1.28 Click OK twice to return to the Type Properties dialogue.

- Change the Wrapping at Ends value to Exterior.

- Click OK.

- In the Element Properties dialogue, set the Top Constraint to Unconnected, the Unconnected Height to **3' 0" [900]**, and the Location Line to Finish Face: Exterior, as shown in Figure 10.21.

- Click OK.

Construction	
Structure	
Wrapping at Inserts	Do not wrap
Wrapping at Ends	Exterior

Constraints	
Location Line	Finish Face: Exterior
Base Constraint	Main Level
Base Offset	0' 0"
Base is Attached	
Base Extension Distance	0' 0"
Top Constraint	Unconnected
Unconnected Height	3' 0"

Figure 10.21   *Type and Instance Properties of the wall you are about to place*

1.29 Place two instances of the new wall type at the edge of the landing and the floor, as shown in Figure 10.22.

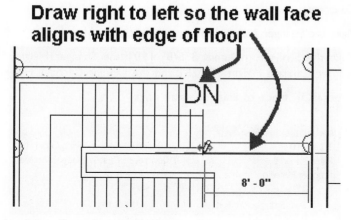

Figure 10.22   *Landing walls in place for this Design Option*

CREATE AN ENLOSED STAIR TYPE

1.30 Open the 3D View Atrium Staircase.

- Select the stair.

- Select the Properties icon on the Options Bar.

- Select Edit/New; and then select Duplicate.

- In the Name dialogue, type **Brick Sides,** as shown in Figure 10.23.

- Click OK.

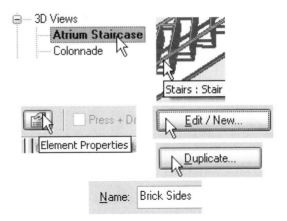

Figure 10.23 *Define a new style for the stairs*

1.31 In the Type Properties dialogue, adjust the values, as shown in Figure 10.24.

- Make the material value for the Tread and Stringer Concrete – Cast-In-Place Concrete.

- Make the Stringer Material Masonry – Brick.

- Make the Stringer thickness **3 5/8" [90]**; the Stringer Height **3' 5" [925]**; the Stringer Carriage Height **1" [25]**; and the Landing Carriage Height **2" [50]**.

- Choose OK twice to finish the stair edits.

Materials and Finishes	
Tread Material	Concrete - Cast-in-Place Concrete
Riser Material	Concrete - Cast-in-Place Concrete
Stringer Material	Masonry - Brick

Stringers	
Trim Stringers at Top	Do not trim
Right Stringer	Closed
Left Stringer	Closed
Middle Stringers	0
Stringer Thickness	0' 3 5/8"
Stringer Height	3' 5"
Open Stringer Offset	0' 0"
Stringer Carriage Height	0' 1"
Landing Carriage Height	0' 2"

Figure 10.24 *Properties for the stairs*

1.32 Select the Rail.

- Use the TAB key to cycle through choices if necessary.

- Select the Properties icon from the Options Bar.

- Choose Edit/New.

- In the Type Properties dialogue, set the values for both Angled Joins and Tangent Joins to No Connector.

- Set Rail Connections to Trim (see Figure 10.25).

- Choose OK twice to finish the rail edits.

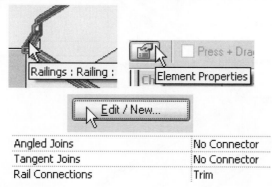

Figure 10.25 *Edit rail properties*

The first option for the stairs is now complete (see Figure 10.26).

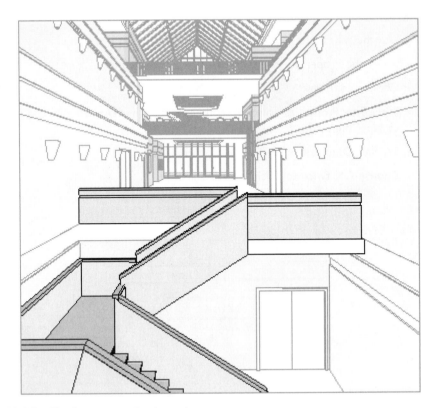

Figure 10.26 *The first stair option is ready*

1.33 Place the cursor anywhere over the Toolbar and right-click.

- Choose Design Options from the list of available toolbars.

- In the Design Options Toolbar, select the Design Options dialogue icon, as shown in Figure 10.27.

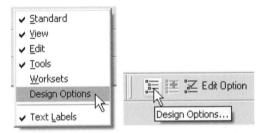

Figure 10.27 *Make the Design Options Toolbar visible and open the Design Options dialogue*

1.34 Choose Enclosed (primary) in the Stairs tree.

- Choose Finish Editing, as shown in Figure 10.28.

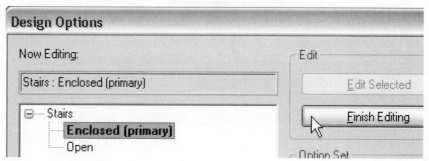

Figure 10.28 *Finish editing the Design Option—you can open it for more edits later if necessary*

The 3D view will change appearance as components of the model become available for editing.

1.35 Select Open in the Stairs tree.

- Choose Edit Selected, as shown in Figure 10.29. The enclosed stairs disappear, and the model becomes gray again.

- Click Close.

- Save the file.

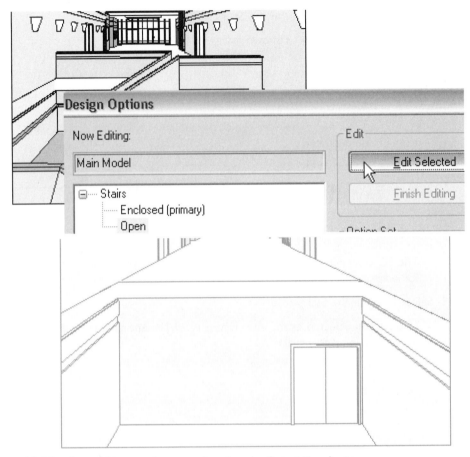

Figure 10.29 *Start editing another named option; the first option disappears*

EXERCISE 2. SECOND DESIGN OPTION

CREATE AN OPEN STAIR WITH CUSTOMIZED BOUNDARY LINES

    2.1 Open or continue working with the file from the previous exercise.

- Make view Lower Level current.

- Zoom into the area you worked in before.

- Select Stairs from the Modeling tab of the Design Bar.

    Revit Architecture will go into Sketch mode.

2.2 Select Stairs Properties from the Sketch tab.

- Set the Stair properties as before: Width is **6' [1800]**, Desired Number of Risers is **23**, Base Level is Lower Level, Top Level is Main Level.

- Click OK.

2.3 Start the run of stairs between the doors.

- Pull the cursor straight up (90°) until the temporary dimension reads 10'-1" [2750] and the riser counter shows 12 RISERS CREATED, 11 REMAINING.

- Click to establish the end of the run.

2.4 Pull the cursor straight up until a horizontal Reference Plane appears, along with the Tooltip Midpoint, as shown in Figure 10.30.

- Click to start the second run, and continue pulling the cursor straight up until the stair outline is complete, as before.

- Click again to set the stair runs.

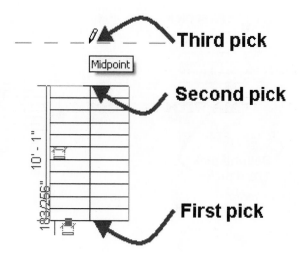

Figure 10.30 *Starting the upper stair run*

The Stair sketch consists of three types of Model lines: green boundaries, black risers and blue runs.

2.5 Select the left-hand boundary of the upper run.

- Copy it 2' [600] to the left.

2.6 Choose Boundary from the Sketch tab.

- Choose the 3-point Arc tool from the Options Bar.

- Select the lower end of the new line for the first point of the arc.

- Select the upper end of the original (copied) boundary for the end point of the arc.

- Pull the cursor to the right of the arc line until the arc bend snaps in place and click to establish the arc sketch, as shown in Figure 10.31.

The arc will display a radius dimension.

2.7 Select the dimension and set it to **26' [7800]**.

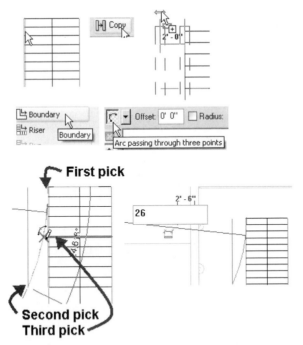

Figure 10.31   *Set the arc radius*

2.8 Click Modify.

- Erase the two straight boundary lines for the upper run of the stairs and the copied sketch line.

- Select the new arc.

- Choose the Mirror tool from the Options Bar and mirror the arc to the right, using the center line of the stairs as the mirror line (see Figure 10.32).

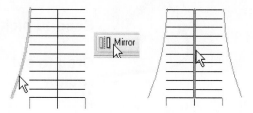

Figure 10.32   *Mirror the arc*

2.9 Erase the boundaries of the landing.

- Move the boundaries of the lower stair run left and right **2' [600]** to line up with the wide ends of the upper runs (see Figure 10.33).

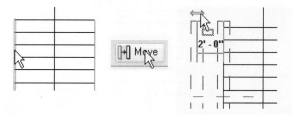

Figure 10.33   *Move the two lower boundary lines*

2.10 Choose Boundary from the Sketch tab.

- Select the three-point Arc tool.

- Select the lower end of the upper-left boundary and the upper end of the lower-left boundary as the start and end points, and pull the cursor to the left to define a 180° arc.

- Click to establish the arc (see Figure 10.34).

- Click Modify.

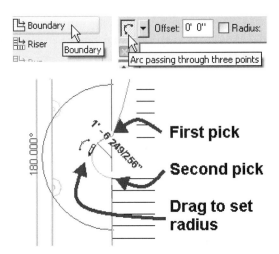

Figure 10.34 *Create a new arc boundary*

2.11 Mirror the arc to the right to create a landing with rounded sides, as shown in Figure 10.35.

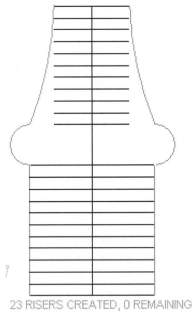

Figure 10.35 *The completed stair sketch*

2.12 Select Stairs Properties on the Design Bar.

- Select Edit/New.

- Select Duplicate.

- In the Name box, type **Open Riser Steel**.

2.13 In the Type Properties dialogue, edit the following values.

- Tread Thickness **1" [25]**

- Riser Type – None

- Stringer Thickness **2" [50]**

- Tread Material Metal – Steel

- Stringer Material Metal – Steel, as shown in Figure 10.36

- Click OK twice.

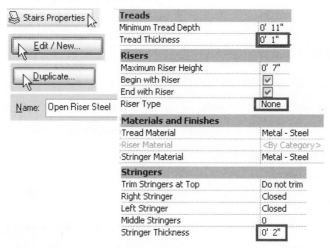

Figure 10.36   *Stair Properties*

2.14 Click Finish Sketch.

- Select the stair and both railings.

- Choose Move from the Toolbar.

- Select the middle of the top of the stairs for the first point.

- Select the middle of the Main Level floor as the second point (see Figure 10.37).

Tip: If you are having difficulty getting Revit Architecture to snap to the middle of the floor edge, type SM (for Snap Middle) to force the Midpoint snap override.

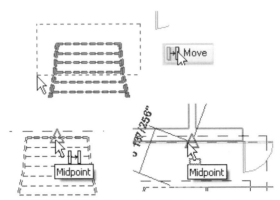

Figure 10.37 *Move the middle of the stair to the midpoint of the floor edge*

ADD A CUSTOM RAIL STYLE

2.15 Open the 3D view Atrium Staircase.

- Select one of the stair rails.

- Change the type in the Type Selector to Railing: Handrail – Pipe.

- Choose the Properties icon from the Options Bar.

- Select Edit/New.

- Select Duplicate.

- In the Name box, type **Handrail – Pipe – Copper and Cherry** (see Figure 10.38).

- Choose OK.

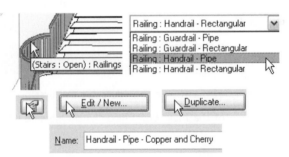

Figure 10.38 *Railing properties*

2.16 In the Type Properties dialogue, select Edit in the Rail Structure value.

2.17 In the Edit Rails dialogue, change Rail I Material to Wood – Cherry.

- Change the Material for Rail 2 to Metal – Trim.

- While in the Materials dialogue, change the AccuRender Texture for Metal – Trim to _accurender\Metals\Copper\Polished, Plain_, as shown in Figure 10.39.

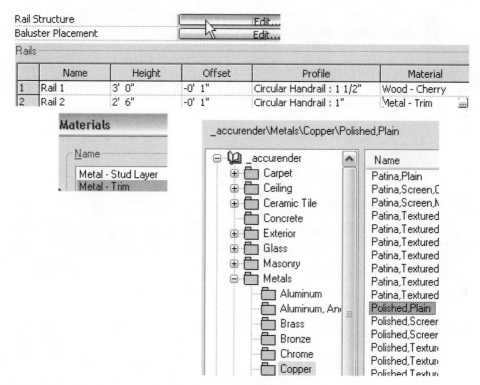

Figure 10.39 *Assign a material and texture to the rails*

2.18 Choose OK twice to exit the Materials Library and Materials.

- Change the Material value for Rails 4 and 6 to match Rail 2.

2.19 Change the Profiles for Rails 3 and 5 to Fascia–Flat: 1" x 4" [Fascia—Flat: 18 x 90].

- Make their Materials Wood – Cherry, as shown in Figure 10.40.

- Click OK.

Rails

	Name	Height	Offset	Profile	Material
1	Rail 1	3' 0"	-0' 1"	Circular Handrail : 1 1/2"	Wood - Cherry
2	Rail 2	2' 6"	-0' 1"	Circular Handrail : 1"	Metal - Trim
3	Rail 3	2' 0"	-0' 1"	Fascia-Flat : 1" x 4"	Wood - Cherry
4	Rail 4	1' 6"	-0' 1"	Circular Handrail : 1"	Metal - Trim
5	Rail 5	1' 0"	-0' 1"	Fascia-Flat : 1" x 4"	Wood - Cherry
6	Rail 6	0' 6"	-0' 1"	Circular Handrail : 1"	Metal - Trim

Figure 10.40   *Rails defined in the new style*

2.20 Choose Edit in the Baluster Placement field.

- Change the Dist. from previous value in Row 2 of the Main Pattern to **3' [900]**, as shown in Figure 10.41.

- Click OK three times.

Rail Structure	Edit...
Baluster Placement	Edit...

Main pattern

	Name	Baluster Family	Base	Base offset	Top	Top offset	Dist. from previous
1	Pattern	N/A	N/A	N/A	N/A	N/A	N/A
2	Regular bal	Baluster - Round : 1"	Host	0' 0"	Rail 1	0' 0"	3' 0"

Figure 10.41   *Edit the baluster spacing*

2.21 Select the rail you have not edited and change its type to Handrail – Pipe – Copper and Cherry.

- Right-click and select Zoom Out (2x).

2.22 Clear Active Option Only from the Options Bar.

This is checked by default in all views, and prevents you from selecting items not in the Design Option that you are editing.

- Select the bottom edge of the 3D View and drag it down to make more of the stair visible, as shown in Figure 10.42.

- Right-click and select Zoom to Fit.

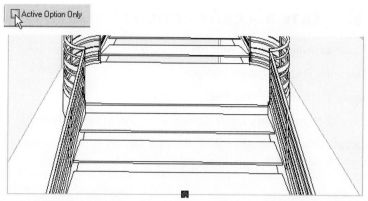

Figure 10.42   *Stretch the view border to see more of the open stairs*

2.23 Select the Edit Option icon from the Design Options Toolbar.

This is a toggle. It is the same as opening the Design Options dialogue and selecting Finish Editing.

The 3D view will show the primary option, the enclosed stairs, as shown in Figure 10.43.

• Save the file.

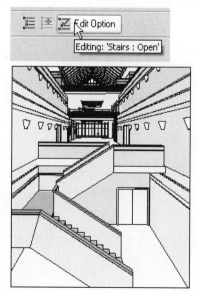

Figure 10.43   *Finish editing the option*

EXERCISE 3. CREATE BALCONY OPTIONS

3.1 Open or continue working with the file from the previous exercise.

- Open the Design Options dialogue.

- Select Open in the Stairs Tree.

- Choose Make Primary under Option, as shown in Figure 10.44.

  The 3D view will change to show the new primary option. A warning box will appear that you can ignore.

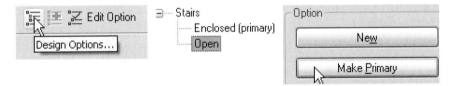

Figure 10.44 *Make the Open stairs the primary option*

3.2 Choose New under Option Set in the Design Options dialogue.

Revit Architecture creates a new Option Set 1 and Option 1 (primary) under it.

- Choose New under Option to create a second Option in the new Option Set.

3.3 Select Option Set 1 and select Rename.

- In the Name box, type **Balconies**.

- Choose OK.

3.4 Select Option 1 and select Rename.

- In the Name box, type **Railings Only**.

- Choose OK.

3.5 Select Option 2 and select Rename.

- In the Name box, type **Twin Balconies**.

- Choose OK.

- Select the Railings Only Option and select Edit Selected (see Figure 10.45).

- Choose Close.

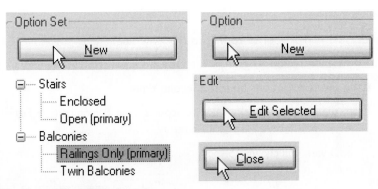

Figure 10.45 *New options for balconies*

3.6 Open the Floor Plan view Main Level.

- Select the Railing tool from the Modeling tab of the Design Bar.

- Select Railing Properties from the Sketch tab.

- Change the Type to Handrail – Pipe – Copper and Cherry.

- Click OK.

3.7 On the Options Bar, change the Offset value to **0' 2" [50]**.

- Draw a line along the edge of the floor left to right from the left wall to the end of the left stair rail, as shown in Figure 10.46.

- Click Finish Sketch.

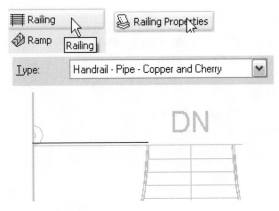

Figure 10.46 *Sketch a rail at the left side*

3.8 Select the new rail.

- Use the Mirror tool to copy it to the right side.

3.9 Click the Edit Options toggle on the Design Options Toolbar.

- Click the Edit Options button a second time.

- From the list that appears, select Balconies: Twin Balconies (see Figure 10.47).

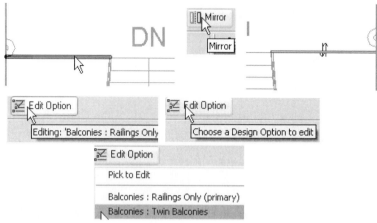

Figure 10.47 *Switch to the other Design Option*

3.10 From the Modeling tab of the Design Bar, select Floor.

- Select Lines from the Sketch tab.

- Select the Rectangle tool from the Options Bar.

3.11 Draw a rectangle **7'-0" [2100]** wide by **6'-0" [1800]** deep starting at the left end of the main floor, as shown in Figure 10.48.

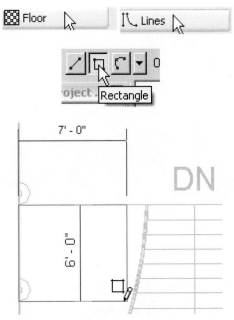

Figure 10.48   *Draw a rectangular floor sketch*

3.12 Select Modify on the Sketch tab.

- Erase the bottom line of the floor sketch.

- Select Lines on the Sketch tab.

- Draw a 180° three-point arc.

- Click Modify.

3.13 Select the entire floor sketch.

- Mirror it to the right, as shown in Figure 10.49.

- Select Finish Sketch.

- Answer No in the question box.

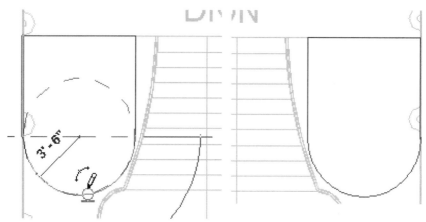

Figure 10.49 *Mirror the balcony floor sketch*

3.14 Select Railing from the Modeling tab of the Design Bar.

- Select the Pick arrow from the Options bar.

- Set the Offset to **0' 2" [50]** as before.

- Select the round edge of the left floor and the right side of that same floor, as shown in Figure 10.50.

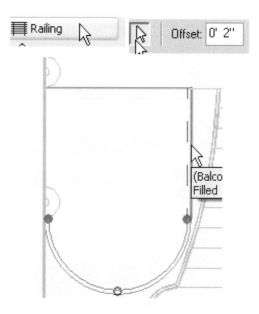

Figure 10.50 *Use Pick and Offset to create these lines*

3.15 Select the Draw (pencil) icon from the Options Bar.

- Keep the 2" [50] Offset.

- Draw a line from the end of the last line you created **1'-6" [450]** to the right. Revit Architecture will offset it from the floor edge.

- Use the Trim tool with the corner option to join the last two lines.

3.16 Select Finish Sketch.

- Mirror the new railing to the right (see Figure 10.51).

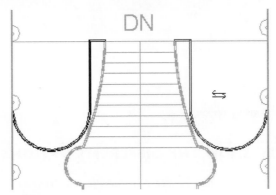

Figure 10.51 *Two new rails*

3.17 From the Toolbar, select the Design Options icon.

- Select Finish Editing.

  The balconies will disappear and the primary option will appear in the view.

3.18 Under Options, select New.

- Rename the new Option **None**.

- Choose Edit Selected.

- Select Finish Editing.

- Choose Close (see Figure 10.52).

  This creates a balcony option with no railings or floor extensions, to be combined with the enclosed stair option.

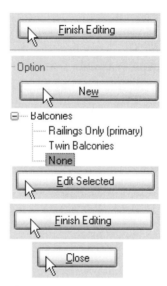

Figure 10.52 *Add a no-railing option*

CREATE VIEWS TO SHOW THE DIFFERENT OPTIONS

3.19 Select the 3D view Atrium Staircase in the Project Browser.

- Right-click and select Duplicate View>Duplicate.

- Repeat to make two copies of the Atrium Staircase view.

3.20 Select View Atrium Staircase, right-click and choose Rename.

- In the Rename View box, type **Atrium Staircase, Enclosed Brick**.

- Choose OK.

3.21 The view name will still be highlighted.

- Right-click and select Properties.

- Choose Edit in the Visibility/Graphics Overrides field.

- The Visibility/Graphics Override dialogue will now have a tab for Design Options.

- Select that tab.

- In the Design Option field for Option Set Stairs, choose Enclosed.

- In the Design Option Field for Option Set Balconies, choose None (see Figure 10.53).

- Click OK twice to end the edits.

  Since that view is not the active view, the screen display will not change.

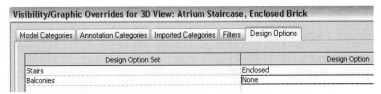

Figure 10.53   *Set the Design Option visibility in this view*

3.22 Select 3D view Copy of Atrium Staircase in the Project Browser.

- Right-click and select Rename.
- In the Rename View box, type **Atrium Staircase Open, No Balconies**.
- Choose OK.

3.23 Right-click and select Properties.

- Select Edit in the Visibility/Graphics Overrides field.
- Select the Design Options tab.
- In the Design Option field for Option Set Stairs, choose Open (primary).
- In the Design Option Field for Option Set Balconies, choose Railings Only (primary).
- Click OK twice to end the edits (see Figure 10.54).

  The view will not change.

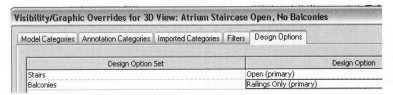

Figure 10.54   *Design Option visibility settings to show open staircase and railings*

3.24 Select view Copy (2) of Atrium Staircase in the Project Browser.

- Right-click and select Rename.
- In the Rename View box, type **Atrium Staircase Open, With Balconies**.
- Choose OK.

3.25 Right-click and select Properties.

- Select Edit in the Visibility/Graphics Overrides field.
- Select the Design Options tab.
- In the Design Option field for Option Set Stairs, choose Open (primary).
- In the Design Option Field for Option Set Balconies, choose Twin Balconies.

- Click OK twice to end the edits (see Figure 10.55).

The view will change to show the balconies.

Visibility/Graphic Overrides for 3D View: Atrium Staircase Open, With Balconies

Model Categories	Annotation Categories	Imported Categories	Filters	Design Options

Design Option Set	Design Option
Stairs	Open (primary)
Balconies	Twin Balconies

Figure 10.55   *Settings to see open stairs with balconies*

3.26 Open each of the Atrium Staircase views in turn to check your work.

- If you wish to print your work, place the Atrium Staircase views on a sheet side by side, as shown in Figure 10.56.

- If you wish to keep the separate design options to experiment with, save a copy of the project file under a different name.

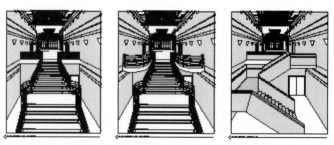

Figure 10.56   *Three options side by side*

ACCEPT THE PRIMARY OPTION

3.27 Open the Design Options dialogue.

- Select Stairs in the left pane.

- Select Accept Primary under Option Set on the right side.

- Answer Yes in the question box (see Figure 10.57).

- Choose Delete in the option dialogue to delete the view dedicated to the eliminated option.

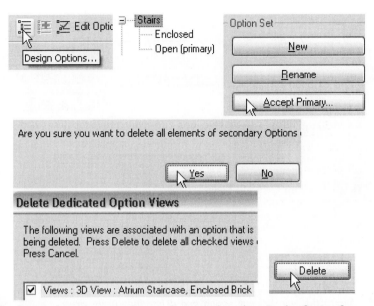

Figure 10.57 *Accept the primary option to eliminate the others in the Option Set*

3.28 Select Twin Balconies in the left pane.

- Select Make Primary under Option.

- Select Accept Primary under Option Set.

- Select Yes in the question box.

- Delete the out-of-date view (see Figure 10.58).

- Choose Close.

- Save the file.

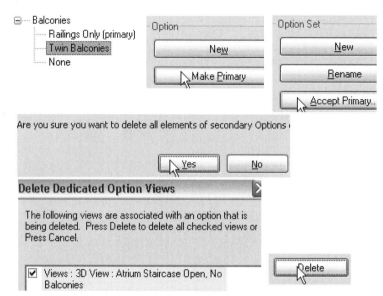

Figure 10.58 *Accept the twin balcony option and delete inaccurate views*

EXERCISE 4. CREATE AND PLACE A NESTED FAMILY WITH CONDITIONAL FORMULAS

Revit Architecture's Families hold and organize definitions for nearly everything in Revit Architecture. Expand the Families section in the Project Browser, as shown in Figure 10.59, to see a list of content by Family. In this exercise, you will create a nested column family using very simple drafting, and then apply Parameters, which you have already studied in previous exercises, to create design rules for types and instances within this family. The Parameters will use Formulas—mathematical, yes/no and conditional (i.e., logical operations such as "if-then-else")—to create smart switches that control the appearance of the column as it is placed.

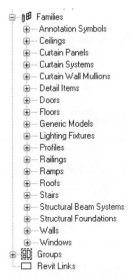

Figure 10.59   *Even a blank Revit Architecture file contains many Families*

CREATE SOME SIMPLE SOLIDS WITH MATERIALS

4.1 Open or continue working in the file from the previous exercise.

- Open view Floor Plans Main Level.

- Zoom to Fit.

- From the Window menu, select Window>Close Hidden Windows.

4.2 From the File menu, select File>New>Family, as shown in Figure 10.60.

- In the New dialogue box, select *Imperial Templates\Generic Model.rft [Metric Templates\Metric Generic Model.rft]*.

- Choose Open.

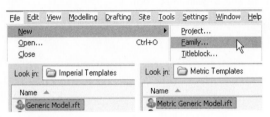

Figure 10.60   *Start a new Family*

4.3 In the new file, open view Elevations: Left.

- Zoom to Fit. Set the View Scale **to 6" = 1'-0" [1:2]**, as shown in Figure 10.61.

Figure 10.61 *Set the view scale so lineweights appear correct*

- Select Solid Form>Solid Revolve from the Family tab (the only one available) on the Design Bar, as shown in Figure 10.62.

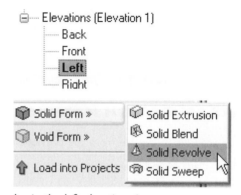

Figure 10.62 *Start a Revolve in the left elevation view*

4.4 Sketch the profile shown in Figure 10.63.

- Do not create the dimensions.

  The profile is basically a 2" [50] wide by 3" [75] tall rectangle set 4" [100] to the right of the origin point, with 1" [25] half-round curves stepped back at 1/4" [6].

- Use the Line and 3-point Arc tool.

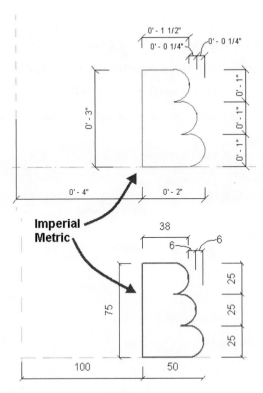

Figure 10.63   *The profile for the trim ring family*

4.5 When the profile is complete, select Axis from the Design Bar.

- Draw a short line vertically from the origin, as shown in Figure 10.64. This profile will create a stepped ring when spun around the axis.

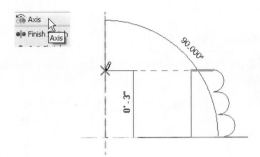

Figure 10.64   *The axis of revolution*

4.6 Select Revolution Properties from the Design Bar.

- Verify that the End Angle is 360° and the Start Angle 0°.
- Select the Material value field to open the Materials dialogue.

4.7 In the Materials dialogue, select Default in the Name panel, then select Duplicate.

- In the New Material name box, type **Chrome**.

- Select the arrow next to the AccuRender Texture field.

- Navigate to _accurender\Metals\Chrome\Satin, Plain_, as shown in Figure 10.65.

- Click OK three times to set the Revolution Properties.

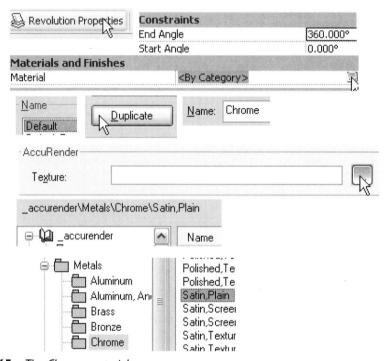

Figure 10.65 *The Chrome material*

4.8 Select Finish Sketch.

- Select the Default 3D view from the Toolbar to see the finished trim ring (see Figure 10.66).

- Adjust the view scale if need be.

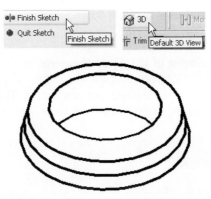

Figure 10.66 *The revolved solid*

4.9 Save the file as **Trim Ring.rfa** in a place where you can find it later.

4.10 Return to the Left Elevation view.

- Select the solid.

- Select Mirror on the Toolbar.

- Clear Copy on the Options Bar.

- Click on the Reference Level to mirror the solid upside down, as shown in Figure 10.67.

- Click File>Save As.

- Save the file as **Upper Trim Ring.rfa**.

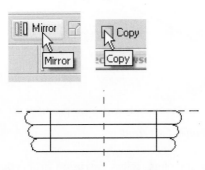

Figure 10.67 *Mirror the solid and save it as a separate file*

4.11 Delete the solid.

- Select Solid from the Design Bar.

- Select Extrusion and choose OK.

- Draw the profile shown in Figure 10.67.

- Do not create the dimensions.

  The profile is a bracket of 1" [25] thick material, 36" x 13" [900 x 325], with a 2" x 1" [50 x 25] flange 1" [25] from the right end and a 1/2" [12.5]-thick arc brace. The exact length and placement of the arc are not critical. The vertical arm of the bracket is centered on the horizontal reference plane, and its left end is aligned with the vertical plane.

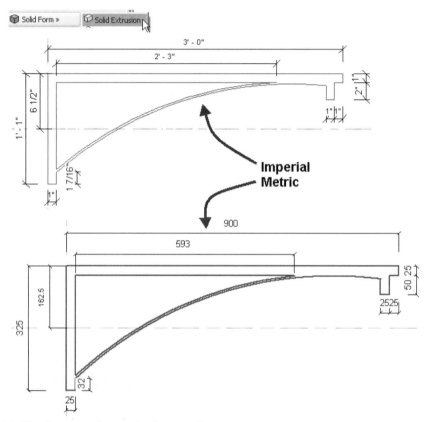

Figure 10.68   *Draw another profile for extruding*

4.12   When the profile is complete, select Extrusion Properties from the Design Bar.

- Set the Extrusion End to **–0 1/2" [-12.5]** and the Extrusion Start to **0 1/2" [12.5]**.

- Select the Material value field to open the Materials dialogue.

4.13   Select the Default material and choose Duplicate.

- In the New Material name box, type **Black Iron**.

- Click the arrow next to the AccuRender Texture field.

- Navigate to *ar2_accurend\matte, black*, as shown in Figure 10.69.

- Click OK three times to set the Extrusion Properties.

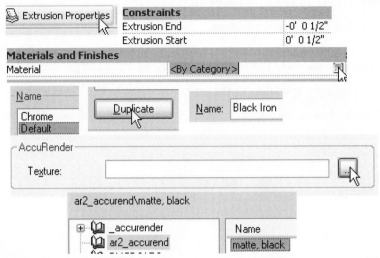

Figure 10.69 *Define the Black Iron material*

4.14 Choose Finish Sketch.

- Open the default 3D view to see the results, as shown in Figure 10.70.

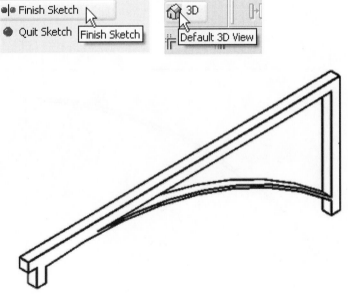

Figure 10.70 *The bracket profile, extruded*

4.15 Return to the Left Elevation view.

- Select Solid>Solid Revolve from the Design Bar.

4.16 Draw the profile shown in Figure 10.71.

- Do not create the dimensions.

  The profile is an arc of approximately 50° from the midpoint of the flange bottom face approximately **1' 9" [525]** to the left by **12" [300]** down, then offset **1/4" [6]**.

- Connect the endpoints of the arc to close the profile.

  Actual dimensions are not critical. The revolved solid will be a translucent light cover.

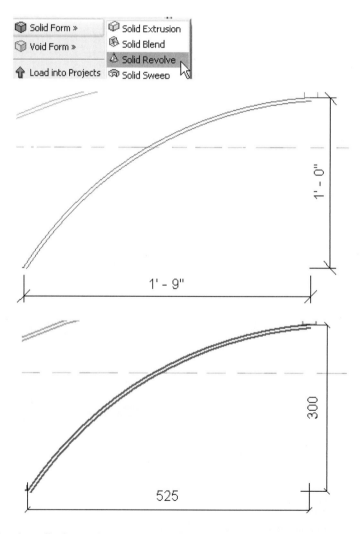

Figure 10.71 *A profile for revolution—this will be a light cover. Imperial above, metric below*

4.17 Select Revolution Properties from the Design Bar.

- Verify that the End Angle is 360° and the Start angle 0°.

- Select the Material value field to open the Materials dialogue.

4.18 Select Glass from the Name list.

- Choose the arrow next to the AccuRender Texture field.

- Navigate to *_accurender\Glass\Tinted\White, Translucent, Frosted*, as shown in Figure 10.72.

- Choose OK three times to set the Revolve Properties.

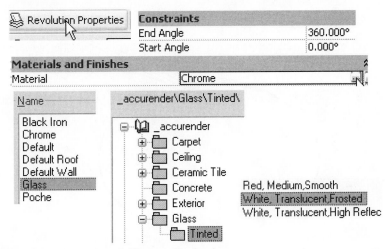

Figure 10.72 *Define the material Glass in this family*

4.19 Select Axis.

- Draw a line down from the right end of the profile, as shown in Figure 10.73.

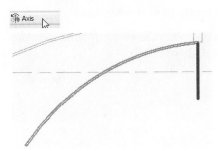

Figure 10.73 *Draw an axis*

4.20 Select Finish Sketch.

- Open the default 3D view (see Figure 10.74).

- From the File menu, select File>Save As.

- Save this file as **Bracket Light.rfa**.

- Close the file.

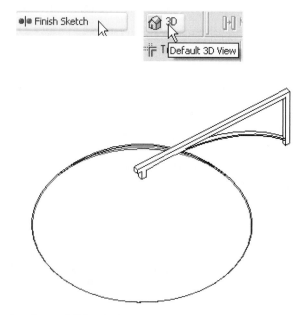

Figure 10.74 *The bracket and light shade*

EXERCISE 5. CREATE AND APPLY CONDITIONAL PARAMETERS IN A FAMILY

5.1 Select File>New>Family from the File menu.

- Select *Imperial Templates\Column.rft [Metric Templates\Metric Column]*, as shown in Figure 10.74, and choose OK.

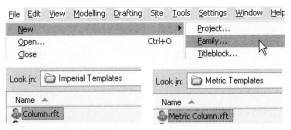

Figure 10.75 *Open the Column template*

5.2 The new file will open to a Plan view.

- Select Solid Form>Solid Extrusion from the Design Bar.

- Sketch a **2' x 2' [500 x 500]** square, as shown in Figure 10.76, using the reference planes around the origin point.

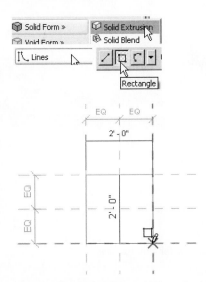

Figure 10.76 *Sketch a square for the first extrusion*

5.3 When the profile is complete, select Extrusion Properties from the Design Bar.

- Accept the default Extrusion Start and End values for now.

- Select the Material value field to open the Materials dialogue.

5.4 In the Materials dialogue, choose Default in the Name panel, and then choose Duplicate.

- In the New Material name box, type **Brick**.

- Select the arrow next to the AccuRender Texture field.

- Navigate to _accurender\Masonry\Brown,8",Running [_accurender\Masonry\Brown,200mm,Running], as shown in Figure 10.77.
- Click OK three times to set the Extrusion Properties.
- Click Finish Sketch.

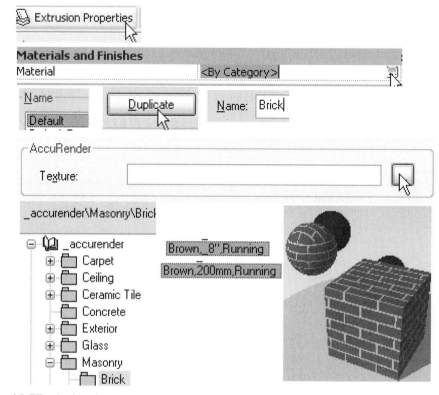

Figure 10.77 *Define the material for this extrusion*

SET UP PARAMETRIC FAMILY TYPES

5.5 Choose Family Types from the Design Bar.

- Select Add in the Parameters section.
- In the Parameter Properties dialogue, verify that Family parameter is checked.
- Name the new parameter **Top Height**.
- Select Length as the Type.
- Set Dimensions as the Group.
- Select Type, as shown in Figure 10.78.
- Click OK.

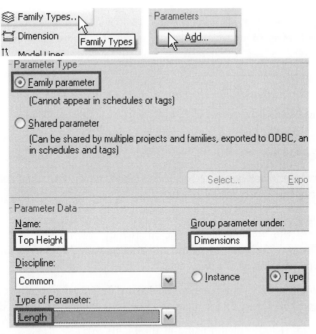

Figure 10.78   *Define the Top height as a Length parameter*

5.6 Select Add in the Parameters section.

- Name the new parameter **Base Height**, and select Length as the Type.

- Set the other options as for the previous parameter: Dimensions, Length and Type.

- Click OK.

5.7 In the Formula field for the new Base height parameter, type **Top Height/6**.

- Choose Apply (see Figure 10.79).

  Revit Architecture will space the text according to its formula syntax. Revit Architecture will notify you if a formula is not correct. Formula text is case-sensitive.

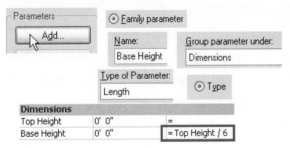

Figure 10.79   *Apply a simple formula to the Base Height parameter*

Tip: This dialogue box will resize. Stretch its borders left and right to give the Formula field plenty of space for typing. Adjust the width of the other columns as necessary.

5.8 Select New under Family Types.

- In the Name Box, type **12' [3600]** and click OK.

- Repeat this step twice more, creating Family Types named **17' [5100]** and **23' [6900]**.

  There are many ways to organize family types. For this column style, you will differentiate the types by height.

  In this project, a 12' [3600] column will fit under the wing roofs at the front (North side) of the building file, and a 23' [6900] column will fit under the third story overhang at the main entry. You will also create a 17' [5100] column, perhaps to use as a parking lot or walkway light.

  Columns of different heights will have different characteristics, which will not be drafted but driven by formulas that use the column height as their basis.

5.9 Put the 23' [6900] column name in the Name list.

- In the Value field for Top height, type **23 [6900]**.

- Choose Apply.

  Revit Architecture will set the units to 23' 0" [6900] and display a calculated value for the Base height, as shown in Figure 10.80.

Name:	23'	
Parameter	Value	Formula
Dimensions		
Top Height	23' 0"	=
Base Height	3' 10"	= Top Height / 6

Figure 10.80 *Create a Family Type and set its Top height. Revit Architecture will calculate the Base height.*

5.10 Repeat this step for the 17' [5100] and 12' [3600] Family Types.

- Enter the correct Top Height value and click Apply each time.

- Make the 23' [6900] Family Type the current Type.

- Click OK.

5.11 Select the extrusion.

- Select Edit from the Options Bar.

- Select Extrusion Properties from the Design Bar.

- Click the small button to the right of the value field for Extrusion End.

- In the Associate Family Parameter dialogue, select Base height and choose OK.

  In the Element Properties dialogue, the Extrusion End value will display a different number, which will be grayed, indicating it is not editable by using this field. The small button to the right of the field will hold an equals (=) sign, indicating that a parameter is driving this value (see Figure 10.81).

5.12 Click OK.

- Choose Finish Sketch.

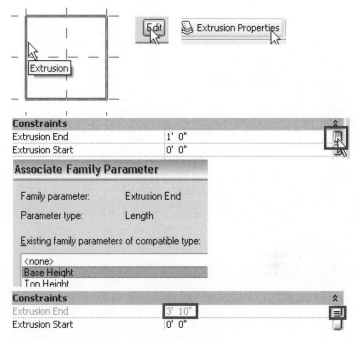

Figure 10.81 *Associating a parameter with an Extrusion*

5.13 Choose Solid From>Solid Extrusion in the Design Bar.

- Select the Circle tool from the Options Bar, and draw a circle of **4" [100]** radius at the origin, as shown in Figure 10.82.

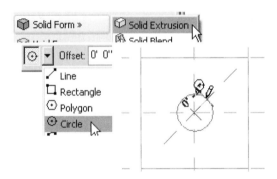

Figure 10.82 *Draw the next extrusion using the Circle tool*

5.14 Select Extrusion Properties from the Design Bar.

- Select the Associate Parameter button as before.

- Choose Top height and choose OK.

5.15 Repeat this step for the Extrusion Start value, using Base height as the Associated Parameter.

- Choose OK (see Figure 10.83).

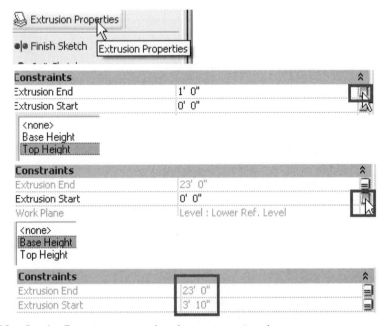

Figure 10.83 *Set the Extrusion start and end to parametric values*

5.16 Select the Material value field.

- Select Brick and choose Duplicate.

- Name the new material **Aluminum**.

- Use the AccuRender Texture field to set the material to _accurender\ Metals\Aluminum\Satin,Plain_, as shown in Figure 10.84.

- Choose OK three times to exit the dialogues.

- Select Finish Sketch.

    Except for placing components, drafting for this family has now finished, and the rest of the work will be with parameters and formulas.

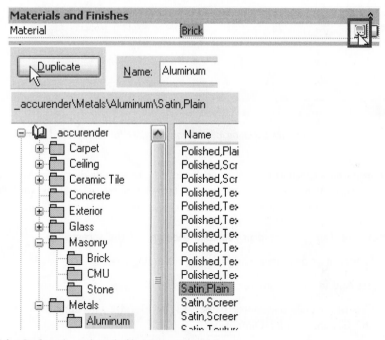

Figure 10.84   *Define the column's Aluminum material*

WORK WITH PARAMETERS AND FORMULAS

5.17 Open the default 3D view.

- Choose Family Types from the Design Bar.

- Cycle the Family Types as you did when you created them, using the Apply button for each Type even though you did not make changes.

- Move the Family Types dialogue box so you can see the column shaft and base change height (see Figure 10.85).

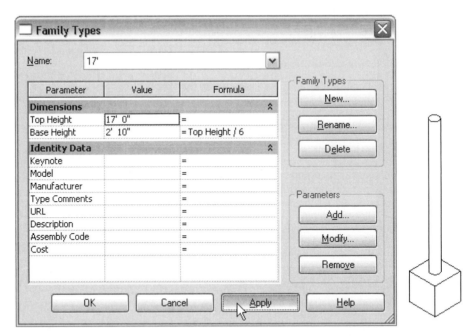

Figure 10.85 *Flex the model by checking all the family types for unexpected behavior*

Note: Checking family types is an important part of creating them effectively. Check your work often when adding parameters, formulas or reference planes with dimensions, to make sure that your families work as expected.

5.18 Move the Family Types dialogue to the center of the screen.

- Stretch its sides left and right to make typing easier.

5.19 Select Add in the Parameter section of the dialogue.

- Create a Family parameter named **Light Height** as a Length value stored by Type and Grouped under Dimensions.

- Click OK.

5.20 Add a Family Parameter named **Trim Height** as a Length value stored by Type and Grouped under Dimensions (see Figure 10.86).

- Click OK.

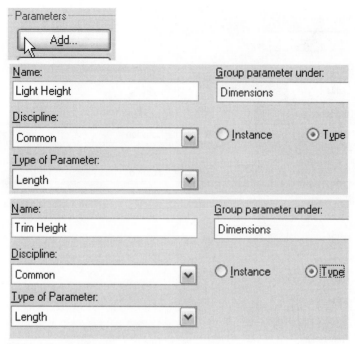

Figure 10.86 *Add height parameters*

5.21 Repeat the Add step to create a Family parameter named **Lights 1** as a Yes/No value stored by Type.

- Click OK.

5.22 Add a Family parameter named **Lights 2** as a Yes/No value stored by Type.

- Click OK.

5.23 Add a Family parameter named **Middle Trim** as a Yes/No value stored by Type.

- Click OK.

The Family Types dialogue will display the five new parameters you have created. They will apply to all the Family Types, not just the type that was active when you started adding parameters (see Figure 10.87).

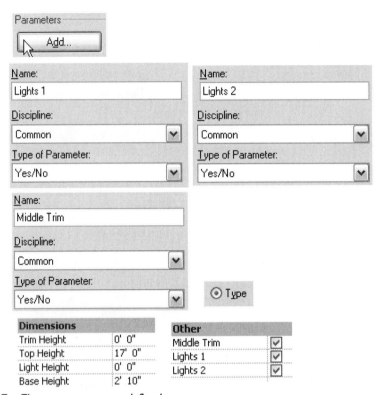

Figure 10.87 *The new parameters defined*

5.24 Enter formulas into the appropriate fields, as shown in Figure 10.88.

- For Trim Height, enter **Top Height/2**.

- For Middle Trim, enter **Top Height>16' [Top Height>4800]**.

- For Lights 1, enter **Top Height>8' [Top Height>2400]**.

- For Lights 2, enter **Top Height>16' [Top Height>4800]**.

  Users in metric units, enter formulas as shown in brackets only. Users in imperial units, do not enter formulas in brackets.

- Choose Apply after entering each one to check your formula.

Name:	12'	

Parameter	Value	
Dimensions		
Trim Height	6' 0"	= Top Height / 2

Other		
Middle Trim	☐	= Top Height > 16'
Lights 1	☑	= Top Height > 8'
Lights 2	☐	= Top Height > 16'

Figure 10.88 *Formulas for the parameters*

The formulas mean the following: The Trim height will always be one-half the height of the column. The Middle Trim value will be Yes when the column height is greater than (>) 16' [4800]. This condition is satisfied in the family types 17' [5100] and 23' [6900] tall, but not in the 12' [3600] type, so any element keyed to the Middle Trim parameter will not appear in that type. The same condition applies to Lights 2.

Lights 1 will appear in all column types taller than 8' [2400]. The three column types you have defined all meet this condition—a shorter column style could be defined, maybe for interior use, that would not have any elements keyed to Lights appear.

5.25 Carefully enter the following text into the formula field for Light Height:

If (Top Height > 20', Top Height –4', If (Top Height < 15', Top Height –2', Top Height –3'))

- Metric users enter the following text:

If (Top Height > 6000, Top Height -1200, If (Top Height< 4500, Top Height -600, Top Height -900))

- Click Apply to check your formula.

 Note: Revit Architecture will supply spaces in formulas, as you have seen before, but be very careful with spelling, case and punctuation. Open and closing parentheses must match.

This formula means: If the column is taller than 20' [6000], the Light Height will be 4' [1200] down from the top. If the column height is less than 15' [4500], the Light Height will be 2' [600] down from the top. For all column heights between 20' [6000] and 15' [4500], the Light Height will be 3' [900] down from the top.

A complete study of conditional syntax is beyond the scope of this book.

5.26 Cycle through the Family Types to see changes, as shown in Figure 10.88.

In the 12' [3600] type, the Middle Trim and Lights 2 value check boxes are empty, since this height is below the threshold value defined by the formula (see Figure 10.89).

- Click OK to finish the parameter and formula setup.

| Name: | 17' | |

Parameter	Value	
Dimensions		
Trim Height	8' 6"	= Top Height / 2
Top Height	17' 0"	=
Light Height	14' 0"	= if(Top Height > 20',
Base Height	2' 10"	= Top Height / 6

Figure 10.89   *Check the Family Types*

ADD NESTED FAMILY COMPONENTS

5.27 Save the file as **decor column.rfa**.

- From the File menu, select File>Load From Library>Load Family.

- Navigate to the folder(s) where you saved the *bracket light* and both *trim ring* files, select them and choose Open.

Tip: If the files are in the same folder, you can hold down the CTRL key and select multiple file names.

5.28 Open the Lower Reference Level plan view.

- Select Component from the Design Bar.

- Put Trim Ring in the type selector.

- Locate an instance of the trim ring at the origin point, as shown in Figure 10.90.

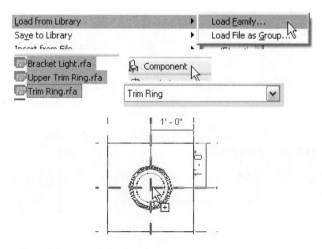

Figure 10.90 *Locate the trim ring at the origin*

5.29 Select Modify to terminate the insert.

- Open the Left Elevation view.

- Select the new trim ring.

- Select the Properties icon from the Options Bar.

5.30 In the Element Properties dialogue, select the Associate Parameter button next to the Offset value field.

- Select Base height and choose OK (see Figure 10.91).

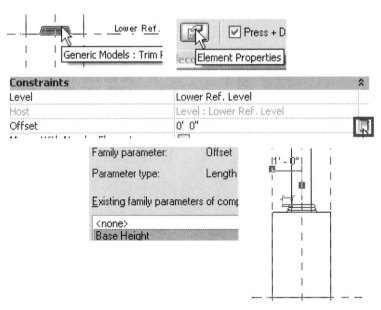

Figure 10.91 *Associate the new component Offset with the parameter Base height*

5.31 The trim ring component will still be highlighted.

- Copy it some distance straight up (the actual distance does not matter.)

- Choose the Properties icon from the Options Bar.

5.32 In the Element Properties dialogue for the new instance of the trim ring, associate its Offset value with the parameter Trim height.

- Choose OK.

- Associate the Visible value for this instance with the parameter Middle Trim.

- Choose OK twice (see Figure 10.92).

The component will change height on the column.

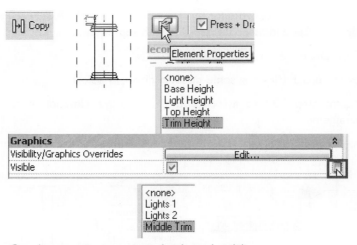

Figure 10.92 *Give the new trim parametric height and visibility*

5.33 Open the plan view again.

- Select Component from the Design Bar.
- Put Upper Trim Ring in the type selector.
- Locate an instance as the origin, as before (see Figure 10.93).

  A warning will appear that you can ignore.
- Choose Modify.

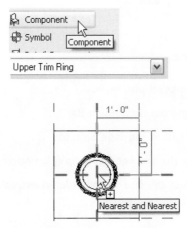

Figure 10.93 *Add a different trim ring*

5.34 Open the Left Elevation view.

- Select the new upper trim ring instance.
- Select the Properties icon from the Options Bar.

- In the Element Properties for this component, select the Associate Parameter button for its Offset and associate its value with Top height.

- Click OK twice.

5.35 Copy the upper trim ring some distance down, as before.

- Repeat Step 5.32 to associate the new copy's elevation with the Trim height parameter.

- Associate its Visible value with the Middle Trim parameter (see Figure 10.94).

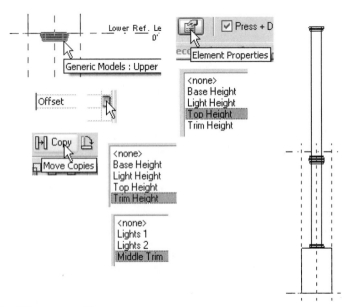

Figure 10.94 *Trim rings in place*

5.36 Open the Lower Ref. Level plan view.

- Select Component from the Design Bar.

- Choose Bracket Light in the Type Selector.

- It will appear with the bracket facing up (plan north).

- Place the instance at the face of the column extrusion, as shown in Figure 10.95.

- Select Modify.

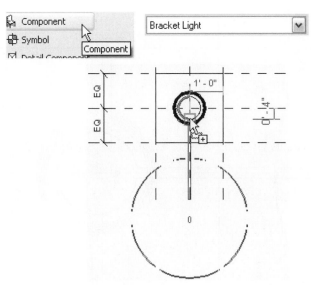

Figure 10.95 *The first light component placed in the plan*

5.37 Select the light.

- Use the Mirror tool and the horizontal Reference Plane to make a copy at the other side of the column.

5.38 Copy the new instance to the left, rotate it 90° and place the new instance at the column face, as shown in Figure 10.96.

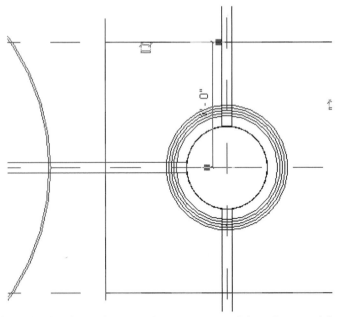

Figure 10.96 *Place the bracket endpoint at the intersection of the column and Reference Plane*

5.39 Mirror the new light to the right, for a total of four lights in place around the column (see Figure 10.97).

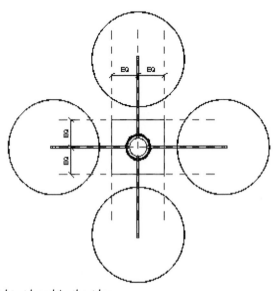

Figure 10.97 *Four lights placed in the plan*

5.40 Open the Left Elevation view.

- Select the four lights and choose the Properties icon from the Design Bar.

- In the Element Properties dialogue, associate the Offset value of the lights with the Parameter Light Height.

- Click OK twice.

  The lights will move up.

5.41 Open the default 3D view.

- Select the front and rear lights, as shown in Figure 10.98.

- Click the Properties icon.

- Associate the Visible parameter of those components with Lights 1.

- Click OK twice.

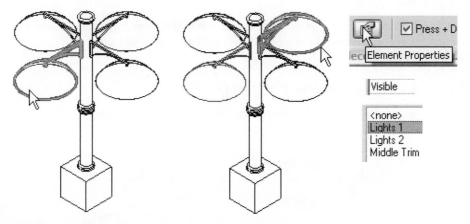

Figure 10.98 *Associate these lights with parameter Lights 1*

5.42 Select the other two lights.

- Associate their Visible parameter with Lights 2.

- Click OK twice.

5.43 Select Family Types.

- Cycle through the Family Types to make sure all components obey the elevation formulas.

- Click OK to exit the dialogue.

  The Visible condition does not show in the family model, so lights and trim will not turn on and off.

- Save the file.

- Close the file.

EXERCISE 6. PLACE FAMILIES IN A PROJECT

6.1 In the Main Level view of the Chapter 10 project file, make the Basics tab of the Design Bar active.

- Select Grid.

  Columns snap to grids, so you will place grid lines to locate columns. For our purposes, the numbering of grid line bubbles does not matter.

6.2 Select the Pick tool on the Options Bar.

- Set the Offset value to **2' 0" [600]**.

- Select the edge of the curved front plaza, as shown in Figure 10.99.

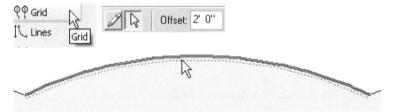

Figure 10.99 *The first curved grid line follows the front walkway*

6.3 Select the Draw and Straight Line tools on the Options Bar.

- Set the offset to **0**.

- Type **SC** to force a Center snap.

- Select the curved curtain wall at the main foyer.

6.4 Pull the cursor straight up at 90° until it crosses the first grid line, as shown in Figure 10.100.

- Click to create the grid line.

- Click Modify.

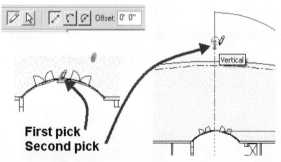

Figure 10.100 *Pull the new grid line vertically*

6.5 Copy or Array the new vertical grid line three times left and right at a distance of 24' [7200], as shown in Figure 10.101, for a total of seven grid lines.

- Click Modify to terminate placement.

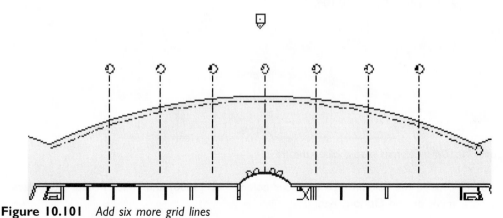

Figure 10.101 *Add six more grid lines*

6.6 Right-click.

- Select View Properties.

- Set the Underlay value to Balcony Level, as shown in Figure 10.102.

- Click OK.

Figure 10.102 *Show the Balcony level as Underlay*

6.7 Make the Modeling tab active on the Design Bar.

6.8 Choose the Column tool on the Design Bar.

- Select Load on the Options Bar.

- Navigate to the folder where you saved the *decor column.rfa* file, and choose Open.

- In the Type selector, make decor column: 23' [6900] the active type (see Figure 10.103).

Figure 10.103 *Load the decor column file*

6.9 Place instances at the intersections of the grid lines along the rim of the front plaza, as shown in Figure 10.104.

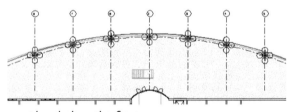

Figure 10.104 *Columns placed along the front*

6.10 Change the column type to decor column: 12' [3600] and check Rotate after placement on the Options Bar.

- Place four instances, rotated 20° or −160°, at the corners and middle of the roof on the front right of the building.

  Precise placement is not critical for this exercise (see Figure 10.105). Note that these columns only have two lights.

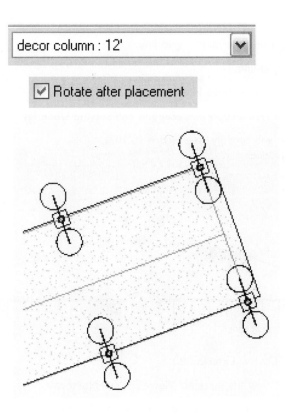

Figure 10.105   *Columns under the right roof*

6.11 Change the column type to decor column: 17' [5100] and place two instances about 24' [7200] apart near the front entrance, as shown in Figure 10.106.

Precise placement is not critical.

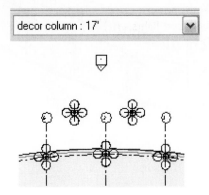

Figure 10.106   *17' columns near the front*

6.12 Mirror the four 12' [3600] (rotated, two-light) columns from the right to the left of the building using the middle grid line.

- Select a grid line.

- Right-click and choose Hide in View>Category.

- Open the View Properties dialogue and set the Underlay value to None.

  The view will simplify (see Figure 10.107).

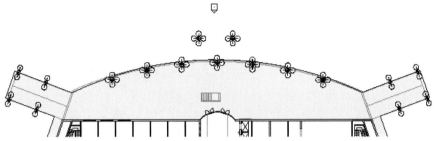

Figure 10.107   *Columns in place*

6.13 Open the 3D view Exterior Iso.

- From the View menu, select View>Orient>Northeast.

- Zoom in a little to study the three types of columns.

  The 12' [3600] columns under the wing roof have two lights and no middle trim, as you defined in the family type formulas (see Figure 10.108).

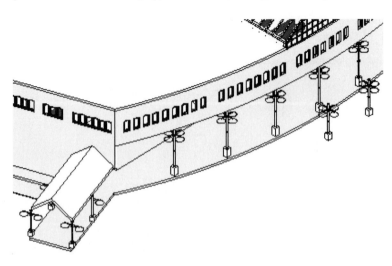

Figure 10.108   *Three types of columns from one family*

There are other 3D views set up in this file to examine the columns and other parts of the model.

6.14 Save the file.

• Close the file.

SUMMARY

This final chapter showed you some of the most powerful and useful features in Revit Architecture—of particular concern to busy designers.

Design Options allow you to create many alternatives inside any project on the fly, and then view them, change them and ultimately resolve them into your preferred options.

Family creation in Revit Architecture is an extremely powerful toolset—practice creating your own component families to suit your purposes. Experiment with writing simple formulas that will take advantage of the design rules inherent in any component so that you can make many versions of the same basic family structure without extensive drafting.

REVIEW QUESTIONS

MULTIPLE CHOICE

1. Family parameter types include

   a. Text, Number, Integer

   b. Length, Area, Volume

   c. Angle, URL, Material

   d. all of the above, plus Yes/No

2. Which of the following is not available when editing a Wall Structure?

   a. Sweeps

   b. Reveals

   c. Balusters

   d. Layers

3. A Design Option Set holds

   a. Family Parameters

   b. As many Design Options as you define, one of which will always be primary

   c. Only the Design Option that is currently being edited

   d. The key to a happy life

4. Stair Parameters that you cannot edit include

    a. Nosing, Thickness and Profile

    b. Riser Type, Thickness, Material and Tread Connection

    c. Stringer Thickness, Height, Material and Trim at Top value

    d. All of the above can be edited.

5. Revit Architecture does not provide family template files (**.rft*) for

    a. Doors, windows, furniture

    b. Roofs, floors, walls

    c. Lights, columns, electrical fixtures

    d. Profiles, plumbing, mechanical equipment

TRUE/FALSE

6. True _ False _ Revit Architecture keeps Design Options as separate files from the main project model.

7. True _ False _ Extrusion Start and End values can be set by parameters.

8. True _ False _ Length parameters cannot use formulas.

9. True _ False _ You can edit a Railing baluster placement value only once if the railing is a Design Option.

10. True _ False _ Stair boundaries can be edited to change the shape of a stair.

 Answers will be found on the CD.

CONCLUSION

Congratulations! You have studied and experimented with the major features of a powerful, efficient building-design application, and you have joined the growing ranks of those able to use the latest technology that the market has to offer. Practice and experience with Revit Architecture will ensure speed in your work and provide the confidence to explore its extensive capabilities.

For those of you in business, it's now up to you and your organization to determine how best to implement what Revit Architecture has to offer. See the Appendix for a discussion of factors that affect the implementation process in design firms. You can blend Revit Architecture into your workflow by combining it with standard CAD applications, or start out with a project planned and executed entirely using BIM (Building Information Modeling). Either way, you can refer back to the concepts and exercises in this text to help you understand and make effective use of Revit Architecture's techniques for massing, modeling, drafting, data management and illustration in your own work process. It's important to match your expectations with your resources. It will be hard to ask too much of the program, but it can be easy to underestimate the time and effort necessary to make it do exactly what you think you are asking.

If you worked through this text on Revit Architecture as part of a school curriculum, you have now explored the only architectural design application in wide use that combines drafting, massing, element modeling, illustration and data extraction in one interface, with one common set of tools throughout. Exposure in the classroom is not proficiency in the workplace and simple understanding is not technical competence, but your expectations heading into the job market should be high.

Those of you who develop expertise with Revit Architecture can and should lead the way in providing guidance, support and training for others in your class, team or company. Patience and sympathy can be key—not everyone learns at the same pace, hits the same stumbling blocks or finds the same solutions to procedural problems. Revit Architecture is well-designed software in that there is more than one way to do just about anything. This can be confusing, both for instructors and students at first, but practice will help everyone develop flexible working methods and the knack for finding answers to questions unaided. Don't forget online resources such as user groups and their ongoing discussion sites. You can engage enthusiastic, experienced, articulate Revit Architecture users worldwide with very little effort.

The authors wish you goodspeed moving on from this point. It is our sincere hope that you will soon find that, just like the fabled Mr. Natural, you can Keep on Truckin' through your design work with your hands firmly on The Right Tool For The Job.

What It Really Takes
to Implement Revit Architecture

 Note: This appendix is taken from a highly rated presentation that co-author Jim Balding gave at Autodesk University in 2003. His audience consisted of executives, managers and engineers from firms seeking to understand what they would face while migrating to Revit Architecture. The term Building Information Modeling (BIM) refers to database systems such as Revit Architecture, rather than vector-based CAD applications.

Readers from educational institutions or other organizations with in-house design departments may readily identify similar concerns that their programs or departments may have when facing upgrade, deployment and training issues.

WHY CHANGE?

There are many reasons to change to the Building Information Model (BIM). In the beginning, the most compelling is the three dimensional aspect of designing and documenting a project. In addition to that, many find the constant, complete coordination of the single model to be a compelling reason. If that doesn't convince you, the fact that firms using Revit Architecture are consistently providing better service and deliverables in less time with less staff—and occasionally both—may sway you. The bottom line is that a totally efficient design process has always been the "Holy Grail" of design and documentation, and only recently have the hardware and software caught up and become able to deliver on this elusive promise.

INTRODUCTION—POINTERS, NOT PRESCRIPTION

We have been asked countless times, "How do I implement Revit Architecture at my company?" The answer to this question is very simple—we have absolutely no idea. That may seem a little odd coming from two people writing a book about the subject, but it is true. All firms and institutions are different in so many ways that there is no way to create a one-size-fits-all, step-by-step instruction book on how to implement a software application such as Revit Architecture in any given firm. With that in mind, we have identified seven factors that will influence the implementation process in just about any office, and have outlined the issues to look for while implementing Revit Architecture.

THE SEVEN FACTORS OF IMPLEMENTATION

While there are many different factors that will affect the success of implementation efforts, most can be grouped into one of seven categories. The following list is in order of importance:

Firm size

Project size

Architectural style

Project type

Scope of services

Firm locations

Firm culture and age

FIRM SIZE

The size of the firm is the number-one issue when it comes to implementing the Building Information Model. While there are many factors that firm size will affect during implementation, the key factors are project size, technology, training/support/R&D, personnel, decision making and standardization. The following sections include *general* observations regarding key factors related to company size. Each category has been rated as holding an advantage toward either larger or smaller firms. This is by no means an absolute system, and the advantage could be great or slight—use your own experience when considering these issues and evaluating your own situation.

PROJECT SIZE—ADVANTAGE: SMALLER FIRMS

Large firms generally have larger projects, and when it comes to Revit Architecture, project size can be viewed as an obstacle with regard to implementation. If you consider that you are adding all of the building data to one file, that file can certainly grow, and becomes exponentially larger than the single referenced files you may be presently using. BIM products including Revit Architecture have strategies to alleviate the issue of large projects. However, many firms feel that they need more time to get to a comfort level using full Revit Architecture implementation on such large-scale projects. It is important to note that this does not count out larger projects; they will simply require a little more planning and more experience.

This is where the smaller firms tend to shine. They typically have smaller projects that are ideal for Revit Architecture. There is not as much information to be communicated in a smaller project, which translates to smaller files and project teams.

One thing to keep in mind when selecting projects based on project size: It is advisable not to base the decision entirely on the floor plan area, but on the "spatial size," including both the amount of mass and detail to be built. For example, designing and documenting a

750,000-square foot (s.f.) warehouse using Revit Architecture can be far easier than creating a 20,000-s.f. gothic cathedral.

TECHNOLOGY—ADVANTAGE: LARGER FIRMS

The fact that Revit Architecture models are larger and more complex than ordinary 2D drafting files means that the software and models require fast machines with plenty of RAM and good video cards. When it comes to technology, larger firms usually have the upper hand. They typically have the latest hardware, fast Local Area and Wide Area Networks, and perhaps even a dedicated Information Systems team to support the systems.

The small firms generally find themselves behind the eight ball in this arena. They may have made a recent, significant investment in connectivity, hardware or software, and can't afford to budget large-ticket items such as these every year. A small firm that has made a substantial investment in technology could find itself in an advantageous position when implementing Revit Architecture.

Firms of any size should pay close attention to the hardware requirements of BIM software, as well as performance on typical firm project types. Maintaining up-to-date hardware and software often pays dividends not only in the speed of deliverables, but also in perceivable pride of ownership, which can provide incentives to learn and maintain knowledge, thus reinforcing a company's commitment to technology.

TRAINING/SUPPORT/R&D—ADVANTAGE: LARGER FIRMS

The larger the firm, the larger the operating expenses. With this in mind, the ability to create and maintain in-house expertise to provide training and support, as well as perform research and development roles for firm-specific tasks, becomes significant. This phenomenon often runs parallel with a firm's CAD manager role. If the firm has a full-time CAD manager, typically it will have a full-time Revit Architecture manager who is often one and the same person.

The smaller firm will tend to have a part-time CAD manager, often an architect who knows the ins and outs of computers better than the rest of the firm. Firms can expect additional overhead and responsibilities when it comes to implementing and steadily using BIM. The return on investment over time is proving to be much larger proportionately when comparing results from Revit Architecture to 2D CAD.

Training and support are two of the keys to success, discussed in greater detail later, and they should be given serious thought when implementing Revit Architecture.

PERSONNEL—ADVANTAGE: LARGER FIRMS

When planning the implementation of BIM, one of the first things you need is a "champion," the person responsible for heading up the change. However, one person cannot do everything required to make the move to Revit Architecture. You will need talented people to carry out the plan and use the software effectively enough to deliver on the promise.

Because larger firms have larger talent pools from which to select, they generally have the luxury of hand-selecting the users who best suit the chosen tool. One thing to note, however—the proportion of talent and roles within a firm will tend to be about the same for both larger and smaller firms.

DECISION MAKING—ADVANTAGE: SMALLER FIRMS

When it comes to making the decision to change, it seems better to have a single entity make that decision rather than a committee. Nowhere does the saying "too many chefs spoil the broth" apply more appropriately than in implementing change. Larger firms with multiple locations will have more challenges in the way of purchasing, planning and organizing an implementation plan.

Smaller firms, generally based in one location, have just one "chef" who makes the decision to implement BIM, and then it is time to move on. When changes in the plan occur, there isn't another round of negotiations, planning, etc., to slow progress.

STANDARDIZATION—ADVANTAGE: SMALLER FIRMS

Here again, the larger firm will tend to have many "chefs," from the top to the bottom. Somewhere in between might be a local CAD manager or two with individual CAD standards and techniques. If your firm is international, you have, potentially, different languages, building techniques, documentation techniques, cultures and infinitely more issues of diversity to face.

Smaller firms get the nod in this arena. With one location, standards can be established, and even voted on, in one location, perhaps at one time. When changes to a standard occur, communicating that change is simple, and most users will understand the reasoning behind it because it undoubtedly arose from issues within the local firm.

FIRM SIZE—CONCLUSION

While there are many factors listed here, there are many more that will be firm-specific. As stated earlier, firm size is the number-one consideration when deciding how to implement Revit Architecture at a firm. And while each subject has been labeled with a score giving an advantage to one size firm or the other, the intention of this section is to make firms aware of the issues regarding firm size and to assist firms in their implementation process.

Firm Size Final Score: Larger firms – 3
                           Smaller firms – 3

PROJECT SIZE

While project size is a major factor in Revit Architecture implementation, it also is influenced more generally by firm size. Refer to the Project Size section under the preceding Firm Size discussion.

ARCHITECTURAL STYLE

When referring to architectural style and the implementation of Revit Architecture, the foundation of the distinction is the size, shape and amount of detail involved. Revit Architecture does not recognize the literal difference, for instance, between Romanesque, Renaissance, and Post-Modern architecture.

Common sense, while not always common, does in fact make sense here. When it comes to modeling a building, straight, square forms are easier to model/build than those of the organic styles of Mr. Frank Gehry. This is not to say that it is impossible to build a sweeping tilted wave of a facade, just that more time and technique are required. If you can imagine the effort required to physically build the building, there is not a great difference to building it virtually using Revit Architecture. Simply put, more detail, shapes and mass equal more time to develop and build.

While modeling all of the components in a project has desirable effects, keep in mind that it is not always necessary. During the planning stages of a project, it should be determined what will be modeled and what can be "represented" with lines, arcs and circles. A simple example: If you have raised wood paneling on a door, there is no harm in drawing lines on the surface of the door to represent the paneling and moving on with the project. Some might argue that it would not render properly or that there is a different material in the panel. The point is that there needs to be an understanding, on a per-project basis, of what is important enough to model and what isn't. Three dimensional modeling in Revit Architecture is not an all-or-nothing proposition.

PROJECT TYPE

Project types often affect different firms in different ways. Generally speaking, however, some project types are better suited for Revit Architecture than others. Again, this is not to say that a certain project or project type cannot be done effectively. Some of the aspects of project types that dictate the ease of implementation include elements that are repetitive, component-driven, area-driven or those that make use of—and find great value in—the different visualization and/or scheduling advantages of Revit Architecture.

The following sections contain a *general* outline of the advantages and challenges typically presented by various project types.

ENTERTAINMENT/RETAIL

Some of the advantages of Revit Architecture in the entertainment and retail industries might be the presentation value gained in color fills for area or room types, or the coordination of area calculations. Perhaps there is a need to schedule and color the tenant spaces by lease expiration, retail type or rent-per-foot. Bi-directional, live scheduling may also have advantages, whether that includes display racks, flooring or parking stalls.

Some of the challenges facing the entertainment/retail architect might include custom furniture and fixtures, or project size and scale. It has also been noted that due to the nature of a retail mall, individual retailers often have architects of their own, so there can be many different designs and ideas flying about at any given time.

HEALTH CARE/HOSPITALITY

Healthcare and hospitality designs, of course, are going to have significant gains using the repetitive, modular or component-driven aspects of Revit Architecture. It could be argued that this is also the case with CAD drawings, where xrefs, blocks or cells are used, but Revit Architecture goes beyond that. When duplicating the data along with graphic representation of that data, you have the benefit of scheduling, coordination and visualization. When a single change to one guestroom sink and vanity can affect the plans, interior elevations, sections, scheduling and specification information for 500 rooms, there is great efficiency in place.

Generally, health care and hospitality projects are large, very complex designs with large project teams. Incorporating and coordinating the necessary data set can become a significant burden. Planning a project of this nature is possible, but requires advanced planning and experienced Revit Architecture-capable staff.

RESIDENTIAL

Residential projects gain tremendously from using Revit Architecture since they are generally smaller in size, as discussed earlier. Beyond that, there is also the fact that residential clientele may not read or understand floor plans, sections and elevations, but certainly would be able to understand a perspective or a rendering.

Challenges facing a residential architect might include the smaller scale of the project as well, surprisingly enough. Residential work tends to be shown at a larger scale, and therefore includes more detailing in each drawing. This tendency can give rise to the notion that all of the information must therefore be modeled. The residential architect would be wise to evaluate and plan the model simply.

PUBLIC/RELIGIOUS

This is a very broad spectrum, and it includes projects large and small, simple and complex. What they have in common is that there is a large body of critics to please. Whether it is the congregation of a church or the citizens of a city, communication is the key ingredient here. This brings the visualization aspect of Revit Architecture to the forefront. City councils can be shown the proposed building model within the proposed setting. The public can readily understand the aesthetics of a rendering and the color-fill diagrams showing the location of the services to be provided.

Challenges facing an architect specializing in these areas might include the fact that many of these building types can be rather large and ornate in nature, requiring additional time spent modeling. The need to provide many different presentation-level images requires additional time working out materials, and possibly significant computer rendering time.

SCOPE OF SERVICES

Due to the nature of Revit Architecture—building a virtual model of the project rather than representing one with lines—it has been noted that more of the effort is front-loaded in the schematic design and design development phases. In other words, you are working out constructability issues earlier in the design process. This generally provides great gains in efficiency in the construction-documentation phase. Architects need to be cognizant of this issue, and perhaps make adjustments in the area. When the scope of services ends at what could be described as the traditional design-development phase, there is a greater level of information within the model than before. Architects are beginning to understand this, and respond in many ways, such as marketing additional services, adjusting fees (front-loading) to maintain projects through their entirety, and using a project information base as a sales tool for current and future work.

FIRM LOCATIONS

It may go without saying, but the more office locations there are in a firm, the more complex the implementation process gets. There are many things to consider, as noted earlier. Training, support and standardization also become more important (and slightly more difficult) when there are multiple office locations. It is not impossible to accomplish widespread implementation efficiently; it will, however, require a little more planning. The planning will need to take into account the different office cultures, personnel and so on.

FIRM CULTURE AND AGE

Firm culture and age may not seem important in the planning stages of implementation However, it will become readily apparent during the implementation itself. How well does the firm accept and anticipate change? Is it a corporate culture? How do the lines of communication operate and what is the management organization? Who will need to back and support the initiative?

When age is considered, generally speaking, younger firms are more open to change, while older firms might tend to be more set in their ways and resistant to a change of this nature.

FOUR KEYS TO SUCCESS

In addition to the seven factors of implementation already outlined, we have developed the following four keys to success in implementation. We have found that when focusing on these steps, the potential for success increases greatly.

Plan

Communicate

Train

Support

PLAN

The first stage of implementation is planning. Of course, each firm is as individual as the employees within the firm. Careful planning is the foundation to a successful implementation. Carefully considering the previously outlined seven factors of implementation can supply a firm with guidelines, but each firm should consider its own unique characteristics at every step. Planning the implementation of any strategy or technology should take into account all of the individuals and groups that the change will affect. It is advisable to discuss these issues with key individuals in these groups.

COMMUNICATE

Once the plan is in place, it is imperative that everyone knows what the plan is. It is very important to share the plan with the entire office. This does not mean inviting the office as a whole to a meeting to show them the latest tool. There are many different roles within any firm; any new tool will affect users in different ways, and therefore users will be interested in different aspects of the tool.

The principals or senior leadership will be interested in how the new technology affects the bottom line—what is the net gain? They will also be interested in what kind of investment in capital and staff it will require. This group will also be interested in why the firm would make the change from the process currently in place. Generally speaking, their question will fall along the lines of "What does this mean to the business of architecture"?

The project managers will want to know how it affects their workflow. Will it really allow them to get work out sooner or with less staff? Revit Architecture's coordination of drawings, details and grid bubbles is greatly appreciated by those who experience it. Managers are also interested in the coordination between consultants and how a team might share the data. What does this mean to the process of architecture?

The architects and designers who will be using the tool will generally be concerned with the user interface, toolsets and learning curve. They will also be asking questions about how to create specific detailed models. What does this mean to the design and documentation of architecture?

TRAIN

When implementing a technology like Revit Architecture, it is advisable to have dedicated training. While on-the-job training can be a great learning experience, it can also be detrimental. When there are deadlines and revisions flying about, trainees tend to want to jump back to their comfort zone, 2D CAD.

The training should be organized and tailored to cover a firm's specific topics and issues. Having general training to cover the basics of the tool can be helpful initially. However, having an expert in the firm covering firm-specific issues will pay dividends later.

SUPPORT

Once the plan is complete and disseminated to your staff, and when training has been finished, the implementation is not yet complete. Firms should plan on maintaining and updating information through ongoing support. It is advisable to have regular meetings to discuss issues as they arise, from topics including new techniques to changes in structure and beyond.

WHAT HAS WORKED

- **Planning**—There is no substitute for good planning. Take the time to consider the issues within your firm and plan on addressing each and every one. It is a good idea to plan on re-planning too, as firm-specific issues arise and need to be addressed.

- **Formal training**—While on-the-job training can be the best-quality training for the real world, real deadlines, revisions and owners can often frustrate users and prompt them to want to give up and go back to good old CAD.

- **On-the-job training**—Occasionally, with the right personnel, project, schedule, team and support, a user can get up to speed while working on an active project. Great care needs to be taken if this is the path you choose to take.

- **Baby steps**—Take a few of the overall benefits of Revit Architecture and focus on those issues on a particular project, master those, and then take on a few more on the next project. Some firms use pilot projects to take these steps.

- **Partial projects**—Consider the "horizontal" approach; use Revit Architecture to design and document the plan view and scheduling while using CAD to document the vertical drawings, like the sections, elevations and perhaps the details.

- **Small projects with experienced users**—It may seem obvious, but overloading the first few small projects with experienced users works well and serves a purpose down the road. The theory here is that these users will gain real-world experience and confidence. These users can then go on to work with others and spread the knowledge.

- **Management buy-in**—It will be important for the entire company to have the management and upper management understand the issues and lend support to the projects and users.

- **Rewards and recognition**—The early adopters of these technologies often have to endure the nay-sayers, critics and the frustration of learning a new software and design process. With rewards and recognition, they will, at the very least, feel that it is worth the effort. Some firms reward new users by upgrading their computers first, or doling out preferred seating in meetings or recognition in newsletters.

- **On-going training**—Many firms, recognizing the fact that trainees cannot learn a large application overnight, organize and support weekly mini-training sessions. These sessions generally last one hour and cover a single topic.

- **Exit Strategy**—On the rare occasion that the BIM is not working out on a particular project, it may be necessary to exit and return to AutoCAD. It will be helpful if that exit is prepared ahead of time.

WHAT TO AVOID

- **Indifferent communication**—As noted, users not only need to know how the tools work, but must know the concepts of Revit Architecture and why the firm is going in this direction. Understanding what BIM is, how it is to be implemented and why, can be just as important as learning how to use it.

- **On-the-job training without full support**—If it is impossible to get formalized, non-project-related training, it will become essential that the new users have a support mechanism available as close to full-time as possible. When it comes down to crunch time with new functionality and there's no one in-house to assist, a deadline can become a breakdown point in implementation. The support team could be an in-house expert or a reseller/training center.

- **Attempting to hit a home run**—Some parts of Revit Architecture can be easier to learn than 2D CAD. It is, however, incredibly comprehensive. Take baby steps into the software. After all, most people did not learn CAD on the first project. Generally speaking, users become very comfortable and highly productive on their second or third project.

- **Isolating users**—This goes back to training and support. Users who are set free on a project without consistent support will tend to lose interest and become increasingly frustrated while they attempt to figure everything out themselves.

- **The oversell**—Avoid selling all of the benefits of Revit Architecture to clients prior to actually having a few projects under your belt. Be certain that the firm can deliver on promises. Avoid repeating vendor marketing promises without in-house verification.

WRAP UP

This appendix is, just that, a little bit of extra information at the end of this book. There are any number of combinations and additional factors that could, and very well will, affect the implementation of Revit Architecture at any given firm. The intent of this appendix is to give general guidance to those on the implementation path. Good luck, and good hunting.

INDEX

IMPORTANT-READ CAREFULLY: This End User License Agreement ("Agreement") sets forth the conditions by which Delmar Learning, a division of Thomson Learning Inc. ("Thomson") will make electronic access to the Thomson Delmar Learning-owned licensed content and associated media, software, documentation, printed materials and electronic documentation contained in this package and/or made available to you via this product (the "Licensed Content"), available to you (the "End User"). BY CLICKING THE "I ACCEPT" BUTTON AND/OR OPENING THIS PACKAGE, YOU ACKNOWLEDGE THAT YOU HAVE READ ALL OF THE TERMS AND CONDITIONS, AND THAT YOU AGREE TO BE BOUND BY ITS TERMS CONDITIONS AND ALL APPLICABLE LAWS AND REGULATIONS GOVERNING THE USE OF THE LICENSED CONTENT.

1.0 SCOPE OF LICENSE

1.1 Licensed Content. The Licensed Content may contain portions of modifiable content ("Modifiable Content") and content which may not be modified or otherwise altered by the End User ("Non-Modifiable Content"). For purposes of this Agreement, Modifiable Content and Non-Modifiable Content may be collectively referred to herein as the "Licensed Content." All Licensed Content shall be considered Non-Modifiable Content, unless such Licensed Content is presented to the End User in a modifiable format and it is clearly indicated that modification of the Licensed Content is permitted.

1.2 Subject to the End User's compliance with the terms and conditions of this Agreement, Thomson Delmar Learning hereby grants the End User, a nontransferable, non-exclusive, limited right to access and view a single copy of the Licensed Content on a single personal computer system for non-commercial, internal, personal use only. The End User shall not (i) reproduce, copy, modify (except in the case of Modifiable Content), distribute, display, transfer, sublicense, prepare derivative work(s) based on, sell, exchange, barter or transfer, rent, lease, loan, resell, or in any other manner exploit the Licensed Content; (ii) remove, obscure or alter any notice of Thomson Delmar Learning's intellectual property rights present on or in the License Content, including, but not limited to, copyright, trademark and/or patent notices; or (iii) disassemble, decompile, translate, reverse engineer or otherwise reduce the Licensed Content.

2.0 TERMINATION

2.1 Thomson Delmar Learning may at any time (without prejudice to its other rights or remedies) immediately terminate this Agreement and/or suspend access to some or all of the Licensed Content, in the event that the End User does not comply with any of the terms and conditions of this Agreement. In the event of such termination by Thomson Delmar Learning, the End User shall immediately return any and all copies of the Licensed Content to Thomson Delmar Learning.

3.0 PROPRIETARY RIGHTS

3.1 The End User acknowledges that Thomson Delmar Learning owns all right, title and interest, including, but not limited to all copyright rights therein, in and to the Licensed Content, and that the End User shall not take any action inconsistent with such ownership. The Licensed Content is protected by U.S., Canadian and other applicable copyright laws and by international treaties, including the Berne Convention and the Universal Copyright Convention. Nothing contained in this Agreement shall be construed as granting the End User any ownership rights in or to the Licensed Content.

3.2 Thomson Delmar Learning reserves the right at any time to withdraw from the Licensed Content any item or part of an item for which it no longer retains the right to publish, or which it has reasonable grounds to believe infringes copyright or is defamatory, unlawful or otherwise objectionable.

4.0 PROTECTION AND SECURITY

4.1 The End User shall use its best efforts and take all reasonable steps to safeguard its copy of the Licensed Content to ensure that no unauthorized reproduction, publication, disclosure, modification or distribution of the Licensed Content, in whole or in part, is made. To the extent that the End User becomes aware of any such unauthorized use of the Licensed Content, the End User shall immediately notify Delmar Learning. Notification of such violations may be made by sending an Email to delmarhelp@thomson.com.

5.0 MISUSE OF THE LICENSED PRODUCT

5.1 In the event that the End User uses the Licensed Content in violation of this Agreement, Thomson Delmar Learning shall have the option of electing liquidated damages, which shall include all profits generated by the End User's use of the Licensed Content plus interest computed at the maximum rate permitted by law and all legal fees and other expenses incurred by Thomson Delmar Learning in enforcing its rights, plus penalties.

6.0 FEDERAL GOVERNMENT CLIENTS

6.1 Except as expressly authorized by Delmar Learning, Federal Government clients obtain only the rights specified in this Agreement and no other rights. The Government acknowledges that (i) all software and related documentation incorporated in the Licensed Content is existing commercial computer software within the meaning of FAR 27.405(b)(2); and (2) all other data delivered in whatever form, is limited rights data within the meaning of FAR 27.401. The restrictions in this section are acceptable as consistent with the Government's need for software and other data under this Agreement.

7.0 DISCLAIMER OF WARRANTIES AND LIABILITIES

7.1 Although Thomson Delmar Learning believes the Licensed Content to be reliable, Thomson Delmar Learning does not guarantee or warrant (i) any information or materials contained in or produced by the Licensed Content, (ii) the accuracy, completeness or reliability of the Licensed Content, or (iii) that the Licensed Content is free from errors or other material defects. THE LICENSED PRODUCT IS PROVIDED "AS IS," WITHOUT ANY WARRANTY OF ANY KIND AND THOMSON DELMAR LEARNING DISCLAIMS ANY AND ALL WARRANTIES, EXPRESSED OR IMPLIED, INCLUDING, WITHOUT LIMITATION, WARRANTIES OF MERCHANTABILITY OR FITNESS OR A PARTICULAR PURPOSE. IN NO EVENT SHALL THOMSON DELMAR LEARNING BE LIABLE FOR: INDIRECT, SPECIAL, PUNITIVE OR CONSEQUENTIAL DAMAGES INCLUDING FOR LOST PROFITS, LOST DATA, OR OTHERWISE. IN NO EVENT SHALL DELMAR LEARNING'S AGGREGATE LIABILITY HEREUNDER, WHETHER ARISING IN CONTRACT, TORT, STRICT LIABILITY OR OTHERWISE, EXCEED THE AMOUNT OF FEES PAID BY THE END USER HEREUNDER FOR THE LICENSE OF THE LICENSED CONTENT.

8.0 GENERAL

8.1 Entire Agreement. This Agreement shall constitute the entire Agreement between the Parties and supercedes all prior Agreements and understandings oral or written relating to the subject matter hereof.

8.2 Enhancements/Modifications of Licensed Content. From time to time, and in Delmar Learning's sole discretion, Thomson Thomson Delmar Learning may advise the End User of updates, upgrades, enhancements and/or improvements to the Licensed Content, and may permit the End User to access and use, subject to the terms and conditions of this Agreement, such modifications, upon payment of prices as may be established by Delmar Learning.

8.3 No Export. The End User shall use the Licensed Content solely in the United States and shall not transfer or export, directly or indirectly, the Licensed Content outside the United States.

8.4 Severability. If any provision of this Agreement is invalid, illegal, or unenforceable under any applicable statute or rule of law, the provision shall be deemed omitted to the extent that it is invalid, illegal, or unenforceable. In such a case, the remainder of the Agreement shall be construed in a manner as to give greatest effect to the original intention of the parties hereto.

8.5 Waiver. The waiver of any right or failure of either party to exercise in any respect any right provided in this Agreement in any instance shall not be deemed to be a waiver of such right in the future or a waiver of any other right under this Agreement.

8.6 Choice of Law/Venue. This Agreement shall be interpreted, construed, and governed by and in accordance with the laws of the State of New York, applicable to contracts executed and to be wholly preformed therein, without regard to its principles governing conflicts of law. Each party agrees that any proceeding arising out of or relating to this Agreement or the breach or threatened breach of this Agreement may be commenced and prosecuted in a court in the State and County of New York. Each party consents and submits to the non-exclusive personal jurisdiction of any court in the State and County of New York in respect of any such proceeding.

8.7 Acknowledgment. By opening this package and/or by accessing the Licensed Content on this Website, THE END USER ACKNOWLEDGES THAT IT HAS READ THIS AGREEMENT, UNDERSTANDS IT, AND AGREES TO BE BOUND BY ITS TERMS AND CONDITIONS. IF YOU DO NOT ACCEPT THESE TERMS AND CONDITIONS, YOU MUST NOT ACCESS THE LICENSED CONTENT AND RETURN THE LICENSED PRODUCT TO THOMSON DELMAR LEARNING (WITHIN 30 CALENDAR DAYS OF THE END USER'S PURCHASE) WITH PROOF OF PAYMENT ACCEPTABLE TO DELMAR LEARNING, FOR A CREDIT OR A REFUND. Should the End User have any questions/comments regarding this Agreement, please contact Thomson Delmar Learning at delmar.help@cengage.com.